DATE DUE

OCT 2 7 1999	

HEREDITY and
HUMAN AFFAIRS

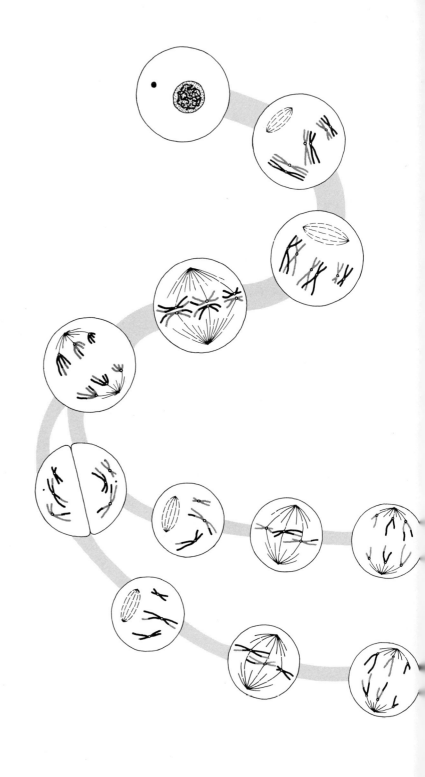

HEREDITY and HUMAN AFFAIRS

JAMES J. NAGLE

Department of Zoology, Drew University,
Madison, New Jersey

With 171 illustrations

The C. V. Mosby Company

SAINT LOUIS   1974

Library of Congress Cataloging in Publication Data

Nagle, James J
    Heredity and human affairs.

    Bibliography: p.
    1. Human genetics. 2. Human genetics—Social
aspects. I. Title. [DNLM: 1. Evolution.
2. Genetics, Human. QH431 N149h 1974]
QH431.N325   573.2′1   73-14547
ISBN 0-8016-3620-5

VH/VH/VH   9   8   7   6   5   4   3   2   1

To the professors who influenced me most
Donald D. Rabb and **Carey H. Bostian**

Preface

One does not have to be a biologist to have an interest in and curiosity about human heredity and the social ramifications of modern genetic technology. This book was developed out of my experiences in trying to relate some of the major biological principles associated with the science of genetics in a meaningful way to students who are not biology majors. I have concentrated on three broad, major areas of biology—evolution, reproduction, and genetics—placed into the context of human heredity. I have attempted to provide a reasonably sophisticated approach to understanding the factual and theoretical foundations of these several areas of biological knowledge which the educated layman should be able to understand in order to act and react, rationally and responsibly to the social, ethical, and moral problems so prevalent today in human affairs as a result of the so-called biological revolution.

The book is divided into three sections: I, Evolutionary processes; II, Reproductive processes; III, Hereditary processes. Capitalizing on our "natural interest" in ourselves, I have chosen to concentrate exclusively on those aspects of the areas under study that relate *directly* to mankind. Any subarea of evolution, reproduction, or genetics that does not lend itself to human example or general layman's concern I have deliberately omitted from the text. This is not because I do not appreciate or acknowledge the value of biological knowledge peripheral to direct human application, but because it does not fit the purpose of *this* text for nonmajors.

A discussion of the evolutionary processes, beginning with the origin of life, is placed first in the book in order to give perspective to subsequent discussions by providing a suitable background to which structure, functions, and mechanisms of reproduction and heredity can be related. For the sake of what I consider to be a coherent approach to fully understanding human heredity in its proper context, I have purposely employed evolutionary theory to set the stage for studying human heredity at the expense of forsaking more in-depth study of the evolutionary processes themselves, which requires a substantial prior foundation of genetic principles. Again, I have done this to suit the purpose of *this* text for nonmajors.

Reproductive processes are discussed next because an understanding of the mechanisms involved here is prerequisite to comprehending hereditary transmission and the important role of the genes in development. Specific attention to human reproduction, and the control of reproduction, also lays the biological foundation for a fuller appreciation and comprehension of the myriad of human interactions that fall into the tangles and nebulous web of "sexual behavior." Far too many persons lack the basic biological knowledge of sexual reproduction and the developmental process given the social controversies within this area in which the average citizen becomes involved today. These range from debates over sex education in the public schools to legislative decisions on abortion and the human status of the fetus.

With the evolutionary setting and reproductive mechanisms established, the final portion of the book addresses the hereditary processes. This section begins with a description of the gene and its function followed by a consideration of the various modes of hereditary transmission. Emphasis here is placed on understanding the mechanisms involved in the production of different types of hereditary patterns. At this point the text becomes more "technical;" however, without delving into the "guts" of hereditary transmission the student would be left with nothing but a set of "interesting facts" about human inheritance. My interpretation of the difference between a popular book written strictly for the reading public and a textbook written for the nonmajor college student is that the latter, which is my objective, is designed to be *studied* with the expectation of gaining fundamental knowledge, while the former is designed to be *read only* with the objective of broadening one's horizons in a more informal and unstructured manner. I firmly believe in the basic difference between a textbook for the nonmajor in any academic subject and a book on the same subject written strictly for the reading public.

Problems are not given at the ends of the chapters dealing with genetics because typical problems are approached and explained in the text. This "working through" of problems is a primary technique employed to explain the hereditary mechanisms and guide the student to an understanding of the basic principles. Problems are also omitted to avoid the temptation to simply assign them as homework and review them in class. For the nonmajors—future shapers of our society—it is much more important to spend class time discussing the principles themselves and the implications, benefits, and problems that their applications present to society and to each individual.

No previous course in college biology is assumed to be necessary for the discussions in this book. Chemistry is not practiced in the text. Only *descriptions* of chemical processes are used, along with the chemical "names" for identification purposes. The mathematics required is hardly more than the arithmetic of fractions and decimals; nevertheless, that is the most common "hang-up" of students I have taught. Recognizing this fact, I have tried to explain all mathematical manipulations fully and in elementary terms. A glossary is also provided to aid in understanding the vocabulary.

So many fine persons have contributed to my development and the development of this book that to list all of them would be impractical—I sincerely thank each one. I do however wish to publicly acknowledge the help contributed by Dr. Lee Ehrman through her constructive criticism of the entire manuscript. Drs. Gerald Stine and E. G. Stanley Baker also read significant portions of the manuscript and offered helpful suggestions. My colleagues at Drew University also deserve a special thank you for the many discussions we had (mostly them listening to me) that helped me to formulate and clarify some of my ideas on the structure of the material presented in the text. It is a pleasure to acknowledge the artistic talent of Mr. Michael Carpenter, whose interest and insight turned my rough draft of the figures in this book into truly outstanding illustrations. Finally, I extend my deepest thanks and appreciation to my wife, Jan, and my sons, Doug and Matt, for their support, endurance, and continued good humor during my preparation of the manuscript.

James J. Nagle

Contents

Evolutionary processes

1

The origin of life

To understand man and his problems from a biological point of view, it is necessary to understand many of the basic processes of the living state. It is often convenient to begin the study of biology by attempting to define life; no simple and concise definition is possible, however. Most attempts to distinguish the living from the nonliving usually amount to listing a set of properties which, taken together, are said to characterize life. This list includes metabolism, growth, reproduction, and responsiveness. These criteria may then be subdivided and elaborated upon to varying degrees in order to study more precisely the vital processes. When life is briefly characterized by a set of properties and followed by a detailed study of the various processes in some sequential order, a problem quite often arises for the novice studying biology. Somehow, life as an integrated whole gets lost in the miry, and sometimes overwhelming, accumulation of facts about the subject. The main objective of this part of the book (Chapters 1, 2, and 3) is to present life as a unified system from its origin on through its evolution to man in the hope that this approach will instill something more than a brief and inadequate definition. The importance of life to man and his society today is not to define but to understand and appreciate life.

In this chapter we shall attempt to gain a fundamental understanding of life by considering its origin and the development of the basic life processes. This will be followed in Chapter 2 by a discussion of evolutionary theory and the processes by which life developed and diversified. This part of the book will be concluded in Chapter 3 with a more specific discussion of the evolution of man.

SPONTANEOUS GENERATION

It is well known today that life arises only from pre-existing life. This is called the principle of *biogenesis*, but where did the first life come from? The experiments of Louis Pasteur in the 1860s disposed of the long-held idea of the spontaneous generation of life. No longer did men believe that maggots could arise de novo from decaying meat, or that bacteria developed spontaneously in spoiling broth, or that earthworms were generated from the soil during heavy rains, or that mice could be produced from sweaty shirts placed in a dark corner and sprinkled with wheat. Pasteur's work was so widely accepted that in the latter half of the nineteenth and the beginning of the twentieth century, spontaneous generation was totally refuted and regarded as scientifically incomprehensible.

It wasn't until the 1920s that A. I. Oparin, a Russian biochemist, and J. B. S. Haldane, a British geneticist, independently proclaimed that life must have arisen spontaneously on the primitive earth. There is a big difference, however, between the modern concept of spontaneous generation and that of Pasteur's day. The modern theory does not suggest that life can arise spontaneously under the conditions that now exist on earth. What it does claim is that life could and did arise spontaneously from nonliving material under the conditions prevailing on the earth at a much earlier time, and that from this origin all present life has evolved. Hence, it is possible today to devise an outline within the domain of modern science that will lead to the gradual development of defined systems that possess the properties characteristic of life (metabolism, growth, reproduction, and responsiveness), without having to incorporate any improbable cataclysmic or supernatural event at any point.

THE TIME ELEMENT

The first factor of major consideration in the origin of life is the time element. Tremendous spans of time would be necessary for the chemical events involved in the process to have any likelihood of occurrence. It has been stated that given enough time, any event, no matter how small its probability of occurrence may be, is certain to occur. As far as the astrophysicists and geologists can tell us, vast spans of time were available.

The earth and our solar system were formed about 4.5 billion years ago. Prior to this time, the basic chemical elements were being formed from free electrons and protons. These elements produced a cosmic gas and dust from which the sun and solar system were formed—a developmental process that may be called *cosmic evolution.* After the formation of the earth, chemical reactions of greater and greater complexity occurred, ultimately producing living material. This period of development may be called *chemical evolution.* The oldest known fossils of living organisms date back nearly 3 billion years. Taking this as the approximate time that chemical systems acquired the properties attributed to living things, *biological evolution* ensued, producing the diversity of life well-documented from the past and the present. In the last 2 million years or so man has evolved, and he has added *cultural evolution* to his biological evolution. Thus we have four major evolutionary periods into which the span of time can be divided (Fig. 1-1). It should be noted that these evolutionary periods are not necessarily mutually exclusive; for instance, man is presently undergoing biological *and* cultural evolution. Considering the universe in its entirety, all of these processes may be occurring at various stages in various places.

The first period, cosmic evolution, is beyond the scope of this text. The period

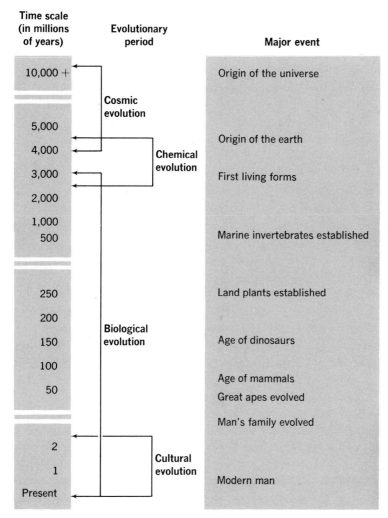

Fig. 1-1. The major evolutionary periods associated with the development of life on earth.

of chemical evolution, particularly the processes leading to the origin of life and collectively referred to as *biopoiesis*, beginning about 4.5 billion years ago and encompassing some billion and a half or more years, is the time span of our immediate concern. Biological and cultural evolution will be discussed in subsequent chapters.

THE PRIMITIVE EARTH

Oparin formulated the key explanation of how life could have been spontaneously generated on the primitive earth, although its continued production in this manner under contemporary conditions is highly improbable, if not totally impossible. He postulated that the atmosphere surrounding the primitive earth was quite different from that which presently envelops the planet. Our present atmosphere contains an abundant amount of free oxygen (21%). This oxygen is supplied and continually replenished as a by-product of photosynthesis from living plants. But on the primitive earth where no life existed, any free oxygen was avidly picked up by a variety of substances, such as iron, which formed the corresponding oxides. As a result, oxygen was lacking in the primordial atmosphere. Instead, hydrogen, the lightest and most abundant element in the universe, was the prevalent gas. Whereas the present atmosphere is an oxidizing one, the primordial atmosphere was a reducing one; that is, hydrogen, as an electron donor for chemical reactions, was available rather than oxygen as an electron acceptor. This difference in the chemical nature of the atmosphere is the basis upon which the theory of the origin of life is built.

As the mass of the protoplanet earth condensed and stratified, carbon, nitrogen, and oxygen were the other elements in abundance in the atmosphere. (Heavier elements like iron and nickel sank to the core.) Each of the lighter elements was combined with the chemically reactive hydrogen in such a way that the nitrogen was probably in the form of ammonia, the oxygen in the form of water vapor, and the carbon as methane. Table 1-1 contrasts the composition of the present and the primitive earth atmospheres.

The significance of a reducing atmosphere as the basis for the origin of life becomes evident in the following explanation. Living material consists chemically of organic compounds. The basis of organic chemistry is the carbon atom in association with hydrogen forming hydrocarbons. It was once thought that organic compounds were the exclusive products of living systems. However, in a reducing atmosphere, such as the primitive earth possessed, the first carbon compounds produced would have been simple hydrocarbons, such as methane, rather than inorganic carbon dioxide, which would have been produced in an oxidizing atmosphere. Carbon first appeared on earth in the reduced state, not in the oxidized state, and this is of great importance because the reduced carbon (methane) represented a simple organic compound not produced by a living organism.

Water was initially present only as a gas, but as the earth cooled, the water vapor condensed to liquid, and torrential rains produced the oceans and rivers. As the waters ran, various inorganic salts and minerals began to accumulate, and some of the methane and ammonia from the atmosphere dissolved in the water. The earth now contained warm seas with an array of chemicals including simple

Table 1-1. The major gases of the primordial and the present earth atmospheres

Primordial atmosphere	Present atmosphere
Ammonia (NH_3)	Carbon dioxide (CO_2), 1%
Hydrogen (H_2)	Oxygen (O_2), 21%
Methane (CH_4)	Nitrogen (N_2), 78%
Water vapor (H_2O)	

hydrocarbons. From this point, the chemical evolution toward more complex organic molecules proceeded.

SIMPLE ORGANIC BUILDING BLOCKS

For life to arise, certain critical organic building blocks are necessary. These include amino acids, simple sugars, and nitrogenous bases. The production of these building blocks requires further chemical synthesis. A mixture of ammonia, methane, water, and hydrogen is thermodynamically stable, however; that is, there is no tendency for these substances to chemically react to form more complex compounds. Further organic synthesis would require an external energy source to drive the chemical reactions.

Several sources of external energy were readily available on the primitive earth. The principal sources were *solar radiation,* especially ultraviolet light, which was not shielded from the earth's surface by an atmospheric ozone (O_3) layer as it is today; *electrical energy* produced by lightening bolts; *thermogenic energy* generated by the numerous volcanic eruptions occurring on the surface of the young planet; and *ionizing radiation* resulting from the disintegration of plentiful radioactive substances.

The experimental breakthrough in support of biopoiesis came in 1953 when Stanley L. Miller reported the results of his now classic experiments on the synthesis of organic compounds from the components and conditions on the primitive earth. Miller constructed an airtight apparatus in which the steam from boiling water was mixed with hydrogen, ammonia, and methane. The mixture was circulated continuously past electrical discharges produced by tungsten electrodes. As the steam passed through the condenser, it liquified, bringing down some of the less volatile products with it. The U-tube at the base of the apparatus served as a collection chamber and also promoted one-way circulation of the gases through the system. Overflow water was revaporized in the boiling flask and recirculated with the other gases (Fig. 1-2). After the apparatus had run for one week, this mixture of simple primitive earth components produced an amazing variety of organic compounds, among which were several important amino acids. Amino acids are the building blocks of proteins, which are among the most important components of living systems. Table 1-2 lists the organic compounds produced in Miller's initial experiments. Miller also conducted control and test experiments of various kinds which unequivocally proved that the organic compounds produced in his apparatus were not from any source of contamination such as microbial synthesis. The Miller experiments thus proved Oparin's most important deduction and the one originally most difficult to accept: organic molecules could be synthesized abiotically from a reducing mixture of simple gases in which methane served as the only source of carbon.

Subsequently, Miller's results were repeated and confirmed by many other scientists using his methods and others. For instance, energy sources other than electrical discharges have been successfully employed. Heat, ultraviolet light,

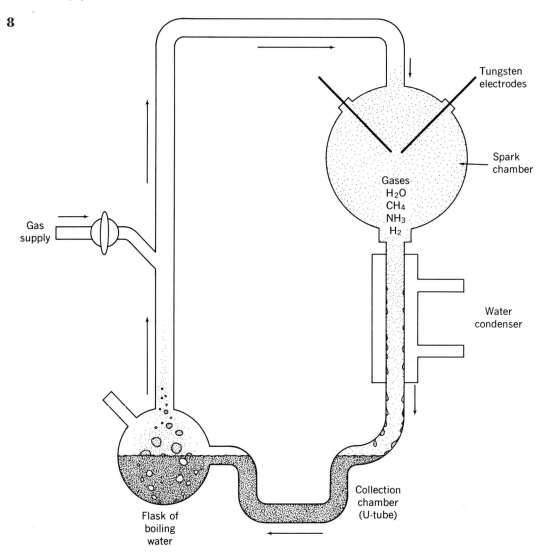

Fig. 1-2. The Miller apparatus used to simulate primitive earth conditions under which abiotic organic synthesis could have occurred.

X-rays, ionizing radiation, and ultrasonic vibrations have all promoted the abiotic synthesis of organic compounds. Most recently, a team of investigators has proposed another natural phenomenon that might have provided the energy for abiotic synthesis. This energy source is *shock waves* produced by thunderclaps or by meteors plunging into the earth's atmosphere. In a relatively simple experiment, a brass-and-Pyrex tube was divided into two compartments by a plastic membrane. One chamber was charged with a mixture of ammonia, methane, ethane, and water vapor. The other chamber contained the chemically inert gas, helium. By increasing the pressure on the helium until the membrane broke, a shock wave was produced that swept into the mixture-containing chamber at high speed, momentarily creating an extremely high temperature. In re-

Table 1-2. Organic compounds produced abiotically in the initial experiments of Miller

Amino acids	Other organic compounds
Alanine	Acetic acid
β-Alanine	Formic acid
α-Amino-n-butyric acid	Glycolic acid
α-Aminoisobutyric acid	α-Hydroxybutyric acid
Aspartic acid	Iminoacetic-propionic acid
Glutamic acid	Iminodiacetic acid
Glycine	Lactic acid
N-Methylalanine	N-Methylurea
Sarcosine	Propionic acid
	Succinic acid
	Urea

peated experiments, at least four amino acids were produced. However, methane is not the only simple carbon source that will produce this reaction. Various experimenters have successfully substituted formaldehyde, paraformaldehyde, formhydroxamic acid, glycol, carbon dioxide, carbon monoxide, and other simple carbon compounds. However, reducing conditions must always prevail for significant organic synthesis to take place. All this demonstrates that abiotic synthesis of organic molecules can occur through so many diverse pathways that even if the conditions on the primitive earth were only roughly similar to those postulated, organic molecules would almost certainly have been formed in the primeval waters. Jupiter and Saturn still possess an atmosphere of methane, ammonia, and hydrogen much like that which shrouded the primitive earth. Data from space probes scheduled to pass Jupiter may help to confirm the biopoietic theory if it can be shown that processes occurring there parallel those postulated to have occurred on earth billions of years ago.

As the amino acids and other complex organic compounds were produced, they would have slowly accumulated in the seas. Even though organic molecules are highly degradable today, their accumulation would have been assured under the conditions on the primitive earth. Degradation of organic compounds occurs primarily through the action of organisms of decay, for example, bacteria, and by oxidation. However, in a reducing atmosphere with no living forms present, neither process would occur.

In addition to amino acids, two other classes of simple organic building blocks are essential to the origin of living systems. The first of these is represented by the simple *monosaccharide sugars:* the pentoses, containing five carbon atoms in their structure and the hexoses, containing six carbon atoms. The pentoses, ribose and deoxyribose, are components of nucleic acids that make up the chemical basis of the hereditary mechanism in living systems. The hexose glucose is the principal raw energy source used to power living systems. Both pentoses and hexoses have been produced abiotically in experiments using the organic compounds that accompanied the initial production of amino acids from the primitive atmosphere.

Nitrogenous bases, purines and pyrimidines, and their derivatives constitute the other class of essential organic building blocks. Nitrogenous bases function in cellular metabolism and are also structural components of nucleic acids. They have also been synthesized experimentally under the abiotic conditions of the primitive earth. Interestingly enough, adenine, one of the most important nitrogenous bases in living systems, was the first formed. This purine is extremely important as a component of adenosine triphosphate (ATP) which is the principle energy transfer molecule in living systems. Adenine is also important in the process of genetic coding.

MACROMOLECULES

A variety of amino acids, sugars, nitrogenous bases, and other simple hydrocarbons cannot of itself produce a living system. Macromolecules, particularly proteins and nucleic acids, are needed. These are polymers (macromolecules) formed from the simpler organic building blocks by further chemical reactions.

It was originally estimated that the primitive oceans had accumulated as much as a 10% concentration of organic building blocks; this concentration provided ample opportunity for their chance meeting and polymerization (chemical aggregation) into macromolecules. However, more recent estimates concerning the rate of abiotic organic synthesis set 1% as a more likely concentration of simple organic compounds in the waters of the primitive earth. Nevertheless, the higher concentrations needed to promote the production of macromolecules could have developed in shallow ponds, estuaries, or beach puddles. In places such as these, the heat of the sun could have evaporated much of the water, thereby increasing the organic concentration and, at the same time, supplying the energy required for polymerization to proceed. The other external energy sources previously discussed would have also been available to drive the polymerization reactions. The polymerizations of amino acids to proteins, nitrogenous bases to nucleic acids, and monosaccharides (simple sugars) to polysaccharides (complex sugars) under primitive earth conditions have all been confirmed experimentally.

Since Miller's classic work 2 decades ago, an overwhelming amount of evidence has accumulated from laboratories all over the world supporting the feasibility of the abiotic synthesis of simple and complex organic compounds under possible primitive earth conditions. Considering the numerous external energy sources available and the manifold pathways of synthesis possible, the question is really not if or how organic compounds could arise spontaneously, but rather how could they have failed to be present?

COACERVATES

Chemical evolution eventually reached the point where the primitive seas, or at least some estuaries, ponds, and puddles, became soupy mixtures of organic compounds, including polymers of various kinds, and dissolved inorganic salts.

But how could this disorganized array of vital constituents become organized into discrete systems with the properties of life?

Proteins, the amino acid polymers, probably played a fundamental role in the further chemical evolution toward living systems. Protein macromolecules are within the critical size range (10^{-7} to 5×10^{-5} cm) to form a colloidal system; that is, they form a semipermanent suspension without noticeable settling but with specific adsorptive properties. The particles of a colloid may remain discrete in their dispersion medium, or they may coalesce and separate from the medium. In addition, proteins are *amphoteric* substances; that is, there are both positively and negatively charged sites on the macromolecule. This property imbues proteins with the tendency to come together and form larger complexes by virtue of their residual valences. The coalescence of colloidal particles in a dispersion medium of water is called coacervation, and the separated complexes are called *coacervates*. The protein macromolecules dispersed in the waters of the primitive earth, therefore, would have had strong tendencies to form complex coacervate droplets consisting of a mixture of various kinds of proteins (Fig. 1-3).

Proteinaceous coacervate droplets have several extremely important characteristics that clearly put them on the pathway toward living systems. The charged

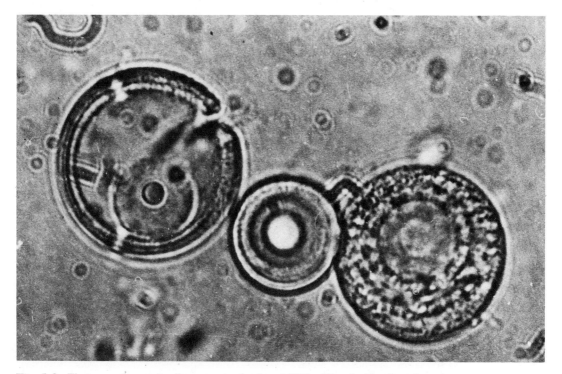

Fig. 1-3. Electron micrograph of coacervate droplets ×1700. (Proteinoid microspheres.) These droplets have internal ultrastructure; they are often mistakenly identified as coccoid bacteria. (Courtesy Dr. Sidney W. Fox, Institute of Molecular Evolution, University of Miami)

nature of the different proteins within a coacervate determines a definite internal organization rather than a random assortment. A definite delimiting "membrane" consisting of a shell of rigidly oriented water molecules plus adsorbed surface-active substances from the surrounding medium forms around the droplet. This boundary does not isolate the coacervate from its environment, however. A very important characteristic of the droplet is its ability to interact with the surrounding medium. This interaction is expressed in the selective absorption of material from the external environment. This ability has been experimentally demonstrated by adding various dyes to the medium surrounding manmade coacervates and observing the droplets gradually become dark while the surrounding solution became correspondingly lighter. Further experiments have demonstrated the selective incorporation of amino acids in coacervates ranging from $100\times$ concentration for tyrosine to $2\times$ for tryptophan and others. In contrast to this, simple sugars and nitrogenous base derivatives were equally distributed inside and outside the droplets. Thus, the coacervate with its outer shell displays the selective permeability that is characteristic of a living cell and its membrane.

With the concentration of selected environmental substances within the coacervates, many chemical reactions that occurred only sporadically in the organic soup could proceed repeatedly. More importantly, the products of these increased chemical reactions could also build up and initiate further reactions within the coacervates which could not even take place in the outside medium because of the extreme dilution of the potential reactants. In this way, chemical sequences not likely to have developed externally could have evolved within the droplets. Among the many simple molecules that coacervates incorporated or adsorbed to their surface would undoubtedly have been some with primitive catalytic properties. *Catalysts* are substances that can speed up chemical reactions but are unchanged themselves when the reaction is over. These simple catalysts could have then been elaborated into highly specific and efficient organic catalysts that in biological systems are called *enzymes*.

Catalysts, or enzymes, introduce an additional selective process to the evolving chemical reaction sequences. They are so specific and efficient that alternate pathways open to the reactants of any particular reaction may be rendered practically inoperative. Thus, coacervates could have developed highly selective chemical sequences occurring at rapid rates, whereas this could not have occurred in the disorganized external medium. The enhancement of the chemical activity within the coacervates would have led to an increase in the mass of the droplet. This would add coordinated internal growth to the vital characteristics of the coacervate droplets. Physical fragmentation of such enlarged coacervates could then produce smaller droplets with a composition similar to that of the original "parental" droplet. As some of these in turn enlarged and fragmented again, a primitive form of reproduction and "family lineage" could have been established. Although coacervates are not alive, they do exhibit many of the properties associated with living organisms, especially molecular organization, a surrounding membrane, selective permeability, internal growth, and reproduction. It should be apparent that the borderline between the living and the nonliving is becoming hazy at this point.

METABOLIC PATHWAYS

A large variety of coacervate types was no doubt formed in the primitive organic soup. Many, if not most, of them would have been chemically unstable and would have disintegrated quickly; others would have been chemically stable and able to persist. All the various coacervates, especially the stable ones with coordinated chemical activity, obtained their resources from the external environment. Therefore, as the most stable coacervates evolved into what might be called proto-organisms, their mode of nutrition was *heterotrophic* (literally, other-feeders). As various types of stable heterotrophs developed, reproduced (by fission), and grew in numbers, competition for the limited organic nutrients in the environment would have arisen, especially since the abiotic synthesis of organic molecules occurs at a very slow rate.

The chemical activities of the proto-organisms were undoubtedly not very complex at first. Most nutrients would have been obtained essentially ready-to-use from the environment. As the primary nutrient supply diminished, any proto-organism possessing, by chance, a catalyst (enzyme) capable of mediating the chemical transformation of some heretofore unused organic compound from the environment into an essential nutrient, would remove itself from direct competition for the primary nutrient and thereby enhance its chances for survival. This selective process represents the mechanism by which elaborate chemical pathways could have developed in a step-by-step manner into metabolic pathways that supply the energy and structural components for living systems.

Probably correlated with the evolution of metabolic pathways to support heterotrophic "life" was the incorporation of nucleic acids (DNA and RNA) as a genetic control system. The structure and function of these important biochemicals will be discussed in Chapter 6. However, it is important to point out here the significance of nucleic acids in the origin of life. The metabolism and reproduction of the earliest proto-organisms was not under any sort of genetic control. But as competition for resources intensified, the chances of survival and persistence would be tremendously increased if a successful metabolic system could be exactly reproduced. The highest selection value would have been possessed by those proto-organisms that could, through incorporated nucleic acids, genetically code for the production of proteins and enzymes that were critical components in their evolving metabolic pathways. Hence, proto-organisms were subjected to an early form of "the survival of the fittest" which will be discussed in the next chapter.

A generalized model, such as that depicted serially in Fig. 1-4, of the evolution of metabolic pathways under genetic control can be developed in the following manner: Suppose that organic compound *A*, which is essential for "life," is

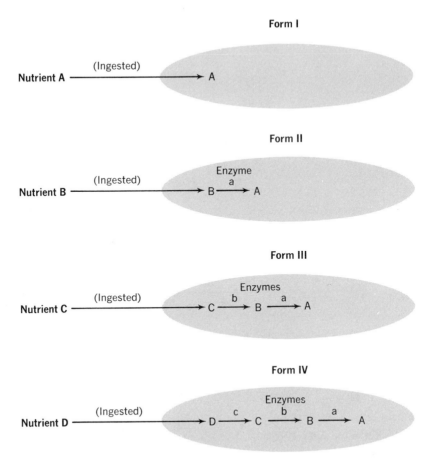

Fig. 1-4. A generalized model of the evolution of metabolic pathways depicted in four simplified steps (*A* through *D*).

initially available in the environment. Proto-organisms of the form I simply have to ingest *A* from the surrounding medium to utilize it (Fig. 1-4, *A*).

As form I multiplies, the slowly produced abiotic supply of nutrient *A* would diminish and competition for it would ensue. Now, if a variant form, II, happened to possess an enzyme, *a*, that could mediate the chemical transformation of some heretofore unused nutrient source, *B*, into *A* as an end product of a chemical reaction, this new form would no longer have to compete directly for *A*; instead it would be able to ingest and utilize *B* as a nutrient (Fig. 1-4, *B*). This would increase the survival value of this form. In addition, if a nucleic acid controlled the production of enzyme *a*, form II would be able to pass this characteristic on to its "progeny."

Form II would now proliferate until nutrient *B* became in short supply, and then competition for *B* would develop. If a variant of form II by chance possessed a second enzyme, *b*, that could mediate the conversion of substance *C*, a new nutrient source from the environment, to substance *B*, then this new form,

III, would no longer have to compete directly for substance *B* (Fig. 1-4, *C*). If nucleic acids directed the production of both enzymes *a* and *b* in form III, then form III and its "progeny" would gain a selective advantage for survival. Substance *B* would now become a chemically synthesized precursor in the emerging metabolic pathway leading to the production of *A*, which is still the essential end product necessary for the stability and survival of the proto-organism.

Continuing this process, competition for nutrient *C* would eventually develop among the form III types. Another variant, form IV, with enzyme *c* added to its repertoire of genetically controlled enzymes could now become the next successful form by utilizing substance *D* as an environmental nutrient and converting it to substance *C*, which leads to substance *B* and then to end product *A* (Fig. 1-4, *D*). Substances *C* and *B* now serve as metabolic intermediaries in the production of the essential end product. This serial process could continue until complex pathways, involving numerous chemical transformations, were established.

While one metabolic pathway was developing as described above, others could be proceeding independently within the same primitive life form. Later on, one pathway might supply a by-product that served the needs of another, or two pathways might merge into one, or long pathways might branch and utilize alternate intermediary compounds. Eventually, a very intricate and interdependent system of metabolic pathways could evolve, and somewhere in this chemical milieu with its concomitant genetic control, the transition to truly living organisms would have been achieved. It should now be apparent that life is dependent on the maintenance of metabolic pathways, and that these are controlled by enzymes that, in turn, are under genetic control. It should also be recognized that the emergence of a variant form, such as form IV in the model, does not necessarily mean the immediate extinction of its predecessors—some could survive for a time unchanged, and others could produce still different forms that might then develop along separate evolutionary lines. Hence, as soon as life arose, it probably expanded into diverse forms from which evolved the multiplicity of types that have characterized life from past forms to those of the present.

The elaboration of metabolic pathways could not have occurred without an energy supply to drive the chemical reactions. It has been previously mentioned that adenosine triphosphate (ATP) is the principal energy transfer molecule of living systems. The universality of ATP in contemporary living systems suggests that its function was very likely established from the onset of the life processes. ATP is composed of the nitrogenous base adenine and the five-carbon sugar ribose, chemically coupled to three units of inorganic phosphate. Since adenine and ribose are among the organic compounds presumed to have been abiotically synthesized on the primitive earth and since inorganic phosphate would have been present as ionic radicles dissolved in the water, it is reasonable to assume that ATP was available to drive chemical reactions and become incorporated as the energy transfer molecule for all subsequent living systems.

16 THE PERSISTENCE OF LIFE

As heterotrophic life forms diversified and became abundant, the abiotic supply of organic compounds could not have been replenished rapidly enough to sustain the forms indefinitely. In addition to the environmental nutrients being consumed faster than they were supplied, the effect of the organisms on the environment would have slowed down abiotic synthesis. For example, all metabolic processes in the primitive organisms would have been fermentative (anaerobic respiration) rather than oxidative (aerobic respiration) in the reducing atmosphere. Fermentation releases carbon dioxide as a by-product into the atmosphere. The carbon dioxide would not have readily contributed to further abiotic synthesis and, as it built up in the atmosphere, it would screen out much ultraviolet light that was probably a major energy source for driving abiotic synthesis. The presence of carbon dioxide in the atmosphere, however, set the stage for the development of the biosynthetic pathways leading to the process of *photosynthesis*. This important chemical evolutionary event made it possible for life to persist through the past 2 billion years or so. In the photosynthetic process, energy from sunlight is used to synthesize organic molecules (carbohydrates) from the inorganic molecules of carbon dioxide and water. Photosynthetic organisms, therefore, became independent of an external source of simple organic compounds and manufactured their own compounds from the abundant inorganic material available; such organisms are called *autotrophs* (literally, self-feeders). Photosynthetic autotrophs took over as the primary producers of organic molecules on earth and assured an adequate nutrient supply for the continued survival and evolution of heterotrophs.

A by-product of photosynthesis is oxygen. The gradual accumulation of oxygen in the atmosphere, coupled with the escape of much of the very light hydrogen into outer space, slowly converted the atmosphere of the earth from a reducing one to an oxidizing one. Once oxygen became plentiful, both heterotrophic and autotrophic forms of life were able to evolve oxidative metabolic pathways through which much more energy could be extracted from organic nutrients than fermentation provided. With the change to an oxidizing atmosphere, the abiotic synthesis of organic compounds probably ceased. Accompanying the spontaneous generation of life was an alteration of the earth's environment which made it incompatible with further spontaneous generation. Life arose at one time; it has flourished since that time, and ultimately Man emerged. Now the future of the whole wondrous process seems to lie in his hands.

2

The principles of biological evolution

Once life had originated, it began to diversify and develop into more and more complex systems that exploited the resources of the earth. The processes by which this occurred represent *biological evolution*. It began some 3 billion years ago and continues today. Man and all contemporary organisms are the representatives of the myriad evolutionary lines that emerged from the humble (or was it majestic?) origin of life.

The development of the theory of biological evolution is an outstanding example of how the educational process operates and how an educated person assumes his responsibilities. No progress is made in any field by the mere transmission of facts from one generation to another; facts and principles are only the building blocks for further advancement. It is the responsibility of the educated person to relate new knowledge to old ideas and, from the conglomerate, to gain new insights for the formulation of more logical theories that can then be subjected to further study and critical testing. Such advancement of man and his institutions through nonbiological processes is *cultural evolution*. The theory of biological evolution is a product of cultural evolution.

EVOLUTION OF EVOLUTIONARY THEORY

Generations of philosophers and biologists have contributed to the evolution of evolutionary theory. From the time man became conscious of himself, he has probably thought and speculated about his origins. The beginning of evolutionary theory dates back to the early Greek philosophers and the origin of Western culture. Their philosophy was one of nature which aimed directly for the development of a theory even before the attainment of any scientific knowledge. Yet, their speculations were very close to the concepts developed from modern science.

Thales (640-546 B.C.) is credited as being the first philosopher to attempt to substitute natural explanations for many previously mythological Greek concepts. He declared water to be the matter from which all things arose and out of which they existed. Fossils were first recognized as the remains of formerly living animals by *Xenophanes* (576-480 B.C.) who interpreted them as proof that the seas had previously covered the earth. And from *Empedocles* (495-435 B.C.) came the first idea of a selective evolutionary process. He believed that there were four great elements of nature—fire, air, water, and earth—which were acted upon by two ultimate forces—love, which was a combining force, and hate, which was a separating force. He visualized all organisms as arising through the fortuitous play of the two great forces of nature upon the four elements. This origin was viewed as a gradual process whereby animals first appeared, not as complete individuals, but as a disorganized array of parts. Out of their random combinations all sorts of extraordinary creatures were produced (animals with the heads of men, men with the heads of animals, or even bodies with two heads). But, as a result of the triumph of love over hate, the unnatural combinations soon became extinct because they were not able to propagate their kind. Thus, in the ancient teachings of Empedocles can be found the germ of the theory of the survival of the fittest.

Aristotle (384-322 B.C.) collected the known facts of nature in his time, studied the ideas of his predecessors, and coupled them with his numerous observations to advance his own philosophy of purposive natural causation. He believed that nature proceeded continually by the aid of gradual transitions from the most imperfect to the most perfect forms. At this point Aristotle became metaphysical. He was so impressed with the marvelous arrangement of the world that he was compelled to assume "Intelligent Design" as the primary cause of things. Nature was viewed as a continuum, with inorganic matter as the lowest stage from which organic matter was produced by direct metamorphosis, and subsequently this matter was transformed into life. Life then ascended from a primordial, soft mass of living matter to the most perfect forms, and even these, he believed, were progressing to still higher forms. If Aristotle had accepted and refined Empedocles' crude ideas of the survival of the fittest rather than an internal perfecting principle as the "force" of evolution, he would have virtually been a prophet of the modern Darwinian concept.

The ideas of Empedocles were not lost, however, and some 4 centuries later *Lucretius* (99-55 B.C.), the Roman poet of evolution, restated the concept of the survival of the fittest in the following stanza from *On the Nature of Things:*

> Verily not by design do the first beginnings of things station themselves each in his right place, occupied by keen-sighted intelligence, . . . but because after trying motions and unions of every kind, at length they fall into arrangements, such as those out of which this our sum of things has been formed, . . . and the earth, fostered by the heat of the sun, begins to renew this produce, and the race of living things to come up and flourish.

Following the establishment of the Christian era, learning in Western culture was under the guardianship of the church. The legacy from the Greeks, especially Aristotle, was expounded by early Christian philosophers of nature such as *Gregory of Nyssa* (331-396), *Augustine* (353-430) and later, *Thomas*

Aquinas (1225-1274). Very little real scientific investigation related to evolutionary theory took place during this period. Finally *Roger Bacon* (1214-1294) began to question the practice of argumentative writing to "prove" conclusions and appealed for searches into nature through the application of the experimental method to determine what was known by nature, what by art, and what by fraud.

The urging of Bacon never really materialized, however, because of an unfortunate set of circumstances that developed within the Church, which greatly affected the advancement of evolutionary thought. The natural philosophy of the Arabs, largely derived from Aristotle, had been gaining influence in Europe for several centuries, especially in Spain. To stem the influence of the Arabic philosophers, the Church Provincial Council of Paris in 1209 forbade the study of the Arabic writers. This interdict included the works of Aristotle, which had been translated into Arabic. Even though Thomas Aquinas and others tried to defend Aristotle, the Church denounced all contrary opinion as heresy, including the developing concepts of evolutionary theory. All scientific work came under the strict censorship of the theologians. Scientific speculation came to a standstill, and no advancement of evolutionary theory was made from this time until the sixteenth century. Even then theology and science became dominated by the idea of *special creation* that was based on a literal interpretation of Genesis and expounded by the Spanish Jesuit *Francisco Suarez* (1548-1617). This became the universal and orthodox theological teaching from the middle of the sixteenth until the middle of the nineteenth century.

Although their work was suppressed, some students of nature still searched for better explanations of life and questioned the sanctity of the dominating theological notions of natural science. The Italian philosopher *Giordano Bruno* (1548-1600) accepted the natural philosophy of Aristotle, Lucretius, and the Arabs and developed his own interpretation of nature, adherence to which cost him his life at the stake. Evolutionary thought slowly continued to develop through the seventeenth century. In England, *Francis Bacon* (1561-1626) recognized variation as a natural phenomenon and pointed out that species could be changed over time by nature acting upon these variations. In France, *Rene Descartes* (1596-1650) attempted to explain all things by the principle of natural law rather than the prevailing dogma of special creation. Once again, the idea of gradual change based on natural laws was considered. Bacon and Descartes set the pattern for the orderly scientific investigation of nature. Slowly the idea that species could change over time in relation to changes in the environment grew through the work of men such as *George Buffon* (1707-1788) and *Erasmus Darwin* (1731-1802), grandfather of Charles Darwin.

By the beginning of the nineteenth century, factual knowledge of the biological world had accumulated to the point where questions concerning the apparent relationships between organisms were receiving much attention from naturalists and philosophers. The fixity of species as objects of special creation had been seriously questioned by Buffon, Erasmus Darwin, and others, and the time was becoming ripe for the formulation of a better theory to explain the structure and order of nature. A threshold had been reached.

Lamarckism

The first truly comprehensive theory of evolution was expounded by Jean Baptiste de Lamarck (1744-1829) (Fig. 2-1). Lamarck specialized in animal classification, and as he worked, he realized that the various species could be fitted into an orderly relationship that formed a continuous progression extending from the simplest little polyp at one end to man at the other. He was the first naturalist to become convinced that animals could be modified in order to adapt to the environment and that species were not constant but were derived from pre-existing species. His theory was not well-received, and he suffered social and scientific ostracism for his convictions; nevertheless he maintained his position to his death.

In 1809, Lamarck put forth a complete theory of evolution in his now classic book, *Philosophie Zoologique*. He first described the three "truths" that were the basis of his theory. First was the fact that species vary under changing external influences; second, the fact that there is a fundamental unity in nature; and third, the fact that there is a progressive sequence of development. Lamarck denied the existence of any "perfecting tendency" and regarded the order in nature as the final and necessary effect of the environment on life. Accordingly, his idea was that animals were not specially created for a particular mode of life; instead, their mode of life had created them. For example, wings were

Fig. 2-1. Jean-Baptiste de Lamarck. He put forth the first comprehensive theory of organic evolution in 1809. (Culver Pictures)

not created to enable birds to fly; rather, birds had developed wings by continually attempting to fly. The processes by which evolution took place were formulated into four propositions, which Lamarck called laws:

The First Law. Life by its own forces tends continually to increase the volume of every body that possesses it, as well as to increase the size of all the parts of the body up to the limit of their function.

The Second Law. The production of a new organ or part results from a new need or want, which continues to be felt. This new need initiates a new movement and causes it to continue. Thus, formation of a new organ (or part) becomes necessary to produce the newly desired movement.

The Third Law. (The law of use and disuse.) The development of organs and their force or power of action are always in the direct relation to the employment of these organs. The frequent and sustained employment of an organ little by little strengthens and develops it in proportion to the length of the employment. On the other hand, the constant lack of use of an organ weakens and deteriorates it and finally causes its disappearance.

The Fourth Law. (The inheritance of acquired characteristics.) All that has been acquired or altered in the organization of individuals during their life is preserved and transmitted to new individuals who proceed from those who have undergone these changes.

Lamarck then applied his laws to explain the adaptation of organisms to their environment. He explained the evolution of the webbed feet of aquatic birds as being initiated by hunger (the need for food) driving the birds to swampy areas to seek food. In this environment the birds would have made efforts to swim by spreading their toes and the continued stretching of the skin between the toes gradually produced the webbed condition as an acquired characteristic that was then passed on to future generations. The possession of horns on the heads of some animals was attributed to the deposition of a secretion of matter upon the forehead stimulated by a need that arose from the habit of butting their heads together during times of conflict. The origin of snakes was explained as a loss of limbs stemming from the habit of moving along the ground and concealing themselves among the bushes. This habit led to continued efforts to elongate the body in order to pass through narrow places and, as a result, the animal acquired a long narrow body. Since long legs would have been useless and short legs would have been incapable of moving the elongated body, continued disuse finally caused total loss of limbs. Probably the most frequently cited example of Lamarckian explanation is that of the long neck and high shoulders of the giraffe. Proceeding from the fact that giraffes browse on the leaves of trees and the assumption that they once had the body proportions of a typical mammal, Lamarck deduced that as the giraffes strained to reach leaves higher and higher on the trees, their shoulders grew higher and their necks longer in response. This increase was passed on and enhanced from generation to generation finally producing the giraffe in its contemporary form.

It is unfortunate that Lamarck had to give such unnatural explanations as these. His illustrations gave critics an opportunity for ridicule that they used to completely mask the sound speculation contained in his theory. It is indeed unfortunate again that Lamarck is best known for his incorrect theory of heredity by acquired characteristics, whereas the value of his pioneering work in the formulation of the first comprehensive biological explanation of evolutionary processes is often greatly underestimated. The processes of heredity were unknown in Lamarck's time, and literally everyone speculated in this area; even Charles Darwin had a totally erroneous concept of heredity. It should also be noted that Lamarck's theory was limited by the scientific knowledge of his day, and his theory was in fact consistent with that knowledge. Lamarck met his responsibilities for cultural advancement. He did not accept the status quo of special creation; for this he is due the respect afforded the other great philosophers of nature.

Darwin and the *Beagle*

Charles Darwin (1809-1882) (Fig. 2-2) was born on February 12, the same day and year as Abraham Lincoln. His father, a well-known physician, expected Charles to follow in his footsteps. However, Charles abandoned the study of medicine while at the University of Edinburgh and subsequently found the

Fig. 2-2. Charles Darwin at the age of 40 (1849). This point in time represents the midpoint of the nearly 25 years Darwin spent gathering evidence and formulating his theory of natural selection. (Culver Pictures)

study of law equally uninteresting. Taking the third respectable profession as the final alternative, his father sent him to Cambridge University to become a clergyman. He completed his degree there but was not ordained. Darwin loved Cambridge most for his association with the sporting crowd, the collection of beetles and, particularly, for his friendship with the Reverend John Stevens Henslow, professor of botany. As a student, Darwin studied the Greeks and encountered the views of Thales, Empedocles, and Aristotle. He also read the views of his grandfather and the comprehensive theory of Lamarck. With this background, it is apparent that Darwin had at least been exposed to much of the past development of evolutionary thought.

Darwin had been interested in nature all his life, and when an opportunity arose to join the H.M.S. *Beagle* as its unpaid naturalist on an extensive exploratory expedition around the world, he managed to overcome the initial objections of his father and finally got him to finance the trip. The voyage lasted for 5 years, from 1831 to 1836 (Fig. 2-3, *A*). The ship had hardly left port when Darwin became seasick, a condition that plagued him throughout the 5 years except for the respites afforded by his explorations on land.

During his land explorations, Darwin collected large quantities of data and recorded many notebooks full of observations. He found cliffs on the Cape Verde Islands embedded with seashells indicating that the entire coast, which was 45 feet above the sea, was at one time the sea bottom. On the Argentine pampas, he found the fossil bones of extinct mastodons and gigantic armadillo-like animals and noted how closely these fossils resembled those known from North America. On the island of Chiloe he witnessed an earthquake and afterwards found dying mussels clinging to the rocks that had been raised 10 feet above the sea level. While crossing the Andes, Darwin found seashells embedded in rock 13,000 feet above sea level. Also in the mountains he realized that the water rushing down the slopes could slowly erode the surface.

All this geology was especially interesting because Darwin had been studying the *Principles of Geology* written by *Charles Lyell* (1797-1875) which his friend Henslow had given him to take on the voyage. In this text Lyell expressed his then revolutionary ideas about the forces of geology. He proposed that the popular belief that catastrophies and cataclysmic events were responsible for producing the geological features of the earth was not correct, but that the action of wind, rain, earthquakes, volcanoes, and other natural forces had altered the earth's surface slowly over time and that the same forces that acted in the past continue to act in the present. This is known as the concept of *uniformitarianism*. In his interpretation of geological strata, Lyell mounted evidence to show that the earth was tremendously older than the then generally accepted age of about 6,000 years. Darwin's firsthand experience convinced him that Lyell was correct, and the uniformitarian idea of progressive change over long periods of time, caused by the gradual action of natural forces, played a key role in the later formulation of his evolutionary theory.

The most impressive biological observations of the voyage were made during Darwin's 5-week exploration of the Galapagos Islands, a volcanic archipelago

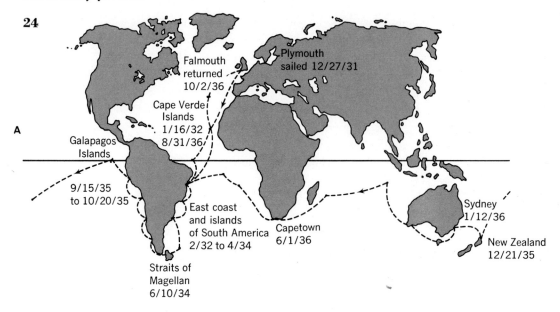

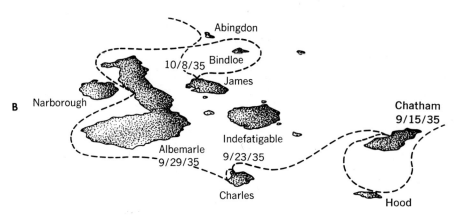

Fig. 2-3. The voyage of the H.M.S *Beagle* (1831-1836). **A,** Route of the entire voyage with stopping points and date of landing indicated. **B,** Enlargement of the Galapagos Islands showing the route and landing dates.

off the coast of Ecuador (Fig. 2-3, *B*). First came his observations on the giant tortoises that gave their Spanish name, *galapago*, to the islands. These huge tortoises weighed up to 500 pounds, and nothing like them existed anywhere else in the world. Ten of fourteen major islands possessed distinct varieties of the tortoise that were easily distinguished by the shape and pattern of their carapace (Fig. 2-4). Equally as interesting were the lizards, or iguanas. One species represented the only seagoing lizard in the world. These marine iguanas grew to 3 or 4 feet in length and weighed about 20 pounds. They had black

Fig. 2-4. Galapagos tortoise. These reptiles roamed at least ten islands before sea captains wiped out entire herds to provision their ships. Today they survive in fair numbers only on Indefatigable and Albemarle Islands. (San Diego Zoo photograph by Ron Garrison)

armor, and when upset they squirted vapor from their nostrils like a dragon. Despite their ferocious appearance, they were harmless vegetarians and were extremely gregarious. In some places large numbers literally covered the rocky coast as they bathed in the sun all day. Another species, very similar to the marine iguanas but brown in color, was completely terrestrial. These land iguanas also congregated in very large numbers. On the sandy beach of one island, Darwin noted, they were so numerous that he could hardly find a spot free of their burrows on which to pitch his tent. Together these two species constituted an entirely new genus found nowhere else on earth. Considering the remarkable reptiles of the Galapagos, Darwin observed that it was not the number of different species that was so great but the number of individuals of each species. Since the only mammal indigenous to the islands was a mouse confined to Chatham Island, he found it fascinating that reptiles were still predominant in this isolated spot of the world. He began to wonder why.

The islands also supported very large bird populations. Twenty-six species of land birds were collected, and all but one were peculiar to the Galapagos. Thirteen of the species were closely related finches that had such diversified habits that they were practically incomparable to finches anywhere else in the world. One of the most unusual is the tool-using finch. There are no woodpeckers on the Galapagos, and this finch has filled the woodpecker's usual niche. Unlike the woodpecker, however, the finch has neither a chisel-like beak nor a long barbed tongue for extracting insects from the bark of trees. To make up for this, the finch uses a twig or cactus spine as a probe to prod its dinner from the bark. Some species became vegetarians and adapted to diets of flowers, fruits, and seeds, while others remained insect feeders but diverged in habitat to reduce competition by eating different kinds of insects (Fig. 2-5).

Recently, another very unusual feeding habit among these finches was discovered by Robert I. Bowman of San Francisco State College. He found that the sharp-beaked ground finch lands on a booby and feeds by stabbing the soft skin at the base of the larger bird's wing feathers and drinking the oozing blood. This is the only known bird species that feeds on blood.

Darwin collected fifteen species of sea fish around the islands, and all were new species. Along the shores, land shells were represented by sixteen species, of which fifteen were previously unknown. Nearly all the insects found belonged to new species, and of 175 flowering plant species 100 were new and confined to the islands. Furthermore, Darwin noted that by far the most remarkable feature of the Galapagos archipelago was that the different islands were each largely inhabited by a different set of species. He wrote in his notebook that the distribution of organisms on these islands would not have been so interesting if, for instance, one island had a mocking-thrush and another had some quite distinct genus (well-defined group), or if one island had its genus of lizard and a second island its distinct genus, or if different islands had totally different genera accounting for the variation in distribution. But this was not the case; instead, several of the islands possessed species of mocking-thrush, finch, tortoise, and plants that were clearly closely related and obviously occupying analogous situations on the islands. This struck Darwin with wonder; if species were fixed entities, why were slightly different but similar ones created for each island?

Darwinism

After his return from the voyage, Darwin began to wade through the voluminous notes and specimens he had collected. By this time the problem of the

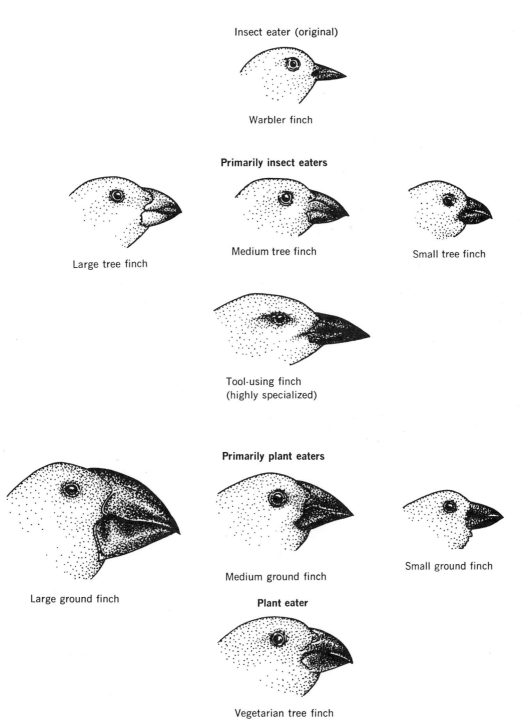

Fig. 2-5. Galapagos finches. The finches on the Galapagos radiated to widely diversified habits, yet they all originated from a common mainland stock that first occupied the islands.

mutability of species had begun to haunt him. All his evidence seemed to indicate that species could gradually become modified. The fossils that were similar to, yet different from, existing animals, the replacement of one species by another closely allied form on a north-south axis down the South American continent, and the remarkable distribution of animals and plants on the Galapagos Islands could only be explained on the supposition that species could gradually change over time. Thus, in 1837, Darwin began the collection of any and all facts that had any bearing on variation among animals and plants in nature and under domestication. He noted that he worked according to true Baconian principle utilizing the inductive method without any preformed theory.

Among the domesticated animals Darwin studied were the pigeons. He found the diversity of the breeds to be astonishing (Fig. 2-6). There were *carriers* with well-developed carunculated skin around their heads, elongated eyelids, large nostrils, and long beaks. The *tumblers* had short faces and a habit of flying at great height and tumbling head over heels. The *runts* were large birds with long massive beaks. The *barbs* were much like carriers but had very short and broad beaks. The *pouters* had elongated bodies, wings, and legs, and well-developed crops that they could inflate to enormous size. The *turbits* had short conical beaks and a line of reversed feathers down their breast. The *Jacobins* had elongated wing and tail feathers, and the reversal of feathers along the back of their neck was so extreme that it formed a hood. The *fantails* had thirty to forty tail feathers, instead of the normal fourteen to fifteen, which they carried so erectly that in certain birds the head and tail touched. Finally, the *trumpeters* and *laughers*, as their names imply, made vocalizations that were quite different from the normal coo of other breeds. The success that breeders had had with pigeons and other domesticated animals and plants led Darwin to perceive *artificial selection* as the force by which the gradual modification of the various breeds was accomplished. He became convinced that the different breeds of pigeons had descended from the common rock pigeon and that all domesticated animals and plants had similar histories.

Darwin commented that the most remarkable feature of domesticated animals and plants was their adaptation, not for their own benefit, but to man's use or fancy. He pointed out that these breeds were not suddenly created as perfect and useful as man wanted them; the history of animal and plant breeding confirmed this fact. The key to man's success, he said, was the power of accumulative selection by which man could take the variations found in nature and add them up over successive generations in directions useful to him. In this sense, man created the domesticated breeds himself.

If man could modify species artificially by a directed additive process, then how might change be brought about in nature? What would be the directing force? Darwin had these questions on his mind when one day in October, 1838, he happened to read *An Essay on the Principle of Population* by *Thomas Malthus* (1766-1834). Malthus pointed out that any natural population of animals or plants, if unchecked, has the capacity to increase its numbers in a geo-

Fig. 2-6. Artificial selection in pigeons. Depicted are some of the pigeon breeds that led Darwin to the conclusion that the directed accumulation of variations found in nature was the key to artificial selection. **A,** The common rock pigeon, **B,** trumpeter, **C,** fantail, **D,** barb, **E,** pouter, **F,** runt, **G,** carrier, **H,** Jacobin, and **I,** turbit.

metric ratio, while food production only increases in an arithmetic ratio. As a result, he said, the difficulty in obtaining food would constitute a strong and constantly operating check on the size of a population. From this Darwin surmised that since more individuals are produced than can possibly survive, there must be a "struggle for existence" either among individuals of the same species, or between species, or with the environment, or any combination of the three. From the large potential number of progeny of any species, only those best adapted to the prevailing conditions of life would be likely to survive and leave progeny of their own. Thus Darwin formulated the cornerstone of his theory. It was *nature* that selected in the wild, rather than man, and as the environmental conditions changed over time, a gradual modification of species would result from the accumulation of those variations that provided the best adaptation to the new conditions.

Darwin also realized that for change to take place, any adaptive variations had to be passed on to future generations; that is, the adaptive characters had to be hereditary. Although Darwin's theory of heredity was incorrect, just as Lamarck's was, he placed the genetic mechanism in the proper context within his theory. Whereas Lamarck's acquired characteristics "became" hereditary and thereby modified the species, Darwin simply required that the adaptive variation that enabled any particular individual to succeed in the struggle for existence have a hereditary basis. It was obvious to him that much of the variation in nature was hereditary. Therefore *natural selection* was the force that determined which characters would be propagated by virtue of their possessors reproducing, and which would be lost, or at least not selected by virtue of their possessors not reproducing (or reproducing at a lower rate). The key here was the transmission of the adaptive traits to future generations. When Darwin spoke of the principle of preservation as the "survival of the fittest," by the fittest he meant those who reproduced most proficiently and passed their adaptive traits on to their progeny, thereby enhancing their chances of survival in the next generation. Hence, natural selection is essentially differential reproduction.

In contrast to the Lamarkian explanation, the evolution of the long neck and high shoulders of the giraffe was explained by Darwin through the action of natural selection. He pointed out that slight proportional differences (variations) in neck length, shoulder height, and so forth would be expected in a population of giraffes. In addition, the giraffes had reduced their competition with other grazing animals by adapting to the habit of browsing on higher growing vegetation than their competitors. If the food supply became inadequate (for example, because of an environmental change involving the growth of fewer trees), the proportional variations in the stature of the giraffe would have served as the adaptations upon which natural selection operated. Accordingly, any individuals that were able to reach even an inch or two higher than others to obtain food could have browsed beyond the reach of competitors. These would have survived in greater numbers than the others and would have intercrossed and left collectively more offspring than the less favored individuals. Individual progeny of the favored types that inherited their parents' adaptive traits or even varied beyond the range of the parental types would have had an increased chance for survival and reproduction in the next and later generations. In this manner the peculiarities of the giraffe's stature could have gradually evolved.

A generalized and very oversimplified model might be helpful at this point. This model is simply designed to impart the "idea" of a selective process operating on a population with high reproductive potential whose growth is held in check by an external limiting factor such as food supply. Only two alternatives, A and B, will be considered, with A having the selective advantage over B. These could be imagined as distinct types of individuals, such as tall and short giraffes, or as single traits, such as long and short neck length. It will be assumed that a constant population size of 1,000 individuals is maintained each generation even though reproductive potential is geometric and that 500 individuals of each type are initially present. Holding the population size at 1,000 and assuming that each individual type tends to produce two like itself each generation, the model will generate 2,000 potential progeny each generation in proportions twice that of the parental ratio. Thus, of the 2,000 potential progeny, half must be eliminated to hold the population size at 1,000. If type A were equal to type B in adaptive characters, survival would be more or less random, and the ratio of A to B would not be expected to change significantly from generation to generation. However, in the model type A has a selective advantage over type B. This advantage will be expressed by having potential A progeny survive at an excess of 10% beyond that which would be expected by simple proportionality related to the parental ratio. Since the population size remains at 1,000, the number of type B survivors among the potential progeny each generation will be 1,000 minus the number of type A survivors. Each succeeding generation will be produced in this manner. Table 2-1 follows the schematism of the model and demonstrates that in eight generations type A would completely take over

Table 2-1. Selective processes in favor of Type A when the selective advantage is 10% in excess of random proportional survival based on the parental ratio*

Generation	Number of potential progeny		Number of selected survivors		Total population size
	A	B	A	B	
0	—	—	500	500	1,000
1	1,000	1,000	550	450	1,000
2	1,100	900	605	395	1,000
3	1,210	790	666	334	1,000
4	1,332	668	733	267	1,000
5	1,466	534	806	194	1,000
6	1,612	388	887	113	1,000
7	1,774	226	976	24	1,000
8	1,952	48	1,000	0	1,000

*See text for further explanation.

in the population. It should be recognized that this model does not consider mating combinations or genetic mechanisms that make the situation much more complex and usually slow down the process tremendously; nevertheless, the process occurs generally as the exaggerated model indicates.

Darwin had worked out his basic theory in 1838, but it was 20 years before he expressed it publicly. He realized how revolutionary his theory was and attempted to build up an impeccable store of supporting evidence lest he be subjected to the same kind of ridicule Lamarck had suffered for his poorly supported examples. Even in 1858 Darwin was not ready to announce his theory, but then he received an unexpected stimulus. For in that year the British naturalist *Alfred Russel Wallace* (1823-1913) sent him a manuscript with a request for comments. In the manuscript Wallace developed essentially the same theory of natural selection as Darwin had been working on for 20 years. Interestingly enough, Wallace had received his impression of the modifiability of species while on a zoological exploration of the Malay Archipelago, just as Darwin had received his on the Galapagos Archipelago. In addition, the idea of the survival of the fittest came to Wallace as he recalled the writing of Malthus on population. Through mutual friends, it was arranged for Darwin and Wallace to present their ideas simultaneously to the Linnean Society of London on July 1, 1858. Darwin had written a draft of his theory in 1839 and had reiterated it in a letter to the American naturalist Asa Gray in 1857. Because of Darwin's much earlier original idea and his much more substantial supporting evidence, Wallace was quite willing to grant Darwin the priority of the theory. Nevertheless, Wallace conceived his theory independently as a logical extension of the cultural achievements of the past coupled with his own observations and data. Such essentially simultaneous, yet independent, development of new concepts has occurred quite often in all disciplines, illustrating the fact that cultural evolution is a continuous human process that generates crucial points in time when accumulated knowledge reaches a "critical mass" that becomes a threshold for major advancements. However, it takes educated individuals meeting their responsibility before the new relationships can be made which provide for the breakthrough. That this occurs independently among two or more individuals at the same time is not really an unexpected coincidence.

At the urging of his friends, Darwin put his theory, which he still considered only an abstract, into book form, and on November 24, 1859, the first edition of *The Origin of Species* was published; its 1,250 copies sold out the first day. This book has proved to be one of the most controversial and influential ever to be produced during man's entire cultural evolution. Following is a paragraph from the summary of Chapter IV, "Natural Selection; or the Survival of the Fittest," in which Darwin states the essentials of his theory:

> If under changing conditions of life organic beings present individual differences in almost every part of their structure, and this cannot be disputed; if there be, owing to their geometrical rate of increase, a severe struggle for life at some age, season, or year, and this certainly cannot be disputed; then, considering the infinite complexity of the

relations of all organic beings to each other and to their conditions of life, causing an infinite diversity in structure, constitution, and habits, to be advantageous to them, it would be a most extraordinary fact if no variations had ever occurred useful to each being's own welfare, in the same manner as so many variations have occurred useful to man. But if variations useful to any organic being ever do occur, assuredly individuals thus characterised will have the best chance of being preserved in the struggle for life; and from the strong principle of inheritance, these will tend to produce offspring similarly characterised. This principle of preservation, or the survival of the fittest, I have called Natural Selection. It leads to the improvement of each creature in relation to its organic and inorganic conditions of life.

In retrospect, Darwin's theory may be summarized by five basic tenets.

1. Many more individuals are born each generation than will be able to survive and reproduce. (The writing of Thomas Malthus influenced Darwin on this point and provided him with the idea of a struggle for existence.)

2. There is natural variation among individuals of the same species. (This had been noted by naturalists from Aristotle through Lamarck; however, Darwin's own observations made the greatest impression of this fact upon him.)

3. Individuals with certain characteristics have a better chance of surviving and reproducing than others with less favorable ones. (This is the concept of the survival of the fittest through favorable adaptations to the conditions of life, which Darwin deduced from the first two tenets but which had been alluded to by Empedocles and even Erasmus Darwin.)

4. Many of the favorable adaptations are hereditary and are passed on to the progeny of future generations. (Darwin, like Lamarck, believed in an incorrect theory of heredity; however, he interpreted the process in the proper context.)

5. Gradual modification of species could have occurred over the long periods of geological time through additive processes occurring in the past in the same manner as they are occurring in the present. (Charles Lyell's geological interpretations provided knowledge of the long span of time necessary for evolution to proceed and supplied Darwin with the concept of uniformitarianism as a modus operandi for the processes of biological evolution.)

Neither Darwin nor Wallace pulled his theory out of thin air—the development of the theory of biological evolution was the product of cultural evolution. Both evolutionary processes continue to operate today, and man must understand both, for each is, or may be, under his control. They must be reconciled to proceed harmoniously or, as far as *Homo sapiens* is concerned, both may stop altogether.

Neo-Darwinism

Charles Darwin's theory did not represent an end point in cultural advancement; rather, it was a milestone marking the path in the proper direction toward man's understanding of himself and his place in nature. Darwin's theories have now been supported and embellished with evidence from nearly every branch of the biological sciences. The Austrian monk *Gregor Mendel* (1822-

1884) elucidated the principles of inheritance in 1865, but his work was essentially unknown until 1900 after which time its elaboration and application to evolutionary theory has been highly productive. Mendel was a contemporary of Darwin and had read and commented on Darwin's theory. It is also known that Darwin had seen references to Mendel's work, but these two great men of biology never linked their ideas. That great union became the responsibility of others. Section three of this book will deal with Mendel's principles, especially with respect to man.

In 1930, the principles of heredity were coupled with evolutionary theory and supported by mathematical explanations in a comprehensive manner for the first time by *Ronald A. Fisher* (1890-1962) in his classic book *The Genetical Theory of Natural Selection*. In this work he demonstrated quite clearly that only natural selection could be the directive agency of progressive adaptation and that Mendelian inheritance was the mechanism by which natural variation was maintained and generated—a point which Darwin was unable to explain. Fisher, along with *J. B. S. Haldane* (1892-1964) and, among contemporaries, *Sewall Wright*, established a firm mathematical basis of evolutionary theory.

The researches of *Theodosius Dobzhansky* with species of the fruit fly, genus *Drosophila*, collectively form much of the experimental basis in support of evolution by natural selection. His book, *Genetics and the Origin of Species*, published in its first edition in 1937, represents the definitive work from which the active area of experimental evolution has developed. Important contributions in the experimental area have also been made by *Edgar Anderson* (1897-1968) on the hybridization of natural species, *G. Leddyard Stebbins* on evolution in the plant kingdom, and *Ernst Mayr* on evolution in the animal kingdom. The work of *George G. Simpson* also contributed greatly to the embellishment of evolutionary theory when he melded paleontologic data into Darwinian interpretation.

Neo-Darwinism represents the modern synthesis of evolutionary theory based on Darwin's original concept but supplemented by evidence from virtually every branch of science. For example, the most recent developments have incorporated biochemical analytic methods of molecular separation according to electrostatic characteristics, a technique known as *electrophoresis*. This technique allows for the analysis of proteins and enzymes, the direct products of the genetic material, and has opened a whole new level of natural variation to analysis with a concomitant re-evaluation of the action of natural selection. This new discovery illustrates the fact that the search for understanding is never-ending; it is always subject to refinement, improvement and advancement. Each generation of man has had the responsibility to perpetuate and direct cultural evolution.

THE ACTION OF EVOLUTION

Darwin correctly recognized the conditions necessary for evolution to occur and explained the general action of the process involving natural variation, the hereditary transmission of adaptive traits, and the differential reproduction or fitness of the selected variants. But it remained for the neo-Darwinists to work out the details of evolutionary action and define the processes or forces that bring it about. According to the modern synthesis, evolution may be simply defined as a directional change in the genetic composition of a population. The local interbreeding population, known as a *Mendelian population,* is considered to be the primary unit of evolution. Individuals do not evolve in a biological sense. An individual has a particular genetic make-up, called its *genotype,* which determines its characteristics or actual appearance, called its *phenotype.* Natural selection then works on the array of phenotypes produced, and those best adapted contribute their genetic material to the succeeding generation. It is the character of the population over generations that changes, not any particular individual. The genetic mechanisms that operate at the population level will be considered in Chapter 13.

The action of evolution occurs at three levels within which various forces operate. The first level of action is that involved in the generation and maintenance of variation. *Recurrent germinal mutations* supply the raw material for evolution. A *mutation* is an alteration of the hereditary instructions which may occur anywhere in the body but is only transmitted if it occurs in the germ line, eggs or sperm; all changes occurring in the nonreproductive cells are called *somatic mutations* and have no evolutionary consequences. In general, mutations are not unique events; they usually occur at a particular rate and are therefore recurrent. Given the long evolutionary history of most living species, nearly all possible mutations have probably occurred repeatedly and have been subjected to selection that either rejected them or incorporated them into the population as part of the "normal" genetic material. The genetic material found in any population, known as its *gene pool,* is generally that which is best adapted to the current conditions of life; all other mutations that have been tested and rejected are therefore usually deleterious when they arise again. When the environmental conditions change, the relative adaptedness of these recurrent mutations may also change. Mutations not only generate variability, but they also form the basis for its maintenance. Not all poorly adapted mutations are rejected immediately; the laws of heredity that will be discussed later in the text explain why this is so. Mutations are not acted upon by natural selection in isolation but as components of the entire genetic constitution of their possessors. The mechanics of cellular reproduction, which will be discussed in Chapter 4, shuffle the genetic material into innumerable combinations upon which natural selection can act. In this way, the force of *recombination* maintains a high degree of variability in most populations. Other sources that generate variation include *immigration,* which introduces genetic material from one local population into another of the same species, and *hybridization,* which involves the occasional interbreeding of different species or Mendelian populations, thereby producing combinations from two separate gene pools.

The second level of evolutionary action is that at which the genetic structure of the population is molded to fit the environmental conditions. This is generally accomplished by *natural selection.* In a well-adapted population, an array of genotypes is produced. These genotypes are expressed as phenotypes that usually form a normal distribution with those having the higher adaptive values occurring in greatest frequency. The predominant, favored phenotypes constitute the *adaptive norm* of the population, and any recurrent mutations that produce phenotypes outside this range will be selected against. When the adaptive norm remains constant for many generations and the population is relatively unchanged, the molding force is called *stabilizing selection.* If the environment is changing (and it always is), the adaptive characteristics of the population will adjust to the change by shifting the adaptive norm. This is accomplished by retesting all recurrent mutations and incorporating or increasing the frequency of those that contribute to new adaptive phenotypes. This is known as *directional selection;* it produces the progressive change that typifies the general concept of evolution by natural selection. A third form of selection is *disruptive* or *diversifying selection* which operates on two or more different sets of phenotypes adapted to two or more different aspects of the environment. This form of selection tends to split a population into subpopulations each of which might, theoretically at least, evolve along separate lines and eventually result in distinctive species. These three forms of selection are illustrated in Fig. 2-7. A second molding force operates in small populations where very few parents produce each succeeding generation. In such small gene pools there may be a chance loss

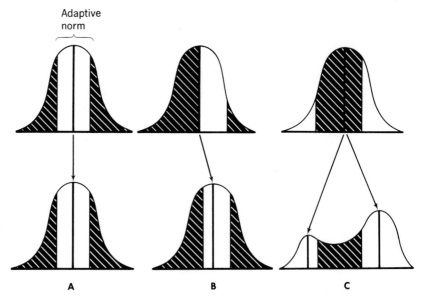

Fig. 2-7. The three basic types of selection: **A,** stabilizing, **B,** directional, and **C,** diversifying. The top row in each case illustrates a normally distributed array of phenotypes making up a population; the unshaded area indicates that portion of the population which is being selected for. The bottom row illustrates the effect in subsequent generations of the respective types of selection on the population structure. The arrows indicate the direction in which the adaptive norm shifts over time.

or fixation of a particular mutation or genetic trait irrespective of the effect of selection simply due to random sampling error in the production of a limited number of gentoypes. These chance effects occurring in small populations constitute the phenomenon known as *genetic drift.*

The third level of evolutionary action concerns the fixation of the adaptive characteristics of a species. This is accomplished through *reproductive isolation* between species. The biological concept of species is based on the criterion of reproductive isolation—good biological species do not freely interbreed. The mechanisms that isolate one species reproductively from others are perhaps the most important set of attributes a species possesses because they make it possible for selection to mold an integrated adaptive norm for each species. This would be impossible if interbreeding between all sorts of related forms took place because the variation generated each generation would be so diverse that little directed progress would be possible. For each species to remain adapted to its niche in nature, or successfully adapt to a change in the environment, its gene pool must remain under the control of the selective forces; reproductive isolating mechanisms keep the gene pool manageable. A classification of the common forms of isolating mechanisms found in nature is given in Table 2-2.

The long-term action of evolution results in the formation of new species adapted to a particular environmental niche and reproductively isolated from all other species. There are three general modes of speciation through which the diversity of past and present living forms have evolved.

When a species persist for a very long time, subtle changes are likely to take place as slow adaptation to environmental change occurs. Viewed from the end

Table 2-2. Classification of the common forms of reproductive isolating mechanisms found in nature

Mechanism	Description
Premating	
Ecological isolation	Potential mates live in different habitats and do not meet.
Temporal isolation	Potential mates have different breeding seasons.
Ethological isolation	Potential mates do not breed because of incompatibilities in behavior.
Mechanical isolation	Potential mates cannot breed because of structural differences in genitalia.
Postmating	
Gametic mortality	Antigenic reaction incapacitates gametes or otherwise prevents the union of egg and sperm.
Hybrid inviability	Egg and sperm unite, but development is arrested or proceeds abnormally.
Hybrid inferiority	Hybrid individuals survive but are intermediate to parental types and cannot compete successfully.
Hybrid sterility	Hybrid individuals survive but have poorly developed or nonfunctional reproductive system.
Hybrid breakdown	Hybrid individuals survive and mate successfully, but their offspring are inviable, inferior, or sterile.

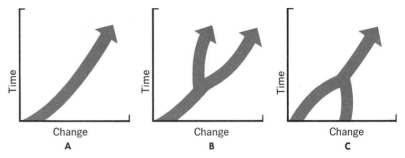

Fig. 2-8. The three general modes of speciation: **A,** phyletic, **B,** divergent, and **C,** convergent.

points of a long period of time, the representative populations may be defined as different species. The past forms, represented in the fossil record, are often designated as one species, a paleospecies, while the modern species has a separate designation. Such gradual change over time is called *phyletic speciation* and is the product of directional selection which, viewed over any relatively short time period, appears to be stabilizing selection. Phyletic speciation does not increase the number of species in existence at any one time, but it does account for much of the diversity observed between contemporary and ancient forms.

The common mode of speciation responsible for increasing the number of species and enhancing the diversity of nature is known as *primary* or *divergent speciation*. This involves the splitting or branching of one species into two or more new species, each reproductively isolated from the others. This is usually accomplished through a fragmentation of a single species into geographically isolated subpopulations that diverge genetically as directional selection works on each independently. If reproductive isolating mechanisms develop, as a byproduct of divergence enhanced by natural selection, the species criterion is established and each population warrants separate species status.

Finally, if two populations normally isolated from each other establish contact and hybridize, the gene pools of each may merge into one. Under these circumstances either directional or stabilizing selection could operate to establish a new adaptive norm. Such hybridization is most likely to occur when ecological isolation breaks down; this very often occurs when man disturbs the environment through his activities. This process is known as *secondary* or *convergent speciation*. It reduces the total number of species if the two original forms completely hybridized. On the other hand, if the original forms remain intact in part of their range and hybridization occurs in a limited area, an increase in the number of species may ensue.

Fig. 2-8 illustrates the three modes of speciation diagrammatically. It should be noted that within any extensive evolutionary lineage, all three modes of speciation might be involved at various times. This is what produces the complex branching in the so-called evolutionary tree used to trace the relationships of organisms over time. We will trace our own ancestry in this manner in the next chapter.

3

The descent of man

The processes of evolution discussed in the previous chapter are important biological concepts to understand because, after all, we are the result of those processes. It is often remarked that man came from the apes or monkeys, but usually in a joking or, even worse, a derogatory way. Why is this remark so often made lightly, while very seldom it is meant in a serious vein? Are people really offended by the implications of a monkey ancestor? A clear understanding of the evolutionary history of man will allow us to put monkey, ape, and man in proper perspective.

In attempts to reconcile the general public to the facts of man's evolution, it has been common to say that men were not descended from monkeys but that both men and monkeys were descended from a common ancestor. This statement is true, but as George G. Simpson has pointed out, there has been too much caution on this subject. Many biologists have apologetically emphasized that man is not a descendant of any living monkey or ape but of some ancestral form. But what was this ancestral form? We shall see later in this chapter that as man's ancestors are traced backward in time, we most certainly reach forms that would undoubtedly be called apes or monkeys by anyone who saw them. Therefore, using ape and monkey in vernacular terms, man's ancestors *were* monkeys or apes or, more likely, both successively. It is simply pusillanimous, if not dishonest, to say otherwise.

The foundation for the placement of man in the evolutionary scheme was laid by Charles Darwin with the publication of *The Descent of Man* in 1871. This book stands second only to *The Origin of Species* as a classic in evolutionary literature and in profound effect on society. It should be noted that Darwin based *The Descent of Man* on abundant evidence and examples from the animal kingdom but without a single subhuman fossil to support his ideas. He admitted that his theory was speculative but thought it undoubtedly true in concept even though some of the details of his views might in time prove to be erroneous. The main conclusion of this exhaustive work is quoted in a paragraph from the final chapter:

> That man is descended from some lowly organised form, will, I regret to think, be highly distasteful to many. But there can hardly be a doubt that we are descended from barbarians. The astonishment which I felt on first seeing a party of Fuegians on a wild and

broken shore [while visiting Tierra del Fuego on the *Beagle* voyage] will never be forgotten by me, for the reflection at once rushed into my mind—such were our ancestors. These men were absolutely naked and bedaubed with paint, their long hair was tangled, their mouths frothed with excitement, and their expression was wild, startled, and distrustful. They possessed hardly any arts, and like wild animals lived on what they could catch; they had no government, and were merciless to every one not of their own small tribe. He who has seen a savage in his native land will not feel much shame, if forced to acknowledge that the blood of some more humble creature flows in his veins. For my part I would as soon be descended from that heroic little monkey, who braved his dreaded enemy in order to save the life of his keeper, or from that old baboon, who descending from the mountains, carried away in triumph his young comrade from a crowd of astonished dogs—as from a savage who delights to torture his enemies, offers up bloody sacrifices, practices infanticide without remorse, treats his wives like slaves, knows no decency, and is haunted by the grossest superstitions.

THE CLASSIFICATION OF MAN

The first attempt to fit man into a taxonomic position in the animal kingdom based on evolutionary relationships according to Darwinian theory was made by the German zoologist *Ernst Haeckel* (1834-1919) who was a contemporary of Charles Darwin and a strong proponent of his theory. Haeckel discussed the genealogy of man and attempted to outline his ancestry in his book *Naturliche Schopfungsgeschichte* published in 1868. Although his genealogical chart, starting with a blob of protoplasm and ending with a modern aborigine, was wrong in fact, its concept was correct (Fig. 3-1). Henceforth, man was viewed as a part

Table 3-1. The classification of modern man

Taxonomic category	Scientific designation	General description
Kingdom	Animalia	Living beings typically capable of spontaneous movement and rapid motor response to stimulation, as distinguished from plants
Phylum	Chordata	Having embryonic elastic rod (notochord) that serves as an internal support
Subphylum	Vertebrata	A supporting backbone (vertebral column) present in the adults
Superclass	Tetrapoda	Four-limbed
Class	Mammalia	Young suckled from the mammary glands of the female
Subclass	Theria	Young born alive from within the body of the female, as opposed to hatching from eggs
Infraclass	Eutheria	Placental mammals; development of young to a relatively mature state in the uterus before birth
Order	Primates	Rather generalized mammals; detailed description given in text
Suborder	Anthropoidea	Resembling man; erect or semierect posture
Infraorder	Catarrhini	Nostrils close together; tail, if present, never prehensile
Superfamily	Hominoidea	Erect and lacking a tail
Family	Hominidae	Terrestrial with bipedal locomotion and a relatively large brain size
Genus	*Homo*	Distinguished in behavior by extensive use of tools, communication by speech, and cultural evolution
Species	*sapiens*	Modern "wise" man

of rather than apart from nature, and the scientific investigation into man's biological past had been launched.

Zoological classification incorporating man into the system has been greatly refined since Haeckel's first attempt. The standard nomenclature of classification proceeds from kingdom through phylum, class, order, family, genus, and species. In addition, these categories may be subdivided into subphylum, superclass, subclass, infraclass and so forth. The genus and species designation constitutes the

THE MODERN THEORY OF THE DESCENT OF MAN.

Fig. 3-1. The ancestry of man as traced by Ernest Haeckel over 100 years ago. This was one of the first attempts to place man in the scheme of organic evolution. (Culver Pictures)

scientific name of an organism, with race, variety, or subspecies names sometimes tacked onto the end. The complete classification of man is given in Table
3-1. Add your own name to the end, and your personal taxonomy is complete.
But what is the rationale for fitting man so neatly into a classification system?

THE EVOLUTION OF MAMMALS

It should be quite apparent that man is a mammal; therefore, we shall begin
to trace his descent from this point. The mammals evolved from an ancestral
reptilian stock that began to develop mammal-like characteristics about 300 million years ago. When the transition to mammals was finally accomplished is
impossible to say; there was no sudden transformation of a reptile into a mammal. The earliest known fossils identified as primitive mammals date back some
180 million years. Some of the major differences between mammals and reptiles
include the following:

1. Mammals are *homoiothermic* (warm-blooded); that is, they are able to
maintain a high and constant body temperature and metabolic rate despite environmental temperature fluctuations. This permits the mammal to remain very
active throughout a wide range of environmental conditions. Reptiles, on the
other hand, are *poikilothermic* (cold-blooded); that is, their body temperature
is determined by the environmental temperature. Since the rate of metabolism is
very closely tied to temperature, the resultant activity of poikilothermic animals
is radically affected by temperature change; for example, they become very inactive in cold weather. The term cold-blooded is misleading, however, because
under very warm conditions a poikilothermic animal may have a higher body
temperature than a homoiothermic one. The advantage of homoiothermy is the
constancy of activity it allows over a wide range of external temperatures.

2. Mammals possess a *diaphragm* that provides a more efficient breathing
mechanism than that in reptiles which relies on rib and abdominal muscles only.

3. The *limbs* of mammals are oriented under the body providing much better
support than those of reptiles which are placed laterally.

4. The mammalian *lower jaw* is composed of a single bone, and the *teeth*
are differentiated into incisors, canines and molars; the reptilian lower jaw is
made up of six or more bones, and the teeth are not differentiated.

5. The *brain*, especially the cerebrum, is larger and more highly developed
in mammals than in any other class of vertebrates.

6. *Hair* forms an insulating coat over the body of mammals, while reptiles
are covered by a dry, scaly skin.

7. Finally, with few exceptions, *embryonic development* occurs in the uterus
of mammals, and the young are born alive; reptiles are egg-layers. After birth,
mammalian young are nourished on milk from the *mammary glands* of the
female.

Modern mammals are divided into three subclasses. The *monotremes* (Proto-
theria) are the only egg-laying mammals. They represent a peculiar offshoot,

which evidently diverged very early from the primitive mammalian stock. The only living monotremes are the spiny anteater and the duckbilled platypus of Australia and New Zealand. The main mammalian line split early in its history into the *marsupials* (Metatheria) and the *placentals* (Eutheria). In the marsupials, embryonic development in the uterus is terminated at a very early stage and is completed in an abdominal pouch of the mother where the young attach to the nipples of the mammary glands. The only surviving marsupial in North America is the opossum; most living forms, such as kangaroos, wombats, and koala bears, are found in Australia. Development proceeds much further in the uterus of the placentals because of the evolution of the *placenta*, which serves as a nutrient, respiratory, and excretory exchange organ for the fetus. About 95% of all living mammals are placental mammals; they include all the familiar animals that suckle their young such as dogs, cats, horses, cows, pigs, lions, tigers, monkeys, apes, and man.

The adaptive radiation of mammals

The mammals originated during the Age of Dinosaurs. The primitive placental mammals were small secretive creatures that fed primarily on insects. Most environmental niches at that time were occupied by the dominating reptiles, and the mammals were relatively unimportant members of the fauna. After the demise of the dinosaurs, the mammals underwent an explosive period of speciation as new forms evolved to occupy the many niches previously held by the reptiles. The evolution of a great diversity of species from one or a few basic prototypes is known as *adaptive radiation*. The mammals adapted to nearly every conceivable mode of life on earth. Most remained terrestrial, but some became aquatic (whales, porpoises, seals), some subterranean (moles), some arboreal (monkeys), and some evolved true flight (bats). The adaptive radiation of the mammals ushered in the Age of Mammals that began somewhat over 60 million years ago and persists today. The living placental mammals are divided into 15 orders, listed and described in Table 3-2.

Success in the trees

Man belongs to the order Primates, which diverged from all other mammals about 60 million years ago. The progenitor of the primate line was an inconspicuous little tree-dwelling insectivore that probably resembled the present-day tree-shrew of Asia. There are nearly 200 living species of primates forming a rather heterogeneous order. No single characteristic can be used to define the whole group, but most are still tree-dwellers, and it is the arboreal mode of life that provided the selection pressures to mold the primate adaptive characters into the blueprint for man.

In terrestrial quadrupedal mammals, the limbs have evolved toward greater stability at the expense of flexible movement. The legs are positioned close to-

Table 3-2. The fifteen orders of living placental mammals

Order	General description	Examples
Insectivora	Small animals with long pointed snouts and sharp molar teeth adapted for insect eating; feet with five toes and claws; most primitive placental mammal	Moles and shrews
Rodentia	Gnawing mammals with two pairs of chisel-like incisors; the largest order of mammals	Mice, rats, squirrels, chipmunks, lemmings, gophers, and beavers
Lagomorpha	Gnawing mammals with two pairs of enlarged incisors plus an extra pair of small upper incisors that lie behind the first pair	Rabbits and hares
Sirenia	Marine herbivores with forelimbs modified as paddles and hindlimbs lost; large horizontally flattened tail used in propulsion	Sea cows and manatees
Cetacea	Large marine mammals with forelimbs reduced to flippers and hindlimbs lost; large tail with horizontal flukes used in propulsion	Whales, dolphins, and porpoises
Chiroptera	Forelimbs modified to wings supported by four elongated fingers; wing membrane attached to hind legs; only mammals to evolve true flight	Bats
Proboscidea	Massive ungulates (hoofed mammals) that have retained five toes, each with a small hoof; two upper incisors elongated as tusks; nose and upper lip modified as a proboscis	Elephants
Hydracoidea	Small guinea pig–like herbivores with four toes on front and three on hind feet, each with a hoof	Coneys
Perissodactyla	Odd-toed ungulates with lateral toes reduced or lost; dentition highly modified for grinding plant food	Horses, zebras, and rhinoceroses
Artiodactyla	Even-toed ungulates with first toe lost and second and fifth toes reduced or lost; highly specialized herbivores	Cattle, sheep, goats, pigs, camels, deer, giraffes, and hippopotamuses
Edentata	Teeth reduced or lost; large claws on toes; tongue long and sticky for feeding on insects	Sloths, anteaters, and armadillos
Pholidota	Teeth lost and long tongue for insect feeding; body covered with overlapping horny plates	Pangolins
Tubulidentata	Teeth reduced and long tongue for insect feeding	Aardvarks
Carnivora	Flesh-eating mammals with large canines and certain molars and premolars modified as shearing teeth; claws well developed	Cats, dogs, bears, skunks, raccoons, seals, and walruses
Primates	Discussed in detail in text	Gibbons, monkeys, apes, and men

gether under the body, providing better support but restricting movement mainly to one plane (backward and foreward, but not sideways). As the fore-limbs evolved to this position, the clavicle (collar-bone) was greatly reduced or lost. In contrast, arboreal primates have retained a prominent clavicle because of its adaptive advantage as a brace for forelimbs attached to the side of the body instead of underneath. This lateral attachment allows greater freedom of movement for swinging from branch to branch in trees. A free-moving shoulder joint and semirotational elbow and wrist joints are also primate adaptations to a tree-swinging form of locomotion. A comparison of the forelimb of a horse with that of a monkey illustrates the basic differences between the forelimb of a terrestrial nonprimate and that of an aboreal primate (Fig. 3-2).

The adaptation of wrapping the digits around a branch instead of driving claws into it, as nearly all other tree-dwelling animals do when they climb, evolved slowly into the grasping ability that characterizes all living primates except the tree-shrew that still digs in with its claws. This grasping ability enabled primates to climb more safely to the extremities of small branches and thereby greatly expanded their food range. The strength and security of a grasping hand (or foot) reduced the risk of falling and also permitted the develop-

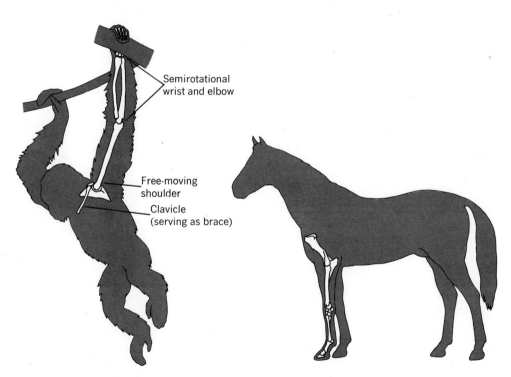

Fig. 3-2. The skeletal structure of the forelimb of an arboreal primate (monkey) compared to that of a terrestrial nonprimate (horse). Note the absence of a clavicle, the inflexibility of the joints, and the placement well under the body in the foreleg of the horse.

ment of larger bodies than could have been possible had their weight depended on claws for support. Accompanying the flexibility of the digits for grasping was the evolution of the thumb and big toe into an opposable position, that is, the displacement of the thumb and big toe from the same rigid plane as the other digits to one that allows the first digit to come under and press against the other four. In addition, the claws were modified into flattened nails, and sensitive tactile pads developed on the digits.

Eventually the tendency toward a sitting posture developed. This released the front feet (hands) from supporting the body and, while they were primarily adaptations to arboreal locomotion, the dexterous hands began to be employed for holding food and other objects. A developmental sequence of primate hands is depicted in Fig. 3-3 and compared to the forefeet of several nonprimate mammals.

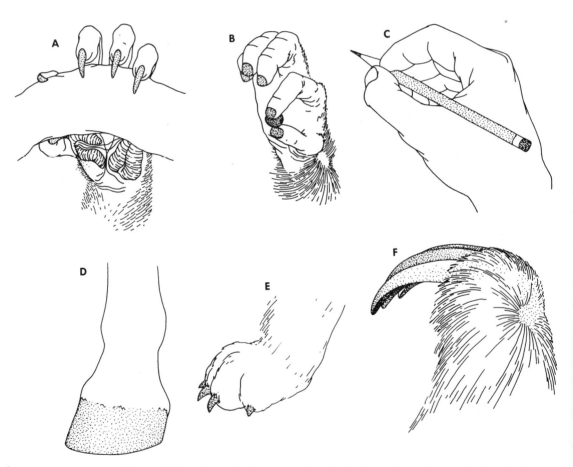

Fig. 3-3. The primate hand compared to the forefoot of nonprimates. **A** through **C** represent a series of primates displaying progressively more advanced hand structure: **A,** tree shrew, **B,** chimpanzee, and **C,** man. **D** through **F** are nonprimates: **D,** the foot of a horse, **E,** the paw of a cat, and **F,** the foot of a sloth. These hand and forefoot structures also represent an adaptive radiation to different modes of life by the various mammalian orders depicted.

Important adaptations in vision also occurred in the primates. To successfully swing from branch to branch or jump from limb to limb, it is very important to accurately determine the position of the next support. Any animal that misjudges a jump or swing through a tree is naturally selected against! Hence, the arboreal mode of life promoted selection in favor of stereoscopic vision for depth perception. This resulted in the slow evolution of a forward placement of the eyes, producing the overlap of their fields of view which leads to three-dimensional vision. In most terrestrial mammals, the eyes are located well to the sides of the head and, although this provides for a wide field of vision, little or no overlap of fields occurs and these animals do not have stereoscopic vision (for example, horses or cows). To facilitate overlapping fields of vision, the typical long snout became modified producing the flattened, forward-directed face characteristic of the most advanced primates.

Besides flexible limbs with grasping hands and stereoscopic vision, arboreal locomotion required great agility and eye-hand coordination. This established strong selection pressures favoring well-developed brain centers for muscular coordination and visual interpretation. Thus, the selective action toward the development of larger brains with greater capabilities was initiated as another adaptation in response to the arboreal habit of the primitive primates. It is interesting to note here that most of the higher mental faculties of the advanced primates have developed alongside the expanded motor centers of the cerebral hemispheres that provide the eye-hand coordination.

The mobility of an arboreal mode of life is not conducive to rearing many young at one time, especially when a prolonged period of infant dependency is involved. A primate mother, with her infant clinging to her hair, must continually move about the trees throughout the day gathering food. As a result, natural selection favored low numbers of offspring per pregnancy and advanced forms evolved to bear one young at a time under normal conditions as part of the adaptation to life in the trees. Concurrently the primates have two mammary glands, while most other mammals, lacking the selection pressure for such a reduction in progeny, possess a larger number of glands, usually corresponding to the average number of offspring per litter. One species of opossum has as many as 25 teats.

If you would pause for a moment to think about these arboreal adaptations, you could observe and demonstrate them on your own body and imagine their value for locomotion through the trees.

From shrew to you

The pathway from primitive primate to man is not straight and narrow. Man was millions of years in the making, and the course of his evolution is broad and complex. It is marked by many dead ends and side branches of various forms. The details of analysis and the wealth of anthropological material on man's evolutionary development lie beyond the scope of this text; nevertheless, we can sketch an outline of the key stages of man's descent.

The most primitive primates are known as *prosimians* (premonkeys). Their surviving forms represent an essentially phyletic progression that constitutes the suborder Prosimii. At the bottom of the living primate ladder are the tree-shrews of Southeast Asia (Fig. 3-4). As their name implies, they resemble the shrews that belong to the order Insectivora and in many ways still straddle the fence between the insectivores and the primates. Tree-shrews seem to be a living relic of the small insectivorous mammal that took to the trees some 60 million years ago and, by so doing, founded the primate order. Other living prosimians include the tarsiers found on a few Southeast Asian islands, (Fig. 3-5), the lemurs, and aye-ayes restricted to Madagascar (Malagasy Republic), and the lorises and pottos of Southeast Asia and Africa. The prosimian line has been distinct from a line that diverged nearly 60 million years ago into the second suborder of primates known as the Anthropoidea (in the shape of man). Thus, today's prosimians are not really closely related to man.

The path must now be taken up from the *anthropoids*. Actually, two distinct lines of anthropoids diverged at about the same time, probably from closely related prosimians. Each line is categorized as an infraorder. The members of the less advanced line are known as the New World monkeys or *platyrrhines* (infraorder Platyrrhini), which literally means broad-nosed. These are the South American monkeys that are characterized by round nostrils widely separated by a broad septum and that have a dexterous prehensile tail strong enough to hang by and sensitive enough to probe about and pick up objects as small as

Fig. 3-4. Tree shrew. A native of Borneo, this creature still resembles the small insectivore from which the primates evolved. Although its thumb and big toe are opposable, note that claws are retained on its digits. (San Diego Zoo photograph)

Fig. 3-5. Tarsier. This prosimian gets its name from the greatly elongated tarsal bones in its feet. Although only about the size of a chipmunk, the powerful leverage provided by its long tarsals enable it to make leaps of 4 to 6 feet through the trees. Its hands possess broad disks on the fingertips that aid in grasping, and fingernails replace claws. Being nocturnal, its night-adapted eyes are the largest of all primates. If human eyes were proportionately as large as those of the tarsier, they would be the size of a softball. (San Diego Zoo photograph by Ron Garrison)

a peanut. This tail is often referred to as the "fifth hand" of these monkeys. Among the platyrrhines are the capuchins (the traditional organ grinder's monkey) and the spider, squirrel, and howler monkeys (Fig. 3-6). In contrast to most prosimians, the platyrrhines are almost all active during the daytime. They are all well-adapted arboreal forms and seldom venture on the ground.

The more advanced anthropoids constitute the *catarrhines* (infraorder Catarrhini) which literally means hooknosed but refers to their comma-shaped nostrils, which are close together and pointed downward. The prototypic catarrhines, the Old World monkeys, inhabit Asia and Africa and are characterized by the absence of a prehensile tail. They have instead developed tough, flattened callous pads on the rump, called *ischial callosities*, as an adaptation for the long periods of time they spend sitting in trees. These callosities are, in fact, survival adaptations. Although the catarrhines are relatively safe in the trees, some of their natural enemies, such as boa constrictors and leopards, are agile climbers. Therefore, these Old World monkeys tend to sleep far from the tree trunk out on the smallest limb that will support their weight. In this way, if a predator approaches, the swaying of the limb will warn the sleeping monkey. To assure balance, the monkeys sleep sitting on their rump with their legs directed upward at a sharp angle grasping another limb for support. The replacement of soft, sensitive haunches by toughened pads enables the monkeys to sit comfortably in this position for hours. Old World monkeys also differ from the New World forms in dentition. The catarrhines, in common with man, possess two premolar teeth in each half of each jaw, whereas the platyrrhines have three. The catarrhines tend to be larger in body size than the platyrrhines and, although primitively arboreal, some of the more specialized types, such as baboons, tend toward a terrestrial life.

The catarrhines are further partitioned into two superfamilies. One is the *Cercopithecoidea* or "apes with tails" (but the tails are not prehensile). This superfamily is represented today by a single family (Cercopithecidae) that encompasses all the Old World monkeys, such as the macaques (Fig. 3-7), baboons, drills, and langurs. This lineage diverged from the other superfamily over 40 million years ago. The living Old World monkeys, then, are not really closely related to man either.

All other catarrhines belong to the superfamily *Hominoidea*. Three families survive today. The gibbons (Hylobatidae) of Southeast Asia represent a lineage that probably diverged from the other hominoids soon after the superfamily itself arose (Fig. 3-8). The gibbons have extremely long arms, and they spend almost all of their time in the trees. These creatures stand no closer to man than 35 million years for a common ancestor. The second family of hominoids is the great apes (Pongidae), which includes the present-day orangutans (genus *Pongo*) (Fig. 3-9) of Sumatra and Borneo, and the chimpanzees (genus *Pan*), and gorillas (genus *Gorilla*) of tropical Africa. Finally, the family of man (Hominidae) probably began to diverge from the pongid line some 25 million years ago. Hence, even though the present-day great apes are our closest living relatives, we have been on separate evolutionary pathways for a very long time. It should be obvious to the point of absurdity that man (genus *Homo*) is not descended from any of today's monkey or ape species. These relationships and the time spans involved are summarized in Fig. 3-10.

Fig. 3-6. Guatemalan red howler. A new world monkey characterized by its prehensile tail, which serves as a "fifth hand." (San Diego Zoo photograph)

52   WHO WERE OUR CLOSEST RELATIVES?

The very closest relatives of man are no longer living. Man is the sole survivor of his entire family. The elucidation of our direct family line has been a slow and laborious process developed by men who have devoted their entire lives to constructing our family tree. Controversies still exist concerning man's predecessors, and fossils are still being gathered and analyzed to develop a clearer

Fig. 3-7. A macaque family. These representatives of old world monkeys, commonly called rhesus monkeys, are used extensively in biological research. Note the comma-shaped nostrils. (San Diego Zoo photograph)

Fig. 3-8. The gray gibbon. A member of the most primitive hominoid family surviving today. Note its extremely long arms. Although awkward on the ground, gibbons travel so gracefully through the trees that they appear to be flying as they swing from branch to branch. (San Diego Zoo photograph)

picture of man's recent geological past. We shall only consider a brief outline of what seems to be the most probable line of ancestory leading to the present condition. It should be noted that the fossil evidence upon which the past is explored, although substantial and mounting every year, is a relatively small sample of the varied forms of apes, man-apes, and men that roamed the ancient countryside. Since it is not uncommon for any genus to be represented by several species, the fossil record may not always contain *the* species that was the direct progenitor of a more advanced form; remember, the pathway is often confounded by dead end branches. The record is complete enough, however, to describe the major steps to man. In some cases, our present state of knowledge must remain at the generic or familial level with no commitment to the particular species involved in the direct evolutionary line.

To begin, some 40 million years ago a generalized hominoid ancestor known

Fig. 3-9. Female and male orangutans. These are members of the great ape family which also includes the chimpanzees and gorillas. Orangutans display marked sexual dimorphism. The male (right) develops peculiar cheek flanges at sexual maturity that, along with a high cranial crest, make the face appear concave. Females lack these features and look quite similar to chimpanzees. (San Diego Zoo photograph by Ron Garrison)

as *Propliopithecus* existed which possessed ape-like characteristics. (The terms hominoid and hominid should not be confused; the former refers to the entire superfamily that includes the gibbons, great apes, and man, while the latter refers specifically to the family of man.) This genus is well-known from a spot in the Egyptian desert about 60 miles from Cairo called the Fayum Depression. It is now one of the driest places on earth, but millions of years ago it lay on the edge of the Mediterranean Sea and was covered with lush tropical forests. Propliopithecines were very likely the progenitors of the hominoid superfamily. The present-day gibbon line (family Hylobatidae) has been distinct from the other hominoids for over 35 million years. Its predecessor was *Pliopithecus* (proto-ape), which is well represented in the fossil record from Europe and Asia. *Pliopithecus* is not in the direct line to man. Near-contemporaries of the plio-pithecines were the dryopithecines, which means oak apes, so called because of

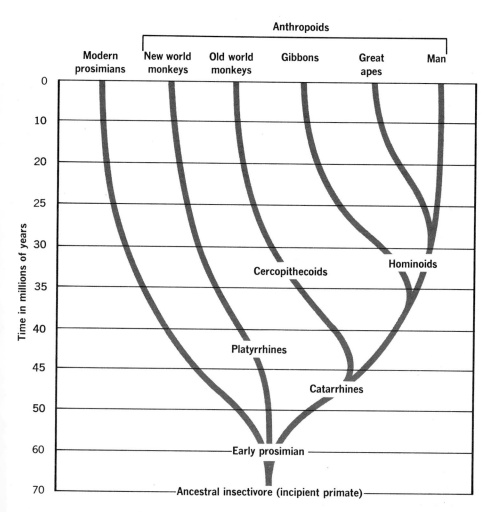

Fig. 3-10. The evolution of the primate order.

the fossil oak leaves found with the first specimens from the Siwalik Hills of northwestern India. The original *Dryopithecus*, described from a lower jaw bone, was considered to be ancestral to man as long ago as the 1920s by William K. Gregory of the American Museum of Natural History; but the fossil inventory was sparse and the evidence inconclusive. In the 1930s L. S. B. Leakey and his wife, working on a small island in Lake Victoria, discovered a 20 million year-old form that was named *Proconsul*. (The name is an inside British joke. Consul was the name of a well-known chimpanzee in the London Zoo, so the generic name *Proconsul* simply refers to the ancestor of the chimp.) *Proconsul* appears to have been an ancestral form of the great apes (family Pongidae). Recent taxonomic investigations indicate that *Proconsul* and *Dryopithecus* were probably members of the same genus. By priority, *Dryopithecus* supercedes *Proconsul* as the generic name, and the latter is now classified as *Dryopithecus africanus*, its species name taken from its place of discovery. It now appears that the dryopithecines constituted a cosmopolitan genus from which the great apes evolved from one species, *africanus*, while man's family (Hominidae) had its origin from some related species, as yet undiscovered. Thus we are back to Gregory's early idea, but the interpretation of the dryopithecines has been greatly expanded.

In a 1966 Yale University expedition to the Fayum region led by Elwyn L. Simons, the oldest true ape skull known to date was uncovered. It dates back between 26 and 28 million years. Studies of this skull plus jaw bones, teeth and limb bones have lead to the placement of this form between the propliopithecines and the dryopithecines in the evolutionary scheme. This genus was named *Aegyptopithecus* (Egyptian ape).

At about the same time that the Leakeys were accumulating the fossil inventory of *Proconsul* in Africa, G. Edward Lewis was leading a Yale-Cambridge expedition in the Siwalik Hills of India. Lewis unearthed a surprisingly human-appearing upper jaw. The form to which this jaw belonged was placed in a new, more advanced genus than the dryopithecines and was named *Ramapithecus* after Rama, one of the three incarnations of the Hindu solar diety Vishnu, the Pervader. The dentition of this form, which roamed throughout Asia and Africa about 15 million years ago, was definitely man-like. In addition, its palate was arched and curved outwards toward the back, like that of man; the palate of all apes and monkeys is flat and U-shaped toward the back (Fig. 3-11). *Ramapithecus* has now been rather firmly established as an early member of the Hominidae, our own family. Thus, well over 15 million years ago (perhaps as long as 25 million years ago), the family to which man belongs must have diverged from dryopithecine ancestors, and since that time the apes and man have had independent evolutionary histories.

A contemporary of *Ramapithecus* was *Oreopithecus*, irreverently referred to as the Abominable Coal Man. In 1958 a nearly complete skeleton, one of the most superb hominoid fossils ever recovered, was removed from a coal mine at Grosseto, Italy. At the time of its discovery, *Oreopithecus* had been considered

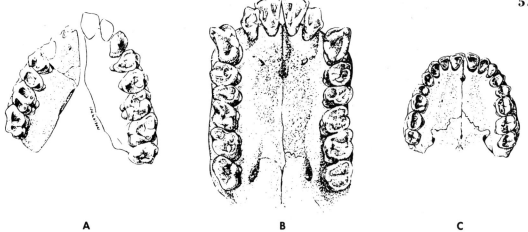

Fig. 3-11. Comparison of the upper jaw of **A,** *Ramapithecus* to that of **B,** an orangutan. and that of **C,** a contemporary human. The dental arcade (curve made by the teeth) of *Ramapithecus* corresponds more closely to the human shape than to that of the apes. The front teeth (incisors and canines) of *Ramapithecus*, like those of man, are small relative to the side teeth (premolars and molars); this is also a distinct contrast to the dentition of apes and monkeys. (From Poirer, Frank E. Fossil man, an evolutionary journey, St. Louis, 1973, The C. V. Mosby Co.)

as a possible precursor of the hominid family, but today it is generally regarded as one of the dead end lines of hominoid radiation.

The evolution of the hominid family is continued in the fossil record by *Australopithecus*. This genus was first described and named in 1924 by Raymond Dart of the University of Witwatersrand in Johannesburg, South Africa. The name literally means southern ape, but it is often called the South African ape man (Fig. 3-12). *Australopithecus* dates back about 2 million years, leaving a gap of some 10 million years between it and *Ramapithecus* from which no fossil evidence is available concerning the transition from the ramapithecines to the australopithecines. There were, of course, doubts as to how man-like *Australopithecus* really was because his appearance was still rather ape-like. Then, in 1948, Robert Broom, a Scottish physician turned fossil hunter, uncovered a nearly perfect pelvis bone of *Australopithecus* from a quarry in Sterkfontein, near Pretoria, South Africa. This and subsequent finds confirmed that australopithecines had an erect posture and walked upright. An erect posture is a necessary development on the road to man, but what type of selection pressure would make erectness and bipedal locomotion favored by natural selection?

Erect posture had always been viewed as an adaptation to a terrestrial habit which, as a by-product, freed the hands and permitted the evolution of tool using. The selection pressure favoring erectness was considered to be related to the search for food on the ground or, because of the drying out of the tropical forests and reduction of arboreal living space, "forced" terrestrial living while searching for suitable arboreal habitats. However, Sherwood Washburn of the

University of California makes a strong case for a reversed situation. He proposes that tool using preceded bipedal locomotion, and, in fact, provided the selective pressure for erect posture. Apes today, and presumably in the past, sit upright and even stand in the trees. Furthermore, they have the manual dexterity to throw stones, use sticks to dig, crack nuts with rocks and sop up water with handfuls of leaves. All of these activities have been observed in the wild state. Hence, Washburn speculates that such tool use in the primitive apes would have placed a selective advantage on free hands, and that this was accomplished through upright posture. Chance success with throwing sticks and stones could have led to their use as weapons for protection and hunting which, in turn, could have allowed their possessors to venture further and further from the safety of the trees. This ground orientation then would have established the selective pressure favoring adaptations for more efficient bipedal locomotion. If this is true, the earliest erect hominids, such as *Australopithecus*, should have been tool-users. Evidence that supports this hypothesis was supplied by Broom's assistant, J. T. Robinson who, in 1957, found australopithecine teeth and bones together with more than 200 stone tools at the Sterkfontein quarry. It has also been established that these ape-men used tusks and teeth as cutting tools, jaw

Fig. 3-12. *Australopithecus.* Known as the South African ape man, he was the earliest user of tools and fire in man's ancestry. (Courtesy the American Museum of Natural History)

bones as saws and scrapers, and leg bones as bludgeons. Thus, the australopithecines were early tool-using hominids.

A contemporary of *Australopithecus* was *Paranthropus*. Broom first discovered *Paranthropus* in a younger stratum than that containing australopithecine fossils and, presuming it to be a closer form to man, assigned its presumptuous generic name which means akin to man. Further investigation revealed, however, that *Paranthropus* was actually more primitive than *Australopithecus*. Finally, Robinson coupled the evidence from South Africa with that collected by Louis and Mary Leakey from the fossil-rich Olduvai Gorge in Tanzania to conclude that *Paranthropus* was not a tool-user. It now appears that *Paranthropus* was another dead end line that became extinct less than half a million years ago.

The immediate predecessor of the species of modern man was *Homo erectus* (erect man). Until recently this form was known by a multiplicity of other names which are listed in Table 3-3. The fossils of Java man are the oldest known *Homo erectus* representatives, dating back over 700,000 years. On the other hand, Rhodesian man lived as recently as 30,000 years ago. The placement of Rhodesian man in the *Homo erectus* category is controversial however; many authorities consider him to be an early form of modern man. In this case, Solo man, who is also considered a modern form by some but whose skull shape is near that of Java man and whose brain size is similar to that of Peking man, would appear to be the most recent *erectus* representative. Solo man dates back approximately 100,000 years. From the accumulated finds that collectively constitute *Homo erectus*, there emerges a picture of men with erect skeletons like ours but with much smaller brains, much thicker and flatter skulls, protruding brows, and larger teeth more primitive than our own. The many forms that make up *Homo erectus* are somewhat analogous to the several races that make up modern man (Fig. 3-13). Heidelberg man, however, seems to have been different from all the others and a little closer to modern man, especially in his dentition; yet he dates back over 500,000 years.

An interesting situation has emerged following a recent find at Vertesszollos, Hungary, (1965) of an occipital bone (back of skull) that indicates a more ad-

Table 3-3. Fossil forms previously categorized as distinct entities, now all considered to be members of the species *Homo erectus*

Common name	Scientific name (s)	Place discovered	Date
Java man	*Pithecanthropus erectus*	Solo River, Java	1891
Heidelberg man	*Homo heidelbergensis* or *Palaeanthropus*	Mauer, Germany	1907
Rhodesian man	*Homo rhodesiensis* or *Cyphanthropus*	Broken Hill, Rhodesia	1921
Solo man	*Homo soloensis* or *Javanthropus*	Solo River, Java	1932
Peking man	*Sinanthropus pekinensis*	Peking, China	1937
Swartkran man	*Telanthropus capensis*	Swartkrans, South Africa	1949
Algerian man	*Atlanthropus mauritanicus*	Ternifine, Algeria	1955
Lantian man	*Sinanthropus lantianensis*	Shensi, China	1963

vanced skull with a larger brain capacity than the typical *Homo erectus*. This Hungarian man has been classified as a form of modern man, *Homo sapiens* (wise man), which dates back nearly 500,000 years, making it contemporary with *Homo erectus*. Before the Hungarian find, the oldest fossils assigned to *Homo sapiens* were the Swanscombe man from England, dated 250,000 years old, and the Steinheim man from Germany, dated 200,000 years old. This faint trail leads to the possibility that *Homo sapiens* emerged in Europe some 500,000 years ago in a line that follows back from Steinheim and Swanscombe to Vertesszollos and finally to Heidelberg man as the incipient *sapiens* species.

Another member of our own species was Neanderthal man. He was first discovered in 1856 in the Neander Valley near Duesseldorf, Germany. This, you will note, was even before Darwin's theory of evolution was published. Thus, attempts were made to write Neanderthal man off as the bones of a Cossack killed during Napoleon's retreat from Moscow, a victim of water on the brain, or an "old Dutchman." An English scholar interpreted the specimen as, "One

Fig. 3-13. *Homo erectus.* The form depicted, known as Java man, represents a species so similar to ourselves that it is classified in the same genus, *Homo.* Sites at which the fossil remains of this "man" have been collected range throughout the European and African continents. These sites indicate that *Homo erectus* led a communal life in bands of 30 or more that cooperated in the hunt for food and their mutual survival. (Courtesy the American Museum of Natural History)

of those wild men, half-crazed, half-idiotic, cruel and strong, who are always more or less to be found living on the outskirts of barbarous tribes, and who now and then appear in civilized communities to be consigned perhaps to the penitentiary or the gallows, when their murderous propensities manifest themselves."

Today Neanderthal man is recognized as a form of *Homo sapiens* who lived from 30,000 to 80,000 years ago. His extinction was very likely due simply to his hybridization with other forms of *Homo sapiens* (races) and the amalgamation of his hereditary traits into the larger modern gene pool. Even today these traits seem to occasionally recombine to produce Neanderthoid-appearing individuals! Neanderthal man had prominent brows connected across the bridge of the nose, large cheeks, and a receding chin. His skull was somewhat differently shaped from that of modern man, but it accommodated a brain of approximately equal, if not larger size. His skeleton marks a powerfully built man standing just over 5 feet tall (Fig. 3-14). The quality of his tools and his culture indicate

Fig. 3-14. Neanderthal man. A member of our own species, *Homo sapiens*, Neanderthal man was widely distributed throughout the Old World. He is no longer thought to have appeared as "primitive" as this restoration indicates. Rather, he was probably quite human looking and not as hunched over as we typically imagine. It is now thought that if he were bathed and shaved, and dressed in contemporary garb, no one would pay him any particular attention as a passer-by on a city street. (Courtesy the American Museum of Natural History)

a mental ability on par with that of modern man. Neanderthal man buried his dead with the corpses carefully arranged in a sleeping position with stone tools and animal bones in accompaniment. This and other evidence indicates that Neanderthal man believed in life after death and had developed primitive religious practices.

The most recently extinct form of *Homo sapiens* appears to have been Cro-Magnon man. This form was discovered in 1868 when contractors widening a road outside Les Eyzies, France, cut into a rock shelter known as Cro-Magnon (literally big-large or great-big). The Cro-Magnon race lived throughout Europe from 35,000 to about 10,000 years ago. These people had prominent chins, reduced brows, high-bridged noses, and other physical features typical of modern man (Fig. 3-15). Cro-Magnons had mastered the art of flint shaping to produce axes and blades, evolved a tribal social structure, and developed their famous cave art, sculptures, and stone engravings. Cro-Magnon man as a distinct form faded about 10,000 years ago, from which time man's cultural and biological evolution have produced the societies and races of *Homo sapiens* as we know them today.

The broad outline of man's emergence from his ancient hominoid ancestors is summarized in Fig. 3-16. To conclude our discussion of man's evolution, it may be of value to also summarize some of the more conspicuous anatomic changes that occurred. The various ape-men and primitive men described previously displayed the progressive transition of these anatomical changes. As erect posture evolved, the points of muscle attachment of the head to the vertebral column shifted from the back of the skull to positions underneath as the

Fig. 3-15. Cro-Magnon man. The most recently extinct form of *Homo sapiens.* Cro-Magnon is best known for his cave art in France and Spain, which seems to have been closely tied to his spiritual life. (Courtesy the American Museum of Natural History)

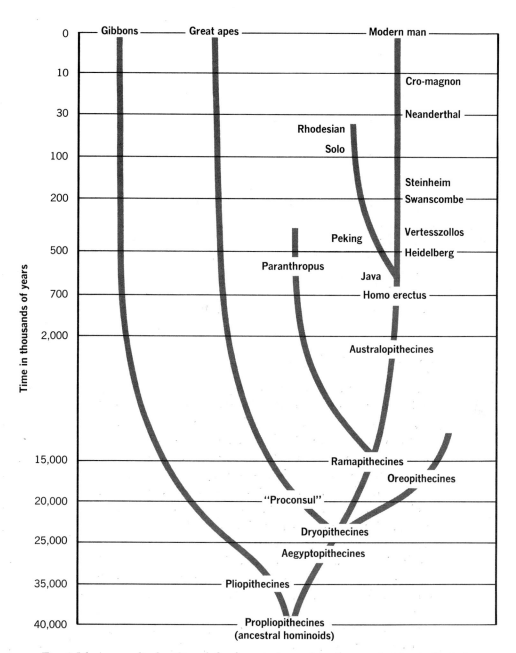

Fig. 3-16. A generalized outline of the descent of man from the ancestral hominoid stock.

Table 3-4. Evolution of the brain as indicated by the average cranial capacities of man, his living relatives, and recent ancestors

Living hominoids	Capacity (cc)	Ancestral hominids	Capacity (cc)
Modern man	1,400	Cro-Magnon man	1,700
Gorilla	500	Neanderthal man	1,400
Chimpanzee	400	Swanscombe-Steinheim man	1,325
Orangutan	375	Vertesszollos man	1,400
Gibbon	100	Rhodesian man	1,300
		Peking man	1,100
		Java man	900
		Lantian man	800
		Australopithecus	600

skull became balanced on top of the column. This resulted in a decrease in the size of the neck muscles. The braincase enlarged as the brain evolved; the cranial capacities of man and some of his relatives and ancestors are given in Table 3-4. A prominent forehead developed as a result of the enlargement of the braincase. The jaw and teeth became smaller as the snout shortened and the nose, with a distinct bridge, grew more prominent. The eyebrows and other ridges on the face were reduced as the muscles attached to them diminished in size. The feet became flattened as the big toe moved back in line with the other toes and ceased to be opposable; the arch of the feet developed subsequently as bipedalism was perfected. The arms became shorter.

FUTURE PROSPECTS

The emergence of cultural evolution in man has made adaptation to changes in the environment by biological evolution (genetic changes) less likely in many respects than in precultural times. Man did not have to evolve a furry body, insulating blubber, or a cycle of hibernation to cope with cold climates; instead he was able to conquer the cold through cultural developments such as warm clothing, insulated shelters, and heating systems. It took millions of years of biological evolution for birds (class Aves) to evolve from their reptilian stock and perfect the faculty of flight, while man accomplished the same feat shortly after his technology supplied sources of power. With far less than a century of flight experience, man now flies from continent to continent in hours, orbits the earth, and lands on the moon. But what kind of long-term biological effects will man's ever-accelerating cultural evolution produce upon him and the rest of the organic world? This question is becoming more critical for our society every year.

Cultural developments have greatly altered the Darwinian fitness values (reproductive differential) in the human population. The common inherited form of sugar diabetes (diabetes mellitus) serves as an interesting and unusual example. Diabetes mellitus is a metabolic disease characterized by the inability to utilize glucose because of the failure of the pancreas to secrete sufficient amounts of insulin. Excessive sugar (glucose) is discharged in the urine. Afflicted persons have very little resistance to infection, and gangrene may set in as a result of simple injuries to the skin. Diabetes is further complicated by the fact that it may strike at any age from childhood through adulthood and may be either very severe or so mild that the individual concerned is unaware of having it. Its onset is usually in adulthood; however, the more severe cases occur during childhood, and death may ensue before the age of reproduction is past. The actual genetically controlled component in diabetes mellitus may be an insulin-inhibiting factor or antagonist.

Persons who have the genetic predisposition for developing diabetes but have not yet manifested the disease are called "prediabetics." Prediabetics reach sexual maturity at an earlier age than nondiabetics. They also appear to have larger than average families. Thus, their fertility rate is high. In addition, the prediabetic child tends to grow at a faster than average rate and the adult prediabetic tends to be fatter than average. These observations suggest that prediabetics make a more efficient use of the food they consume. The disadvantage of the prediabetic genotype appears to be the eventual overproduction of the insulin antagonist that brings on the symptoms of diabetes. An interesting and plausible attempt has been made to explain this seemingly confused situation. It is thought that starvation was an important agent of natural selection in early man and his ancestors. Various environmental factors, such as droughts or floods, undoubtedly operated in the past to produce periods of sparse food supply; in fact, famines have been a factor in most of man's development extending into the present. The metabolic efficiency of the diabetic gentoype (in prediabetics), sometimes called the thrifty genotype, would have been an important energy-conserving mechanism and would have imparted a definite advantage to its host in time of famine. The obesity of prediabetics today may in fact be a result of stored surpluses despite a normal diet because their metabolic efficiency is such that "normal" is "excessive" for them. Greater fertility coupled with the ability to subsist on minimal amounts of efficiently metabolized food may well have been a distinct selective advantage in the past. Thus, the prediabetic would have had a high Darwinian fitness, and consequently the genes for diabetes would have spread and increased in the population. The disadvantage of prediabetics becoming diabetics would have been minimal because these early men had life expectancies of 30 years or less, and the onset of diabetes most often occurs later than that age. (Today most cases appear between the ages of 45 to 65.)

With cultural improvement, man's environment became less hostile, and, in general, his diet improved, and starvation became less commonplace. Because of medical science, sanitation measures, and a less rigorous physical existence, man's life expectancy has risen to a high of about 70 years. Coincident with this, diabetes has become an ever more debilitating disease. It might be expected that under these circumstances the genes for diabetes would lose their selective advantage, and natural selection would eventually lower their frequency in the population. However, the frequency of the genes for diabetes has not diminished. The reason for this is that medical treatment (cultural development) has enabled diabetics to survive and reproduce. Diabetes is now controlled by insulin injections, and diabetics can live a relatively normal life. Even the more severe cases, which some authorities believe carry a double dose of the genes for diabetes (homozygotes) as compared to the milder forms which carry only one dose (heterozygotes), can now pass their genes to future generations—a feat they rarely accomplished in the past because of their early death. Thus, cultural evolution, has, in this case, essentially negated biological evolution by natural selection.

Many other instances of cultural evolution superceding the action of natural selection could be cited. Optical defects with a genetic basis, such as nearsightedness or farsightedness, are easily corrected by lenses today, but in our ancestors who swung through trees or hunted on the ground visual defects would have conferred distinct disadvantages for survival, and affected individuals were probably selected against. Today there is no particular selection pressure operating on these defects. Likewise, many genetic diseases that once were lethal at an early age are now compensated by medical treatment, and the Darwinian fitness of the afflicted individuals is little different from that of the "normal" population. We will discuss many examples that fit into this category, such as phenylketonuria (PKU) and galactosemia, in later chapters. The point of all this is that natural selection no longer operates to effectively keep the frequency of these genetic defects low (essentially near their recurrent mutation rate). This leads to another form of pollution to worry about—polluting the human gene pool with defective genes. Although many biologists discount the detrimental effects of this pollution, we must remember that as medical science treats more and more genetic defects, this pollution will be compounded, and ultimately every individual could become dependent on society to provide the treatment for his particular defect or defects. True, this seems a long way off, and genetic manipulation may someday correct or eliminate most genetic defects. But for the present, pollution of the human gene pool is a distinct possibility, and therefore man and society must assume the responsibility for the preservation of man's genes for future generations. It has been remarked that the problem is negligible, citing as an example the calculation that it would take 40 generations, or more than 1,000 years, to double the incidence of PKU from its present 1 in 10,000 to 1 in 5,000 births. This might seem negligible but doubling all other genetic defects in like manner changes the picture drastically. More importantly we have seen that man has a history of millions of years; therefore, what kind of time span should we really consider when we think of man's future?

Despite its power and speed, cultural evolution has not prevented natural selection from operating. Biological evolution consists of adaptation to a changing environment through differential reproduction favoring the best-adapted types. Cultural evolution does not make man's environment uniform and stable.

In fact, cultural evolution, through air, water, chemical, and thermal pollution, has drastically altered the conditions under which man and all other living forms must live and has created a very unstable environment. Social pressures and psychological pressures accentuated by urban crowding and the complex and rapid pace of life are bearing down on man's nervous system. Perhaps the ability or inability to cope with pressures such as these is becoming a contributing factor affecting differential reproduction in man today. Environmental pollution tolerance may very well be, or become, a determining factor in fertility. Continued pollution of our environment may even become the determining factor in man's survival as a species. It has been pointed out that throughout man's evolutionary history he had never been subjected to "pollution pressure" until very recently. Natural selection never had to operate on adaptations to the myriad of chemicals that now enter our bodies through the air we breathe, water we drink, and food we eat. Therefore, practically no biological mechanisms exist to cope with many of these pollutants. For instance, we cannot metabolize or degrade DDT in our bodies; rather, it accumulates and is stored primarily in our fatty tissues. Such storage of foreign chemicals in the body can reach levels that induce genetic damage or become toxic and produce illness or death. Environmental pollution has created a tremendous selection pressure that challenges natural selection to remold our biological systems to accommodate. But natural selection operates slowly and, without even rudimentary biological processes to enhance or reshape, it seems very unlikely that man will be able to biologically adapt to increased and continued pollution. The necessary adjustments will have to be made by man through his cultural processes. He can do it by controlling and abating pollution, thereby allowing his biological systems to function normally, or he can perhaps do it through more and more "advanced" technology—producing synthetic foods, limiting human activities to environmentally controlled urban enclosures with relatively clean air and water supplies, and so forth. But this would create a much different society from any that man has ever experienced, and these conditions would probably place additional, yet unknown, stresses on his biological systems. Furthermore, what man does today affects all the rest of the organic world. If his pollution kills him, it may also eliminate or drastically alter all life on earth. The personal pronoun has a changed meaning in this context, but the song is very true, "He's got the whole world in his hands."

It has also been speculated that man through his cultural advancement will overcome all obstacles and proceed toward human perfection. Natural selection under these circumstances is viewed as favoring ever-increasing intellectual capacity which, in due course, would lead phyletically to a more advanced man who might be named *Homo superior*. But, if we are speculating on future hominid species, let us consider two other alternatives. The pollution factors discussed above, if unabated, may not wipe man and beast completely off the the face of the earth. It is likely that a poisoning of our environment would leave survivors, albeit most would probably be carriers of genetic damage from the

68   mutagenic effects of the pollutants. On the other hand, the constant threat of a nuclear holocaust may be realized, leaving again some survivors, many genetically damaged this time from radiation. In either case, sparse numbers of men and other living forms would be left to pick up the pieces. Natural selection would have to operate on a genetically impoverished population living in an even more impoverished environment. Man in this condition might be reclassified as *Homo inferior.* Finally, pollution or a nuclear holocaust might totally wipe out man, or our resources may become depleted (the sun will eventually burn out), and all life on earth would become extinct. In evolution, as illustrated by the dead end lines of even the predecessors of man, the ultimate fate of any entity is extinction. At this point, classification is strictly superfluous, but I will propose a null-entity to be called *Nomo homo!*

Suggestions for further reading on evolutionary processes

Appleman, Philip, editor. 1970. Darwin.* W. W. Norton & Company, Inc., New York. This book presents some background to Darwin, selected excerpts from his works, and many essays concerning Darwin's influence on science, theology, and society.

Bajema, Carl Jay, editor. 1971. Natural selection in human populations.* John Wiley & Sons, Inc., New York. An extensive collection of papers emphasizing human evolution as an ongoing process. The operation of natural selection on the human population is discussed in relation to physical characteristics, disease, behavior, and our future genetic composition.

Darwin, Charles. 1960. The voyage of the Beagle.* Bantam Books, Inc., New York. The account of Darwin's explorations and adventures on his 5-year voyage around the world.

DeBeer, Gavin. 1964. Atlas of evolution. Thomas Nelson and Sons, London. A broad view of evolution beautifully illustrated with over 500 plates.

Dillon, Lawrence S. 1973. Evolution: concepts and consequences. The C. V. Mosby Company, St. Louis. A stimulating treatment of the current state of evolutionary theory, designed to present the conceptual aspects and biological facts associated with Neo-Darwinian thought.

Eimerl, Sarel, Irven DeVore, and the Editors of Life. 1965. The primates. Time Inc., New York. A very well illustrated book on primate evolution from tree-shrew to Man.

One of the volumes in the Life nature library.

Keosian, John. 1968. The origin of life.* Second edition. Reinhold Publishing Company, New York. A clear and concise account of the theory of the origin of life.

Kerwin, Carlotta, series editor. 1972. The emergence of man. Time Inc., New York. A five-volume series on the evolution of man with numerous illustrations and novel-like text.

Moore, Ruth and the Editors of Life. 1962. Evolution. Time Inc., New York. Another book in the Life nature library series illustrated with outstanding photographs.

Oparin, A. I. 1968. Genesis and evolutionary development of life. Academic Press, Inc., New York. This book is written in a nontechnical manner by the foremost scientist in this field for over 40 years. It is designed to inform the layman of the state of knowledge concerning the origin of life.

Stebbins, G. Ledyard. 1971. Processes of organic evolution.* Second edition. Prentice-Hall, Inc., Englewood Cliffs, New Jersey. The principles of biological evolution are explained with supporting evidence from the field of experimental evolution clearly interpreted.

Young, Louise B., editor. 1970. Evolution of man.* Oxford University Press, Inc., New York. A collection of short papers and excerpts dealing with various aspects of human evolution ranging from evolutionary theory to the ecological and technological problems confronting mankind today.

*Denotes paperback.

Reproductive processes

4

The mechanisms of reproduction

In living organisms, growth and development, tissue repair and regeneration, propagation, and the transmission of hereditary information are nearly always accomplished through cellular reproduction. Two basic types of cellular reproduction are involved. The first, called *mitosis*, is the process by which like begets like and two genetically identical cells are produced from one. The second type of cellular reproduction occurs only in the germ line of sexually reproducing organisms and results in products that contain only half of the genetic information of the original cell rather than being identical. These represent the gametes, eggs and sperms, involved in propagation. This process, called *meiosis*, also provides the underlying mechanism of hereditary transmission.

In this chapter we shall discuss the processes of mitosis and meiosis, placing special emphasis on the maneuvers of the chromosomes that are the carriers of the genetic material. Comprehension of the mechanisms of cellular reproduction is essential to understanding human reproduction and development (Chapter 5) and the hereditary processes (Section three).

THE CHROMOSOMES

The cell is the fundamental structural and functional unit of living systems. The cytoplasm of the cell contains several important organelles (cytoplasmic structures) of which the *nucleus* is of primary importance in cellular reproduction. The nucleus is a spherical body usually appearing more dense and granular than the surrounding cytoplasm in a nonreproducing cell. The chromosomes (colored bodies), so named because of their staining properties, are located within the nucleus. They are composed of deoxyribonucleic acid (DNA) and proteins and serve as the carriers of the genetic material (Chapter 6). When the cell is not reproducing, the chromosomes are in the form of extremely thin and long threads less than 100 angstrom units in diameter (one angstrom is 1/254,000,000 inch) dispersed throughout the nucleus and producing its characteristic appearance. Prior to cellular reproduction, these threads replicate and then condense by coiling into tightly massed bodies that are readily distinguished. The two coiled threads of the replicated chromosome, each called a *chromatid* (chromosome precursor), are held together at some clearly discernible point along their length; this juncture is called the *centromere* or primary constriction (Fig. 4-1).

The condensation from long, fine threads to compact bodies enhances the maneuverability of the chromosomes for the movements they will perform when the cell reproduces. It is clear that the centromere directs the movement of the chromosomes even though the manner in which this is accomplished is not understood. Thus, when chromosome movements occur during mitosis or meiosis, the centromere leads the way; chromosomes lacking a centromere have no directed movement. Depending on its position, the centromere divides each chromosome into two arms of varying lengths. The location of the centromere is a very constant feature of each chromosome which can be used for identification purposes. If the centromere is centrally located, the chromosome is *metacentric;* if it is off-center, the chromosome is *submetacentric;* and if it is very near to one end, the chromosome is *acrocentric.* During movement the arms of the chromosome trail behind the leading centromere, thereby producing a characteristic shape: a metacentric chromosome becomes V-shaped, a submetacentric becomes J-shaped, and an acrocentric chromosome becomes I-shaped (Fig. 4-2).

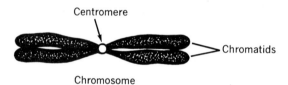

Fig. 4-1. The general structure of a chromosome.

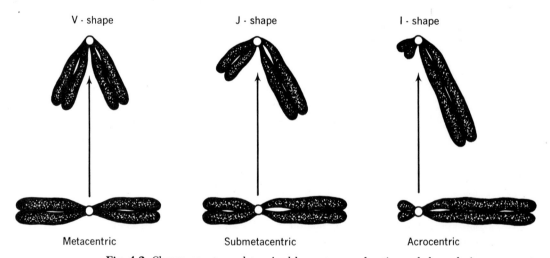

Fig. 4-2. Chromosome types determined by centromere location and shape during movement.

In addition to centromere location, chromosomes are classified according to total length and, in some cases, the possession of *satellites,* which are a distal portion of the chromosome arm separated from the proximal portion by a narrow region known as the secondary constriction.

Furthermore, the chromosomes typically occur in matched pairs with the members of each pair referred to as *homologous chromosomes.* There are characteristically 46 chromosomes (23 matching or homologous pairs) in each human cell. One pair is specifically involved in sex determination, and its members are called *sex chromosomes.* These are not necessarily a matching pair; rather, two types of sex chromosomes exist. One is a medium sized submetacentric type called the *X chromosome,* and the other is a much smaller acrocentric type called the *Y chromosome.* A female possesses two X chromosomes, whereas a male possesses one X and one Y (Chapter 9). The other 22 chromosome pairs are alike in both sexes and are referred to as *autosomes.*

An international standard of nomenclature for human chromosomes was established in 1960 by a small study group of human cytogeneticists convened in Denver, Colorado under the auspices of the American Cancer Society. According to the so-called Denver classification, the autosomes are serially numbered, 1 through 22, in descending order of length. These are further classified into seven lettered groups, A through G, according to similarities in length, centromere location, and satellite presence or absence. This grouping is convenient because it is relatively easy to place a chromosome within the proper group even though it is often difficult to distinguish accurately among chromosomes within a group. A conspectus of the Denver classification of human chromosomes is provided in Table 4-1.

When the chromosomes of a cell are prepared for microscopic observation, photographed, cut out, and arranged in homologous pairs by descending order

Table 4-1. Conspectus of the Denver classification of human chromosomes

Group	Number	Description
A	1–3	Large metacentric chromosomes
B	4–5	Large submetacentric chromosomes
C	6–12	Medium-sized submetacentric chromosomes. The X chromosome closely resembles pair number 6, the longest in the group
D	13–15	Medium-sized acrocentric chromosomes. All three pairs possess satellites on their short arm
E	16–18	Smallest of the medium-sized chromosomes. Pair number 16 is metacentric, others submetacentric
F	19–20	Short metacentric chromosomes
G	21–22	Very short acrocentric chromosomes. Both pairs with satellites on their short arm. The Y chromosome is similar to these pairs but is typically larger, although its size varies, and it is not satellited

of length, the montage is called a *karyotype*. Normal human female and male karyotypes are depicted in Figs. 4-3 and 4-4 respectively. The existence in pairs is the typical condition of the chromosomes in most cells of sexually reproducing animals. Thus, the 46 chromosome condition, 23 pairs, in human cells is referred to as the *diploid* (after *diploos* meaning double) or (2n) number. On the other hand, in sexual reproduction the fusion of the gametes, egg and sperm, to produce a new individual represents a combination of half of the genetic material from each parent. These gametes, produced by meiosis, possess only one member of each chromosome pair found in the typical body cells (somatic cells). This is called the *haploid* (after *haploos* meaning single) or (n) number.

It should now be apparent that cellular reproduction, whether for body growth and maintenance or for propagation, must incorporate a mechanism for replicating and portioning the genetic material (chromosomes) to all progeny-cells in a very precise manner. With this mechanism as our prime objective, let us discuss the processes of cellular reproduction.

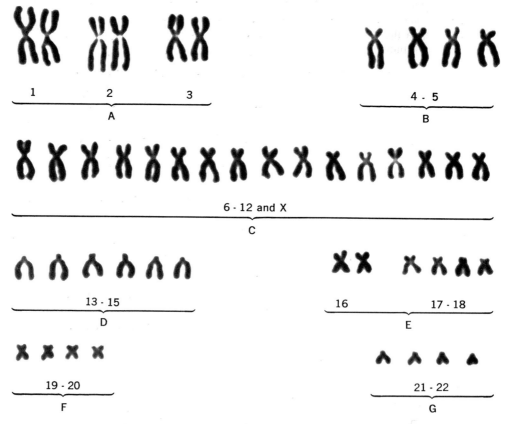

Fig. 4-3. The human female karyotype. Note that specific pairs within groups cannot always be distinguished. The two X chromosomes are included in group *C*. (Courtesy Dr. D. S. Borgaonkar, Johns Hopkins School of Medicine)

THE MITOTIC CYCLE

Mitosis accomplishes the production of two diploid cells from one in a very precise manner through the replication of the chromosomes and the distribution of one copy of each into two identical groups that form the nuclei for each new daughter cell. Exactly what stimulates a functioning cell to reproduce itself is not known; however, it is well known that as the chromosomes replicate prior to cellular reproduction, the DNA content doubles. Whatever the control mechanism is, it apparently operates by initiating DNA replication which, in turn, initiates the subsequent mitotic events.

Mitosis is a continuous process in which a series of events take place that ensure a precise distribution of the genetic material (chromosomes). For convenience of study, the cycle has been divided into stages that describe the major events.

A typical functioning cell that is not actively dividing is said to be in *interphase*. At this time the cell is metabolically active; that is, the genetic material characteristically operative in the particular cell type is doing its job. Observa-

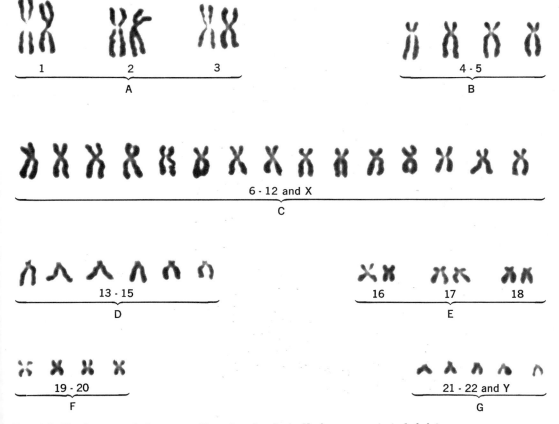

Fig. 4-4. The human male karyotype. Note that the single X chromosome is included in group *C*, while the *Y* chromosome is a member of group *G*. (Courtesy Dr. D. S. Borgaonkar, Johns Hopkins School of Medicine)

tion of this stage would reveal the structure of a generalized cell (Fig. 4-5, *A*) with the chromosomes in the attenuated condition.

Following DNA replication, the chromosomes condense into readily discernible bodies (see Fig. 4-1) randomly dispersed within the nucleus. Concomitantly the centriole (a tiny body outside the nucleus in animal cells) replicates, and the daughter centrioles migrate toward opposite ends (poles) of the cell; the nuclear membrane begins to break down. This constitutes the first recognizable stage of the mitotic cycle and is called *prophase* (Fig. 4-5, *B*).

As the centrioles move apart, a spindle-shaped arrangement of contractile protein fibers forms between them. When the centrioles reach the poles, the chromosomes, which by now have reached their maximum condensation, randomly line up on an equatorial plane of the cell at right angles to the spindle, with the centromere of each chromosome attached to the spindle fibers. This stage is called *metaphase* (Fig. 4-5, *C*). This arrangement of chromosomes forms a more or less two-dimensional *metaphase plate* which, viewed from above (polar view), presents the best opportunity for chromosomal count and study (Fig. 4-5, *D*).

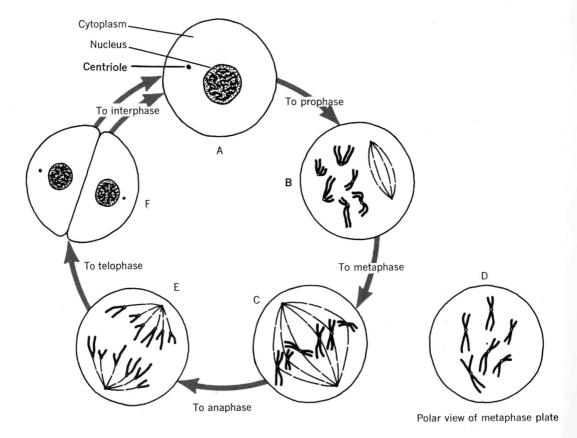

Fig. 4-5. The mitotic cycle. **A,** Interphase (chromosomes in attenuated state) ; **B,** Prophase (duplicated chromosomes condensed) ; **C,** Metaphase (chromosomes arranged on equatorial plane) ; **D,** polar view of Metaphase (best time to count and study chromosomes) ; **E,** Anaphase (chromatids move toward poles) ; **F,** Telophase (chromosomes uncoil; cycle complete).

As we have noted, each replicated chromosome consists of two chromatids joined by the centromere. Soon after establishment of the metaphase plate, the centromeres split along a line parallel to the long axis of the chromatids, and the chromatids travel toward opposite poles as if drawn by the spindle fibers. During this movement, the centromere leads the way, and the chromatids assume a V, J, or I shape, depending on the position of the centromere. This stage is called *anaphase* (Fig. 4-5, *E*).

When the chromatids arrive at their respective poles, they are considered to be full-fledged chromosomes, and the final stage of the cycle, called *telophase*, begins. At this time, a furrow forms around the equatorial plane, which grows continually deeper until the cell is cleaved in two. This process that divides the cytoplasm in two is called *cytokinesis*. At the same time, the chromosomes uncoil, and a nuclear membrane is re-established around them. Eventually the chromosomes reach their fully attenuated condition, and each daughter cell becomes a cell in interphase carrying out its particular function in the organism until it either dies or is stimulated to reproduce itself some time later.

The life span of a typical cell has been divided for convenience into four descriptive periods which are symbolized as M, G_1, S, and G_2. The M stands for

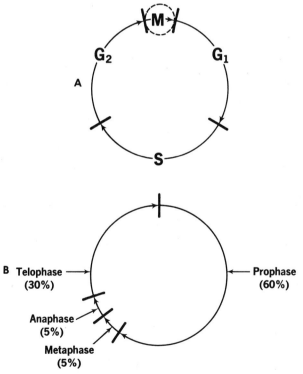

Fig. 4-6. A, Life cycle of a human cell, from tissue culture, divided into periods proportioned by their relative durations. *M* is the mitotic cycle; *G_1* is gap between end of mitosis and next DNA synthesis; *S* is DNA synthesis; *G_2* is gap between end of DNA synthesis and beginning of mitosis. **B,** The relative time durations (in percent) of the stages of the mitotic cycle.

mitosis, from prophase through telophase, while the other three periods occur during interphase. The S represents the period of DNA synthesis (chromosome replication), and G_1 and G_2 represent the nonsynthetic gaps prior to and subsequent to the S period. The M period is of short duration in relation to the complete cycle, and all periods vary in duration from cell type to cell type and from organism to organism. In human cells, studied in actively growing tissue culture, it takes 18 to 19 hours to complete a cell cycle, but only 45 minutes is spent in mitosis. The other 17 hours or more are spent in interphase during which the middle 6 to 8 hours constitute the S period (Fig. 4-6, *A*). The time durations of the various stages in the mitotic cycle are also variable but in general, prophase is the longest stage, usually occupying over half the total time; metaphase and anaphase are very short, and telophase is relatively long (Fig. 4-6, *B*).

Cancer: uncontrolled mitosis

Cancer, one of man's most dreaded diseases, is basically a problem of cellular reproduction gone out of control. The term cancer encompasses a wide variety of diseases, but all are characterized by cells that reproduce when they should not. Although the control and regulation of mitosis is not understood, it is apparent that the loss of control over DNA synthesis underlies cancerous growth, and this loss probably results from a genetic change in the cell. Almost any kind of cell—skin, bone, liver, kidney, and so forth—can become cancerous. The transformation of a normal cell to a cancer cell, whether in a whole animal or a tissue culture, can be brought about by exposure to carcinogenic chemicals, high energy radiations, or certain viruses. Common to all transformations to cancer is a change in the chromosome make-up of the cell. The chromosomal change may be very slight, and tremendous variability exists from one transformed cell to another. Nevertheless, such changes reflect a change in the DNA which is the genetic material, and this change leads to uncontrolled cellular reproduction. The change is heritable, for all cell progeny of a cancerous cell possess the property of uncontrolled mitosis. Thus, it should be apparent that a complete understanding of the cell cycle, especially the events governing DNA synthesis (S period), is fundamental to "controlling" cancer.

THE MEIOTIC PROCESS

Sexual organismic reproduction is also a function of cellular reproduction. In this case, however, a specialized form of cellular reproduction, called *meiosis,* has evolved which affords a commingling of the genetic material of two parents in the production of offspring. Meiosis is the process that shuffles the genetic material into endless combinations and is the underlying mechanism for understanding the principles of heredity. An entire organ system, the reproductive system, has evolved to accommodate, transmit, and service the specialized cells of the meiotic process; this system will be discussed in the next chapter. Presently, we shall discuss the cellular mechanics of meiosis.

Meiosis accomplishes the production of haploid reproductive cells, the *gametes,* from specialized primordial diploid cells known as *germ cells.* Meiosis, like mitosis, involves a mechanism of precise chromosomal distribution to new daughter cells. The meiotic process is an elaboration upon the mitotic cycle whereby two sequences of mitotic-like events are required to (1) reduce the chromosome number from diploid to haploid and (2) to return each replicated chromosome back to its unreplicated, functional form. Hence, two series of prophase, metaphase, anaphase, and telophase, designated as I and II respectively, constitute the meiotic process. As in mitosis, the process is continuous, and the stages provide a convenient reference to the major events; however, meiosis is not truly cyclic because the end products, the gametes, do not repeat the process. Meiosis can therefore be viewed as a terminal process with respect to any further meiotic activity of its immediate products.

Meiosis, like mitosis, is initiated after DNA replication in an interphase cell (Fig. 4-7, *A*). Prophase I follows with events similar to those of mitotic prophase—the chromosomes condense, the centrioles replicate and establish the spindle and poles of the cell, and the nuclear membrane breaks down. However, Prophase I is characterized by two other events that are not common to mitosis and have very important genetic consequences. The first is the attraction and actual pairing of homologous chromosomes in a very intimate point-for-point association to form units known as *bivalents* (Fig. 4-7, *B*). This process is called *synapsis.* A little later in Prophase I, the tight association of the synapsed chromosomes is relaxed somewhat, but it can be seen that the homologous chromosomes are still in contact at several places. These contacts form crosslike patterns called *chiasmata* (singular: *chiasma*) (Fig. 4-7, *C*) within the bivalent. The chiasmata are apparently evidence of an actual exchange of a portion of a chromatid of one chromosome with a corresponding chromatid portion of the paired homologous chromosome. Such an exchange of genetic material between the so-called nonsister chromatids of a bivalent is called *crossing over.* (Sister chromatids are the matched replicates that make up a single chromosome; nonsister chromatids are members of the two separate homologous chromosomes making up a bivalent.) The phenomenon of crossing over has important genetic implications that will be discussed in Chapter 8; suffice it now to note its occurrence here in Prophase I.

At Metaphase I the bivalents line up at random on the equatorial plane of the cell (Fig. 4-7, *D*). For convenience, Fig. 4-7 depicts the same three pairs of chromosomes as Fig. 4-5, but in Fig. 4-7 one member of each pair is gray, whereas the other is black. This is done to facilitate identification of chromosomes (genetic material) of maternal and paternal origin. Since each individual in a sexually reproducing species is itself the product of the fusion of two haploid gametes, we can acknowledge the fact that one member of each pair of homologous chromosomes was contributed from each parent by distinguishing one set in gray (maternal) and one set in black (paternal) for the purpose of determining how the genetic material is shuffled and recombined during the process of meiosis. The random alignment of the bivalents on the metaphase

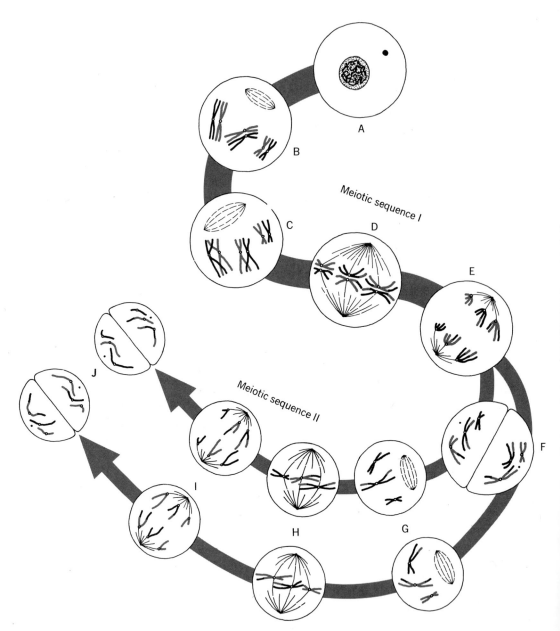

Fig. 4-7. Diagrammatic scheme of the meiotic process. **A,** Interphase; **B,** Prophase I (showing synapsis); **C,** later Prophase I (showing crossing over); **D,** Metaphase I (bivalents on metaphase plate); **E,** Anaphase I (univalents going to poles); **F,** Telophase I (end of first sequence); **G,** Prophase II (univalents show effects of crossing over); **H,** Metaphase II (univalents on metaphase plate); **I,** Anaphase II (unduplicated chromosomes approach poles); **J,** Telophase II (haploid cells formed; genetic material shuffled with one member of each homologous chromosome pair represented; process complete).

plate produces many combinations of end products, and it should be definitely recognized at this point that the arrangement and orientation of maternal and paternal chromosomes toward one pole or another depicted in Fig. 4-7, *D* is only one of the many possibilities that could occur.

At Anaphase I the paired homologous chromosomes separate in such a way that one member of each pair goes to each pole (Fig. 4-7, *E*). It is important to note that the centromeres do not split as they do in mitosis and that entire duplicated chromosomes, each called a *univalent,* go to the poles. This separation of bivalents into univalents assorts the maternal and paternal chromosomes randomly; the particular pole where any univalent ends up is simply a function of the metaphase orientation. Thus, the bivalents are said to assort independently with respect to the maternal and paternal genetic material that now forms a random mixture at each pole, but this mixture contains precisely one representative of each chromosome pair. The importance of synapsis is now apparent—it ensures that one member of each of the matched chromosome pairs will go to each daughter cell.

Telophase I is an abbreviated stage in meiosis (Fig. 4-7, *F*). Cytokinesis forms two daughters cells, but chromosome uncoiling and nuclear reorganization does not take place. Instead, the second series of meiotic stages follows directly. It should be noted that the two cells formed at this point are actually haploid, that is, they contain half the number of chromosomes of their diploid progenitor. Thus, the reduction in chromosome number is accomplished in the first sequence of meiotic stages. However, since the centromeres did not split at Anaphase I, the chromosomes are still duplicated. The primary function of the second meiotic sequence is to reduce the duplicated chromosomes to the unduplicated functional condition. This sequence very closely resembles an ordinary mitotic reproduction (Figs. 4-7, *G* through *J*).

At Prophase II the chromosomes are essentially still coiled from the previous sequence; the centriole replicates, and a new spindle forms. At Metaphase II the chromosomes line up singly to form a metaphase plate. Anaphase II is initiated by the splitting of the centromeres and, just as in mitosis, the chromatids of each chromosome move to opposite poles. Upon reaching the poles, each chromatid becomes a full-fledged, functional chromosome, and Telophase II is initiated. The chromosomes uncoil to their attenuated condition; the nuclear membrane is re-established; cytokinesis occurs, and the process is completed. Since two daughter cells of the first sequence each enter the second sequence to produce two more daughter cells each, the total number of haploid products produced from a single germ cell undergoing meiosis is typically *four.*

Looking at the genetic constitution of the four products, one can see that all four contain different combinations of maternal and paternal genetic material (chromosomes or chromosome segments). This variability is generated in two ways. Initially, the alignment of bivalents at Metaphase I determines the chromosomal orientation to the poles. It can be readily seen that if the alignment at this stage were different (and it occurs at random), the combination of maternal and paternal genetic material (gray and black chromosomes) in the end products would be changed. Secondly, the phenomenon of crossing over recombines maternal and paternal genetic material *within* the chromosomes; therefore, the location and frequency of crossovers generates a considerable amount of genetic recombination. All this means that, in total, the genetic material found in any gamete is not likely to be identical to that in either of the gametes that initially produced the individual in whom the meiosis takes place, nor are any two gametes produced by any individual likely to be absolutely identical to each other. This is the genetic significance of meiosis—it shuffles the genetic material into endlessly new combinations which, when united in the sexual process to another gamete similarily produced by a member of the opposite sex, produce a new and genetically unique individual. Considering only the shuffling of intact chromosomes in the independent assortment of man's 23 pairs, the number of different combinations any one individual could produce in his gametes is mathematically 2^{23} or 8,388,608. When the effect of crossing over is added, the number of different gametes becomes astronomical. Even with each of two parents having 2^{23} different combinations of gametes possible, they could theoretically produce more than 64×10^{12} ($2^{23} \times 2^{23}$) genetically different offsprings, and again crossing over increases this figure tremendously. It is no doubt safe to say that, with the exception of identical twins, the sexual process never has and never will produce two identical human beings!

Human spermatogenesis and oogenesis

The production of sperm through the process of meiosis is known specifically as *spermatogenesis* (Fig. 4-8). The sperms are produced from germ cells, called *spermatogonia* (singular: *spermatogonium*), that line the numerous tiny tubules that make up the sperm-producing organ, the *testis*. Spermatogenesis technically begins with a series of mitoses leading to the production of cells, called *primary spermatocytes*, that then undergo meiosis. Meiosis in the male is quite typical; four haploid products are produced from each diploid primary spermatocyte. The first meiotic sequence produces two cells called *secondary spermatocytes*. Each of these contains 23 univalents (duplicated chromosomes). The second meiotic sequence produces two functional haploid cells from each of the secondary spermatocytes. These cells, a total of four, are called *spermatids*. During a period of maturation, the spermatids are transformed into *mature sperms* consisting of a head, which contains the 23 chromosomes and very little cytoplasm, a neck or midpiece, which houses the mitochondria that provide the energy to the cell, and a long tail whose wagging propels the cell (Fig. 4-9).

The time required for meiosis to occur is about 32 days, 16 days for each meiotic sequence. Another 16 days is required for maturation from spermatid to sperm. Thus, the entire meiotic process in males, from Prophase I in the primary spermatocyte to a fully functional sperm cell, takes about 48 days. Spermatogenesis in total, including the mitoses leading from spermatagonia to

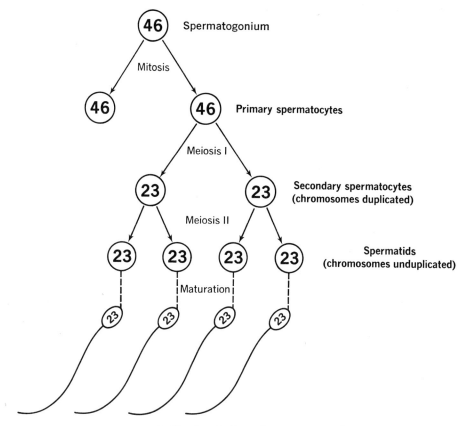

Fig. 4-8. Human spermatogenesis. Numbers indicate chromosome complements.

primary spermatocytes in addition to meiosis *per se,* takes at least 64 days. This 64-day process is carried on by large numbers of cells, and at any given time all stages of the process are occurring so that sperms are constantly produced. Their production normally begins at the sexual maturity of the organism and is continuous throughout the life of the male.

Egg production in the female is call *oogenesis* (Fig. 4-10). The eggs are produced from germ cells in the female egg-producing organ, the *ovary*, called *oogonia* (singular: *oogonium*). In a fashion similar to spermatogenesis, the oogonia produce *primary oocytes* through mitosis, and the primary oocytes then undergo meiosis. However, the process is altered from this point on. Instead of the typical four functional haploid products, each primary oocyte produces only *one functional* egg. This altered pattern has evolved as a mechanism to conserve a food and energy supply of large enough quantity to carry development, should the egg be fertilized, through its initial stages until a food supply can be established through the mother. Since the sperm contain little or no stored food (they have evolved a streamlined design adapted for swimming to the egg), the egg is the sole source of nutrient supply immediately after fertilization.

Thus, meiosis in the female occurs somewhat differently than in the male.

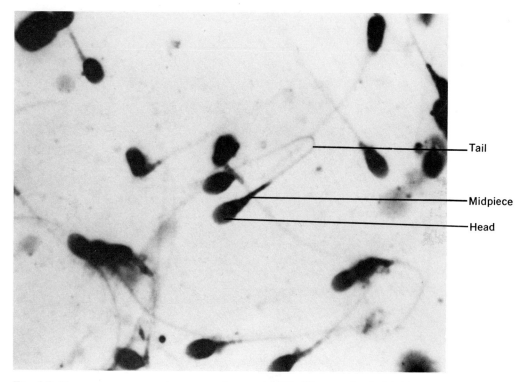

Fig. 4-9. Human sperms as seen in a smear of semen ×2,000. (Courtesy Dr. Judith Ramaley, University of Nebraska Medical Center)

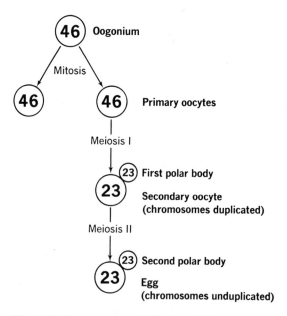

Fig. 4-10. Human oogenesis. Numbers indicate chromosome complements.

The first meiotic sequence produces two cells, of quite disproportionate size, each having 23 univalents. The entire spindle is set off to one side of the cell in such a way that at Telophase I, one set of univalents remains in the central cytoplasmic mass, while the other set is literally pinched off from the edge during cytokinesis. The large central portion retains most of the original cytoplasm of the cell and is called the *secondary oocyte;* the other cell contains a minimal amount of cytoplasm with its set of univalents and, since it was pinched off from one pole of the cell, is called the *polar body.* One secondary oocyte and one polar body represent the products of meiosis I in the female. The secondary oocyte goes on to the second meiotic sequence, while the polar body disintegrates without further reproduction. The second meiotic spindle is also set off to the side of the cell, again producing one large cell and a second polar body which also disintegrates. The large cell is the haploid *egg* with 23 unduplicated chromosomes and a large amount of cytoplasm (Fig. 4-11). The first and second polar bodies simply dispose of the excess chromosomes that must be shed during the production of the single haploid egg. No maturation or further cellular differentiation is required to form a functional egg.

The timing of egg production is also very different from the timing of sperm production. In females the germ cells multiply rapidly during prenatal life, and by about 3 months *before birth,* the oogonia cease mitosis, and all the primary oocytes the female will ever have are already established. Hence, at birth the fe-

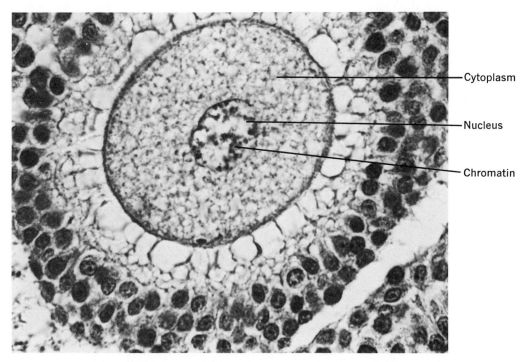

Cytoplasm

Nucleus

Chromatin

Fig. 4-11. A potential human egg (secondary oocyte) in the follicle of the ovary prior to ovulation ×2,000. (Courtesy Dr. Judith Ramaley, University of Nebraska Medical Center)

Table 4-2. Comparison of the timing of spermatogenesis and oogenesis

Spermatogenesis		Oogenesis	
Spermatogonium to	⎱ 16+ days	Oogonium to	⎱ By 6 months after conception
Primary spermatocyte to	⎱ 16 days	Primary oocyte to	
Secondary spermatocyte to	⎱ 16 days	Secondary oocyte to	Anytime from 12 to 50 years after establishment in the ovary
Spermatid to	⎱ 16 days		
Mature sperm		Mature egg	
Total time	64+ days	Total time	12 to 50 years

male has her life-time supply of primary oocytes numbering about 400,000, all suspended in Prophase I of meiosis. The meiotic process remains arrested from birth until puberty at which time hormonal changes occur, and eggs are produced at a rate of one about every 28 days. (This hormonal control of the egg cycle will be discussed in the next chapter.) Even at ovulation (release of an egg from the ovary) the egg is a secondary oocyte, rather than a fully haploid cell since only the first meiotic sequence is completed in the ovary. Without the stimulus of fertilization, meiosis will not be completed before degeneration of the secondary oocyte occurs. Upon fertilization, oogenesis is completed, producing a functional haploid egg to unite with the sperm nucleus. About 400 potential eggs are ovulated during the reproductive life span of a normal female, and only a very few, those actually fertilized, ever really complete meiosis. Hence, complete oogenesis may take anywhere from 12 to 50 years with each egg being older than the female producing it—according to the common practice of counting age from time of birth! A comparison of the timing of spermatogenesis and oogenesis is given in Table 4-2.

5

Reproductive systems, development, and birth control

Two tracts lead to the bridge of life that spans the generations. They are the male and female reproductive systems through which the gametes are transported and by which both gametes and developing offspring are housed and serviced. A new life quite literally begins in the tubes of these systems, and it is here that the new genetic combinations, similar to their parents yet different, are made and nourished until they emerge from the female tract to take up life on their own—at first still highly dependent on extrinsic support but growing evermore independent with age.

We will discuss the reproductive systems of man in some detail in this chapter, not only for the sake of placing human heredity and development in the proper context for the subsequent discussion of more specific inherited traits and genetic defects, but also because these systems receive a great deal of attention associated with cultural and behavioral traits by many people who are woefully uninformed about the basic biological function of these systems. Biological illiteracy often leads to unwanted pregnancies, unwarranted fears and guilts, or sometimes physical harm. Man's cultures have also devised many myths, superstitions, and dogmas associated with these systems, many of which reek of biological ignorance. Finally, the practice of contraception involves the manipulation of these systems in some way or another so that they will "work but not function" or provide "sex" without babies. This is viewed as necessary if man is to control his population explosion yet maintain the myriad of behavioral and psychological characteristics and needs other than baby production that have evolved (mostly through cultural evolution) in an inseparable association with sex. Whether one fears sex or feels guilty about it, has a special superstition or conforms to a particular dogma concerning sex, or is for or against certain methods of contraception, cognizance of the biological function of the reproductive systems should not be neglected—after all, that's why they evolved!

87

88   THE SPERM TRACT

The sperm tract is essentially one long tubular system with one opening to the outside (Fig. 5-1). The tract begins in the *testes*, which are composed of large numbers of tiny tubes, the *seminiferous tubules*, surrounded and separated from each other by cells known as *interstitial cells*. The essential reproductive function of spermatogenesis occurs in the seminiferous tubules, whereas the interstitial cells are responsible for the production of the primary male sex hormone *testosterone*. The seminiferous tubules collect into an interconnected, highly coiled network that is attached to the upper surface of the testis and is called the *epididymis*. Spermatid maturation occurs in the epididymis, and the mature sperms are stored there until ejaculated. The tract continues as the coiled epididymis more or less straightens out into a thick-walled *sperm duct* (vas deferens), which goes to a point below the bladder and joins another tube, the *urethra*, coming from the bladder. The urethra passes to the outside, serving as a common duct for the urinary and reproductive systems.

It should be noted that the sperm duct transports the mature sperm from the testes, which are located *outside* the body proper, to the lower abdominal area where accessory glands and the urinary system are located. The pouchlike sac in which the testes are suspended is called the *scrotum* (bag). It is an evolutionary adaptation found in most mammals for lowering the temperature of the

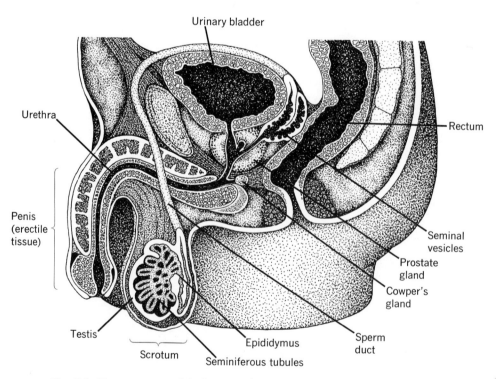

Fig. 5-1. The sperm tract of the human male.

testes. Spermatogenesis will not occur normally at internal body temperature, but the scrotal temperature, about 4° C lower, is optimal. The testes develop in the abdominal cavity at a position near the kidneys and shortly before birth descend into the scrotum through an opening in the abdominal wall called the *inguinal canal*. Failure of the testes to descend normally is called *cryptorchidism*, which results in sterility; it is usually corrected by surgery. The inguinal canal is also quite prone to rupture, making men highly susceptible to hernias in this region.

The three sets of glands associated with the sperm tract are essential to sperm transport and survival. Proceeding from the testes, the first glands, found toward the end of the sperm duct, are the *seminal vesicles*. They produce a thick yellow alkaline secretion that contains large amounts of the sugar fructose. Sperm motility is adversely affected by the acid environment typical of the urinary system to which the sperm tract joins at the urethra. The alkaline nature of the seminal vesicle fluid reduces this acidity, while the sugar provides the nutrient required by the sperm. Next is the *prostate gland,* which secretes a thin milky alkaline fluid that aids sperm motility by acting as a pH buffer. Finally, at the beginning of the urethra are the *bulbourethral glands,* also known as *Cowper's glands.* These glands secrete an alkaline mucous that coats the urethra prior to ejaculation; this secretion provides an advance neutralization of the urethra. The secretions of these glands are collectively called the *seminal fluid,* and the combination of sperms and seminal fluid is the *semen.*

Delivery of the sperm

It was mentioned previously that spermatogenesis is an essential male reproductive function. After the sperms are produced, the other essential male function is their delivery to the proper location inside the female reproductive system. The evolutionary adaptation for this form of fertilization was the *penis.* This structure actually evolved in the reptiles, who were the first group of animals to master a fully terrestrial life-style. The amphibians, predecessors of the reptiles, evolved from fish stock and, although they lived on land, still had to return to the water for reproduction; water was the medium into which sperms and eggs were released to subsequently combine. Reproduction on land requires *internal fertilization,* and the penis is an adaptation for sperm delivery from the male tract directly into the female tract. Both the birds and the mammals have inherited this mode of fertilization from their reptilian ancestors.

The penis, which constitutes the male external genitalia, consists of three cylindrical masses of tissue bound together and covered with skin through which the urethra passes to the outside of the body. At the end of the penis, the bottom cylindrical tissue is expanded into a blunt cone-shaped head called the *glans penis.* The glans is covered by a fold of skin called the *prepuce* or foreskin; *circumcision* is the surgical removal of the prepuce. Most of the penile mass is *erectile tissue* that enables the penis to become hard and erect and facilitates penetration of the female tract for sperm delivery. This tissue is a spongelike network of vascular spaces. Normally the blood vessels (arterioles) supplying these spaces are constricted; the network contains very little blood, and the penis is limp. During sexual stimulation, however, the blood vessels relax and open, and blood rushes into the spongy spaces and distends them into a rigid condition (like filling a balloon with water). As the erectile tissue expands, the veins that normally drain it are compressed; this minimizes the outflow and contributes to the distention. The whole process occurs rapidly, with complete erection sometimes taking only 5 to 10 seconds. Erection is the result of nervous stimulation that initiates a reflex reaction. As a result, thoughts or emotions can cause erection in the complete absence of physical stimulation; conversely, failure to attain erection (impotence) is most often a psychological rather than a biological problem. The interference of erection by alcohol, for example, is probably caused by its effect on the higher brain centers that control the erection reflex.

The final step in sperm delivery is *ejaculation* (sudden ejection). This is also basically a nervous reflex in which the muscle of the prostate gland and the muscles surrounding the urethra rapidly and rhythmically contract, pushing the semen out of the sperm tract. A simultaneous nervous stimulation of the whole body associated with the feeling of great pleasure also occurs, the entire event constituting the *orgasm*. The average amount of semen ejaculated is about 3 milliliters (slightly less than a teaspoonful) and each milliliter normally contains about 100 million sperms. Thus, of the 300 million or more sperms that enter the female, only one will fertilize the egg. A fairly large quantity is necessary at the onset, however, because the sperms have a long way to travel after being ejaculated into the vagina of the female. We will resume their journey when we discuss the egg tract.

Regulation of the sperm tract

Spermatogenesis is under hormonal control. Adequate amounts of the male hormone *testosterone* are essential to maintain normal fertility; deficiency invariably results in sterility. Testosterone also exerts widespread bodily effects, specifically on the secondary sex characteristics (pubic and axillary hair, beard, voice, skin texture, and pattern of muscle and fat distribution) and generally on all types of growth. Small amounts of testosterone or testosterone-like substances, generally called *androgens*, are also necessary to promote normal growth in women. In females the primary site of production is the adrenal gland.

Testosterone alone does not control spermatogenesis; other hormones produced by the pituitary gland located at the base of the brain are also necessary. The pituitary hormones, known as gonadotrophins (hormones that affect the *gonads* or reproductive organs), are follicle-stimulating hormone (FSH) and luteinizing hormone (LH). They were named for their effects in the female, which we will discuss later, but they exist in both sexes. LH in the male is fre-

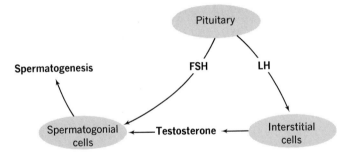

Fig. 5-2. Hormonal control of spermatogenesis.

quently called interstitial-cell-stimulating hormone (ICSH). As its male syn-onym implies, LH stimulates the interstitial cells of the testes whose function is the production of testosterone. FSH has a direct stimulation on spermatogenesis but cannot sustain it alone. Fig. 5-2 illustrates the hormonal control of spermato-genesis.

Exactly how LH and FSH production are stimulated in the male is not clear; however, it is important to note that both hormones seem to be produced at a rather fixed and continuous rate during the adult life span. Accordingly, tes-tosterone production and spermatogenesis also occur at relatively constant rates. Such constancy is unusual for hormonal systems and is completely different from the cyclic nature of the regulation of oogenesis, which will be discussed subsequently.

THE EGG TRACT

Like the sperm tract, the egg tract is essentially a tubular system with a single opening to the outside. The tract begins at the *ovaries*, but the ovaries are not an intact part of the tubular system like the seminiferous tubules of the male testes. Rather, the ovarian tissue consists of many discrete clusters of cells known as *primordial follicles* (Graafian follicles). Each follicle is composed of one primary oocyte surrounded by a layer of cells. At birth the ovaries consist of some 400,000 such follicles, one for each preformed primary oocyte. Recall that a female releases only about 400 mature eggs during her reproductive life time; thus, 99.9% of the primordial follicles are destined to degenerate. Be-ginning at puberty, certain follicles mature at varying times in such a way that follicles in different stages of development would characterize the ovarian tissue of an adult female. Development of the follicle entails an increase in the size of the primary oocyte, a proliferation of the surrounding follicle cells, and the for-mation of a fluid-filled central space. The mechanisms that stimulate the develop-ment of certain primary follicles while leaving the majority unstimulated are not known; however, the hormonal regulation of this process will be discussed in the following paragraphs.

The remainder of the female tract consists of two *oviducts* (Fallopian tubes),

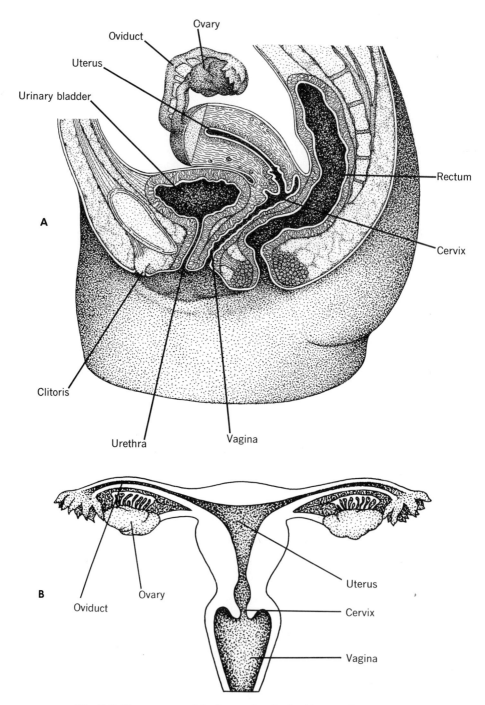

Fig. 5-3. The egg tract of the human female, **A,** side view, **B,** front view.

one from each ovary, which unite to form the *uterus,* which then passes to the *vagina,* which opens to the outside (Fig. 5-3). The urethra coming from the bladder opens to the outside directly above the vaginal opening. The ovarian end of the oviduct is broad and fringed (fimbriated) and is in intimate contact with but does not penetrate the ovary. When an egg is released, it literally bursts through the ovarian wall and falls into the oviduct to begin its journey through the egg tract.

The release of the egg from the ovary into the oviduct constitutes *ovulation.* Once in the oviduct, the egg moves rapidly for several minutes, propelled by cilia and muscular contractions of the duct, but soon it slows down and takes several days to reach the uterus. The egg remains fertile for only 10 to 15 hours after ovulation. This means that fertilization must occur in the oviduct. Since the sperms are ejaculated into the vagina, they must travel about 7 inches (a tremendous distance for a microscopic sperm) through a rather harsh acid environment, buffered only by the seminal fluid and supplied only with the energy from the sugar in the fluid. Sperms can live about 48 hours in this environment, but many succumb along the route. Whether or not any number below the average of 300 million sperms per ejaculation could still ensure a high probability of success at reaching a viable egg is questionable; only a few thousand of the initial number reach the oviduct. Although the life span of eggs and sperm varies from person to person, in general, to maximize the likelihood of fertilization, coitus (sexual intercourse) should occur no more than 48 hours before or 15 hours after ovulation. Even these are broad limits for, although still fertile, the older the egg is, the greater the likelihood of continued egg degeneration, even after fertilization. Once an egg is fertilized, a chemical transformation of its outer membrane occurs, making it impenetrable to other sperms.

Externally the vaginal opening (orifice) is bordered by a thin fold of membrane called the *hymen*; it usually forms a circular border around the orifice, but great variation in shape occurs from person to person. Occasionally the hymen may cover the entire vaginal opening; this is called an imperforate hymen and requires surgical opening. The other external genital organs, collectively known as the *vulva,* are illustrated in Fig. 5-4. The vaginal opening is surrounded by two sets of protective coverings. The outer set is the *labia majora* (major lips). These two folds of tissue lie longitudinally on either side of the opening; the skin of the labia majora contains sweat glands and sebaceous (oil secreting) glands and is hair-covered. Two smaller folds of skin known as the *labia minora* (small lips) are situated between the labia majora. The labia minora are hairless and lack sweat glands; at their anterior juncture they form a small hood of skin called the *prepuce* that partially covers the *clitoris* located underneath. The clitoris corresponds in structure and origin to the penis of the male. It becomes erect and very sensitive upon sexual stimulation. Located above the labia majora and minora is a rounded cushion of fatty tissue padding the pubic bone called the *mons pubis.* At puberty it becomes covered with dense hair that enhances its cushioning power; it is then called the *mons veneris.*

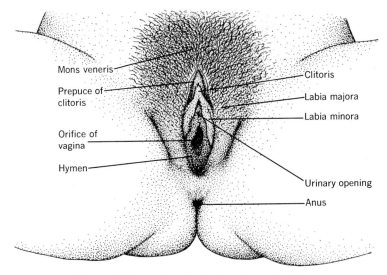

Fig. 5-4. External genitalia of the human female.

Regulation of the egg tract

Regulation of the egg tract includes the hormonal control of oogenesis plus the associated preparation of the uterus in anticipation of fertilization. The same two gonadotrophins from the pituitary gland that regulate the male tract, FSH (follicle-stimulating hormone) and LH (luteinizing hormone), also stimulate the female tract. In addition, two hormones produced by the ovary, *estrogen* and *progesterone,* are involved in the female regulatory cycle. The cycling of these hormones is in distinct contrast to the stable continuous rates of male hormone secretion. As mentioned in the previous chapter, oogenesis produces a single egg from one primary oocyte, and this event occurs about once every 28 days during the reproduction life span of the female in the following manner.

FSH and LH are both secreted by the pituitary gland and, as its name implies, FSH stimulates a certain follicle to begin developing; the selective mechanism here is not known. The follicle cells manufacture and secrete *estrogen,* the primary female hormone responsible for the development of the female secondary sex characteristics (breast development, hair distribution, muscle and fat distribution, and so forth). As the follicle cells multiply (by mitosis), estrogen secretion progressively increases. This secretion, however, requires the presence of the other pituitary hormone, LH. The meiotic process, which was arrested in Prophase I at birth, resumes in the primary oocyte of the maturing follicle. This development proceeds for about 14 days, and then a brief but rapid increase in the quantity of LH occurs. This produces a shift in the FSH-LH balance and provides the stimulus for ovulation. The follicle bursts, and the egg (technically, the secondary oocyte) is released into the oviduct. Exactly what causes the dramatic increase in LH secretion is not known; one of the latest theories is that a

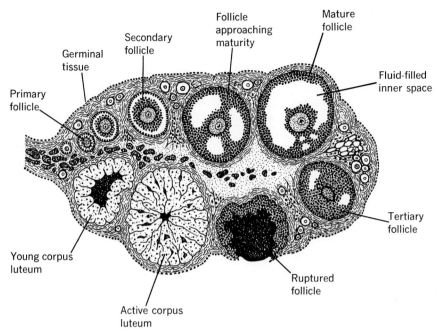

Fig. 5-5. Schematic diagram of the ovarian events in the female reproductive system.

"biological clock" mechanism in the brain may control it. This first half of the regulatory cycle is called the *follicular phase.*

Upon rupture of the follicle, estrogen secretion decreases slightly, but other ovarian cells maintain its production. The ruptured follicle, which has now lost its egg, almost immediately undergoes a transformation into a new structure called the *corpus luteum* (yellow body). The corpus luteum manufactures the other ovarian hormone *progesterone* and also estrogen in large quantities. The increased concentrations of estrogen and progesterone inhibit production of LH and FSH by the pituitary gland, thereby lowering their concentrations and preventing the development of another follicle during the lifespan of the corpus luteum. If fertilization does not occur, the corpus luteum degenerates after 10 to 14 days of functioning. This constitutes the second half of the regulatory cycle, called the *luteal phase.* When it ends, LH and FSH production resume as the progesterone and estrogen concentrations drop and a new follicular phase initiates another cycle. A schematic diagram of the ovarian events is given in Fig. 5-5.

Profound changes in the uterus also occur during the hormonal cycle, especially in the mucous membrane called the *endometrium,* which lines the inside. If the egg is fertilized in the oviduct, the female tract must be prepared for the ensuing pregnancy. The uterus, in which further development takes place, receives the fertilized egg that becomes attached to the endomentrium, which is in a thickened actively glandular condition conducive to *implantation,* as the process is called. At the beginning of the follicular phase in the ovary, the endometrium is quite inactive. As a matter of fact, it is disintegrating from its previous preparation. By about the fifth day of the follicular phase and until ovulation, the endometrium grows because of the influence of increasing estrogen concentrations in the blood. This period is called the *proliferation phase* of the uterine cycle. Following ovulation and formation of the corpus luteum, the influence of progesterone acting with estrogen stimulates the endometrium to become an actively secreting glandular tissue; much glycogen, a form of stored sugar energy, is secreted into the tissue. The blood vessels become more numerous, and various enzymes accumulate in the endometrial tissue. All these changes create an environment ideally suited for implantation. This is called the *secretory phase* of the uterine cycle, and it corresponds to the luteal phase of the ovarian cycle. When the corpus luteum degenerates, the fall in progesterone and estrogen concentrations deprives the highly developed endometrium of its hormonal stimulation, and disintegration of the tissue ensues. The entire lining breaks down and is sloughed off; the rupture of tiny blood vessels in the lining produces minor bleeding, and some blood and endometrial debris pass to the outside through the vagina. This process is *menstruation,* and the whole regulatory system involving the ovarian cycle and the uterine cycle is commonly called the *menstrual cycle.* The menstrual flow occurs during the first 4 to 6 days of the next ovarian follicular phase. Thus, the beginning of menstruation marks the first day of a new regulatory cycle in the egg tract. Table 5-1 summarizes the hormonal changes and correlated ovarian and uterine changes that occur during the regulatory cycle of the egg tract.

Eventually the hormonal cycling in the female diminishes. Estrogen secretion drops even though the pituitary gonadotrophins are secreted in increasing amounts. Finally ovulation and menstruation cease completely. This ending of the reproductive life span of the female is the *menopause;* it usually occurs between the ages of 45 and 50. Some estrogen secretion generally continues beyond menopause, but it gradually diminishes until it can no longer maintain the estrogen-dependent tissues (the breasts and genital organs), and they gradually atrophy (degenerate). Emotional disturbances frequently accompany menopause, in part because of the direct effect of estrogen deficiency and also because of the disturbing awareness that the reproductive potential is ended. It may take a year or two until a new physiological balance is finally established and the female adapts to her changed life.

Male changes with age are much less drastic. Testosterone secretion does decrease with age, but it usually remains in high enough concentrations to maintain reproductive capacity throughout the lifetime of most men.

DEVELOPMENT

Upon fertilization, the egg's 23 chromosomes combine with the 23 from the sperm to reconstitute the diploid number of 46. This 46-chromosome cell is the *zygote.* The zygote undergoes cellular reproductions (mitosis), producing a number of cells. After its 2- to 3-day trip down the oviduct, the cell mass floats in the fluid of the uterus for several days and finally, about 7 days after fertilization, implantation occurs. By this time the zygote has developed into a ball of 60 to 100 cells with a central fluid-filled space and is called a *blastocyst*. The cells of the blastocyst form an outer layer, which will produce the embryonic membranes, and an inner compact mass, destined to become the *embryo*. The outer cell layer makes contact with the uterus and secretes digestive enzymes that break down the thickened endometrium, and the embryo becomes embedded in the uterine wall by literally eating its way in. The breakdown of the nutrient-rich endometrium provides the initial food and raw material supply to the tiny embryo.

Meanwhile, embryonic blood vessels develop in the outer cell layer, and this vascularized structure becomes the membranous *chorion*. A second membrane, the *amnion*, forms between the growing embryo and the chorion. It is connected with the chorion at the stalk supporting the embryo; this stalk eventually develops into the *umbilical cord*. The amnionic membrane encloses the embryo in a chamber filled with fluid, which is very much like sea water with respect to its salt concentrations. Thus, even in man, development occurs in an environ-

Table 5-1. Summary of hormonal, ovarian, and uterine changes during the regulatory cycle of the egg tract

Days	1 2 3 4 5 6	7 8 9 10 11 12	13 14 15 16	17 18 19 20 21 22 23 24	25 26 27 28
Ovarian phase	Follicular		Ovulation	Luteal	
Uterine phase	Menstrual	Proliferation		Secretory	
Hormonal changes					
FSH	Increasing	High, then declining	Low	Low	Increasing
LH	Low	Low, then large increase	High	Decreasing	Decreasing
Estrogen	Low	Increasing	High	Decline, then increase to high	Decreasing
Progesterone	None	Very little	Very low	Increasing to high	Decreasing
Ovarian changes	New follicle begins to develop Old corpus luteum degenerating	Growth and maturation of follicle Meiosis occurring in primary oocyte	Follicle ruptures, and egg is released into oviduct	Corpus luteum formed and secreting actively	Corpus luteum begins to degenerate
Uterine changes	Degeneration and sloughing off of thickened and glandular endometrium	Development of new endometrial layer	Continued	Active secretion by glandular endometrium; high vascularization	Endometrium begins to degenerate

ment similar to that of the primitive sea where life arose and evolved. The amniotic fluid also provides a protective cushion for the delicate embryo.

More recently the amniotic fluid has been used to detect various genetic abnormalities long before birth. Cells from the embryo break off and float in the amniotic fluid, and by a technique known as *amniocentesis* a needle can be inserted through the pregnant woman's abdomen and into the amniotic chamber to extract some embryonic cells. These cells can then be cultured and tested to determine various biochemical abnormalities that may have been inherited, or the chromosomes can be karyotyped and analyzed for aberrations. If a severe abnormality is discovered, an abortion may be performed to terminate the pregnancy.

The chorion forms finger-like projections, called *chorionic villi*, which protrude into the uterine wall and firmly attach the embryo and its membranes. The villi are abundantly supplied with blood vessels, as is the uterine tissue, the endometrium, into which they penetrate. This highly vascularized area, called the *placenta*, serves as the embryo's respiratory, nutritive, and excretory organ from about 5 weeks after implantation until development is completed. The endometrium contains maternal lakes or sinuses of blood adjacent to the chorionic villi. Oxygen and nutrients, provided by the mother, readily diffuse from the endometrial sinuses through the walls of the villi into the embryonic bloodstream that flows through the umbilical cord into the embryo. At the same time, waste products of the embryo move in the opposite direction, diffusing from the embryo's bloodstream into the endometrial sinuses to be ultimately eliminated by the mother. Fig. 5-6 diagrammatically illustrates the embryonic membranes and the placenta. It should be noted that *no direct connection* between maternal and embryonic bloodstreams exists. Exchange is by diffusion, and any substance of small enough molecular size can readily be transmitted from mother to child.

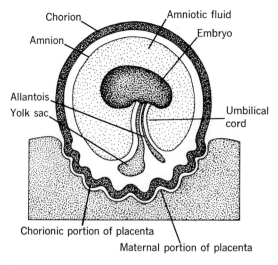

Fig. 5-6. Diagram of human embryonic membranes and placenta.

Drugs and chemicals in the mother's bloodstream also enter the embryo, as do viruses but not most bacteria, which are too large.

The prenatal period of development lasts about 40 weeks, or 10 lunar months (28 days each), approximately 9 calendar months. It can be characterized in three phases. The *egg phase* lasts from day zero (the moment of fertilization) until the end of the second week. Implantation occurs about midway through this phase, and during the second week primitive chorionic villi begin to form. Weeks 3 through 8 constitute the *embryo phase*. The chorionic villi develop into a fully functioning placenta about midway through this phase. By the end of the eighth week, the rudiments of all the adult organs are established, and the embryo is about an inch long. From the end of the eighth week (second lunar month), the embryo is called a *fetus*, and the remaining time of development to birth is the *fetal phase*. This phase consists mainly of the growth and maturation of the rudiments set down during the embryo phase; few, if any, major structures are newly formed. The fetus is distinguishable as human in form at the beginning of this phase, and by the end of week 16 (fourth lunar month) it has grown to 6 inches with sex discernible by the external genitalia (Fig. 5-7). By the end of week 24 (sixth lunar month), the fetus is about 12 inches long, and the first fat deposits are made beneath its wrinkled skin (Fig. 5-8). Weak cries and energetic movements of the limbs occur by the end of week 32 (eighth lunar month). The fetus is now about 16 inches long and has more than doubled its weight since week 24. A fetus born from this time on may survive; the probability of survival increases with age. By week 40 (tenth lunar month) the fetus is normally about 20 inches long, and pregnancy has reached its full term. The single egg cell fertilized by a single sperm cell has developed into a new individual ready to begin life on its own in the much harsher environment of man's external world.

To maintain the pregnancy, the endometrium must retain its thickened, glandular condition. If it should break down, as it usually does within 2 weeks after ovulation, implantation would not occur, the placenta would not be established, and development would not proceed. Continued estrogen and progesterone secretion is required to sustain the endometrium and prevent menstruation. It is not known just what controls the degeneration of the corpus luteum during a nonpregnant cycle or its persistence during pregnancy. It is known, however, that a new hormone, very similar to LH, is produced by the chorionic villi almost immediately after implantation. This hormone, called *chorionic gonadotrophin* (CG), strongly stimulates estrogen and progesterone secretion by the corpus luteum. Recall that the increased concentrations of estrogen and progesterone inhibit the pituitary hormones; therefore, LH and FSH levels remain extremely low during pregnancy. This ensures that further follicle development, ovulation, and menstruation will be suspended for the duration of the pregnancy. Placental CG, even though similar to LH in activity, is not inhibited by increased estrogen and progesterone concentrations, thereby permitting its continued stimulation of the corpus luteum.

Since CG is produced only during pregnancy, determination of its presence is the basis of most clinical pregnancy tests. By late in the first month of pregnancy, CG is excreted in the urine. A urine sample is mixed with a solution of antibodies against CG and added to a suspension of CG-coated red blood cells or latex particles. The antibodies typically react with CG to chemically bind it. Normally the antibodies would react with the CG coating of the suspended cells or particles, producing an easily visible clumping. In the case of pregnancy, however, the antibodies react with the CG in the urine before the suspended particles are added. Thus, when the CG-coated particles are added, the antibodies are already bound to the CG in the urine and the suspended particles

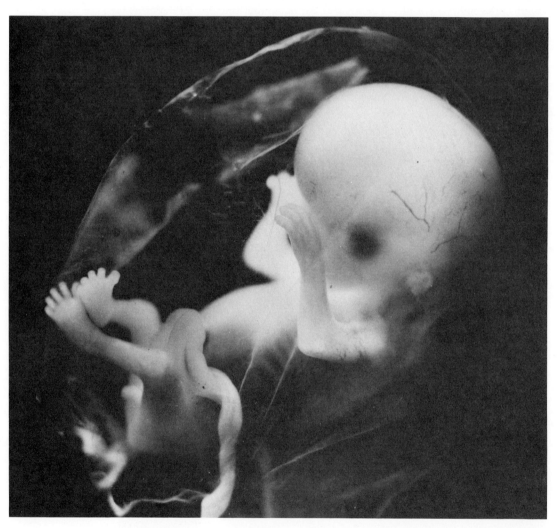

Fig. 5-7. Human fetus at eleventh to twelfth week. The safest forms of abortion are performed up to this stage of development. (From Nilsson, L., Ingelman-Sundberg, A., and Wirsén, C.: A child is born, Stockholm, 1966, Albert Bonniers Förlag.)

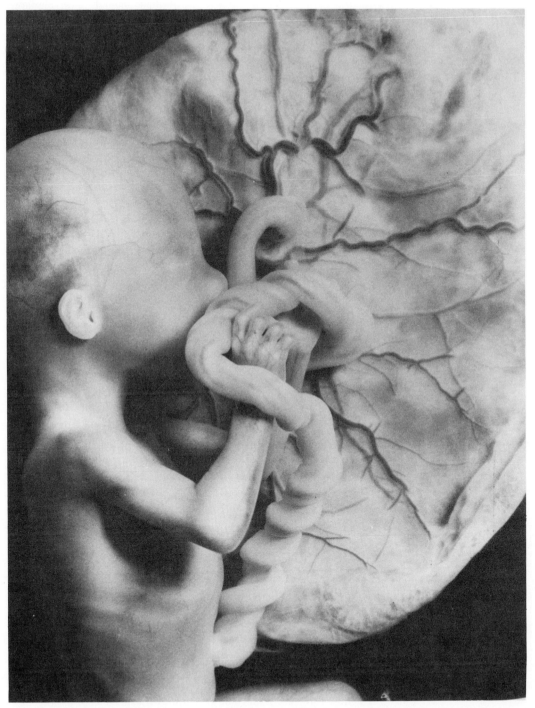

Fig. 5-8. Human fetus at twenty-fifth week. (From Nilsson, L., Ingelman-Sundberg, A., and Wirsén, C.: A child is born, Stockholm, 1966, Albert Bonniers Forlag.

float freely. This constitutes a positive test for pregnancy. Earlier pregnancy tests relied on the LH-like activity of CG to stimulate ovulation in immature mice, rats, or rabbits when injected with urine of a pregnant woman. These tests were much slower and more complicated than the technically simple test-tube method that can give the answer within an hour.

The secretion of CG increases rapidly during early pregnancy, reaches a peak in week 9 or 10, then rapidly falls to a low by week 14, and remains relatively constant for the duration of the pregnancy. As CG secretion falls, the placenta begins to produce large quantities of estrogen and progesterone and from this point on sustains the pregnancy itself. In humans, removal of the ovaries (which includes the corpus luteum) during the first 3 calendar months causes immediate loss of the fetus through breakdown of the endometrium; however, during the last 6 months their removal has no effect, indicating placental control of the endometrium.

What is the relationship of development to heredity and to human affairs? Development is under genetic control. The progression from a fertilized egg to a fetus with the rudiments of all the adult tissues and organs is not simply the growth and elaboration of a tiny preformed being into an increasingly larger product that finally bursts forth into the world. Neither the zygote nor the blastocyst has eyes, limbs, or internal organs, but the fetus has all of these. The formation of these organs comes about through the *differentiation* of what were at first similar cells into cells with more and more specialized structures and functions. This specialization is complex and not fully understood, but it results in the construction of tissues and then organs and organ systems. Throughout the process of differentiation, different genes on the chromosomes perform different functions at different times. The regulation of this process is the subject of much current research. (Gene function and the basic concept of gene regulation will be discussed in the next chapter.) The important concept here is that a very precise regulation of gene function, turning genes on and off, operates during embryonic development to literally create the new individual from the raw materials supplied by the environment. Development proceeds from the zygote through a series of patterns based upon previous patterns. It is because the later patterns are different from, but dependent upon, the previous patterns that development is said to be *epigenetic;* that is, development begins in a relatively structureless and homogeneous mass and gradually the organs and parts differentiate to produce the adult body. If any gene in the blueprint is defective, a defective product may result. Likewise, if the environment is defective or in any way inhibits the proper functioning of the genes, a defective product may also result.

Many of the genetic traits and abnormalities that will be discussed in subsequent chapters have their initial effect on the individual during embryonic development. It is therefore very important to retain the basic concept of epigenesis under genetic control.

Much of man's behavior not only affects the external environment but also alters the internal fetal environment. It has been noted that drugs, chemicals, and viruses pass through the placenta and into the embryo. In the case of drugs, both legal and illegal, we usually know little or nothing of the effect they may have on any particular critical stage of embryonic development. The classic example of this situation occurred in the early 1960s, involving the sleep-inducing drug *thalidomide*. This drug was found safe and effective for adults, but its effect on a critical stage of embryonic development was tragic. When thalidomide was used during weeks 3 to 5 of the pregnancy, the normal development of the limbs, whose rudiments were forming at this time, was altered, thus producing babies with seriously deformed arms or legs (Fig. 5-9). This condition of flipperlike limbs is called *phocomelia* (seal limbs). In Germany, where the drug was most prevalently used, thousands of deformed babies were born before the cause was determined. A similar condition is known to be produced as an inherited defect. Thus, it is evident that during development when differentiation of the limb buds is taking place, the controlling genes must operate correctly. A defect in the genetic mechanism produces phocomelia, or an environment (thalidomide) that apparently inhibits proper gene function may have the same result. Specific environmental changes that modify development in such a way that the resultant phenotype simulates the effect of an inherited trait pro-

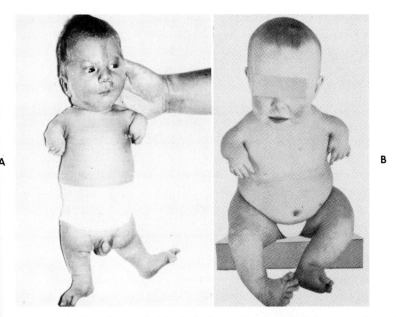

Fig. 5-9. A, An infant with inherited phocomelia (seal limbs), a recessive trait; and **B,** a thalidomide baby. The thalidomide infant has normal genes, nevertheless its limbs developed abnormally because of poor environmental conditions at a critical stage of development. Such environmentally induced mimics of inherited traits are known as phenocopies. (**A,** From Raney-Brasher, B.: Shand's handbook of orthopaedic surgery, St. Louis, 1971, The C. V. Mosby Co. **B,** from Stevenson, R.: The fetus and newly born infant, St. Louis, 1973, The C. V. Mosby Co.)

duce what are known as *phenocopies* (Fig. 5-9, *B*). Phenocopies do not have defective genes, and their aberration is not inherited. Thalidomide produced phenocopies of phocomelia, but if these individuals should reproduce, there is no danger of their passing this trait on to their children. Unfortunately we do not know what subtle physical changes or genetic mutations are being produced by the myriad of drugs and chemicals that deliberately and inadvertently enter our bodies and, in pregnant women, pass through the placenta into the embryo.

German measles (rubella) is a virus infection that also causes severe abnormalities in the developing embryo. If a pregnant woman contracts the disease during the early weeks of pregnancy (the first trimester), there is a 60% chance that her child will be abnormal in some way. The main abnormalities caused by rubella are deafness, congenital heart disease, eye defects (especially cataract), and mental retardation. These are also phenocopies that resemble known hereditary conditions.

Heredity and environment, environment and heredity, one cannot exist without the other. Development and, for that matter, life cannot proceed beyond the limits set genetically, even in a "perfect" environment. On the other hand, genetic potential cannot be realized without the proper environment. This relationship between heredity and environment will be emphasized in particular cases in subsequent chapters. However, it should be noted here that the heredity-environment interaction is *always* occurring, and when it is not specifically mentioned, we simply assume that the genes are operating in a "proper" environment.

CONTROL OF BIRTH

Birth control, in general, includes all methods by which the frequency of actual childbirth is regulated. All birth control measures relate to a knowledge of the male and female reproduction systems, pregnancy, and development. One cannot rationally expect to make social decisions or moral judgments concerning birth control, and more importantly population control, without a clear understanding of what is happening biologically and how any particular method works physiologically.

Birth control methods fall into two categories. *Contraception* includes all temporary and permanent measures designed to prevent coitus from resulting in pregnancy. This may be achieved by preventing ovulation, fertilization, or implantation. Interference with embryonic survival beyond implantation by physical or chemical methods is considered *postconception*. Until very recently all contraception was based on preventing the sperm from reaching the egg to fertilize it. Current and prospective methods, on the other hand, operate on the ovulatory cycle in the female and the hormonal control of reproduction in both sexes. Table 5-2 lists the various methods that account for virtually all birth control practiced in the world today.

Table 5-2. Methods of birth control used in the world today

Contraception			Postconception
Folk methods	**Traditional methods**	**Modern methods**	
Coitus interruptus	Condom	Intrauterine devices	Abortion
Postcoital douche	Vaginal diaghragm	Oral contraceptives	
Prolonged lactation	Spermicides	Surgical sterilization	
	Rhythm		

Coitus interruptus

Coitus interruptus involves withdrawal of the penis from the vagina prior to ejaculation. It is probably the oldest contraceptive procedure known to man. It is referred to in the Old Testament and is known to primitive tribes throughout the world. In Europe during the Middle Ages and beyond, where marriages occurred late and unmarried adults frequently had intercourse, the strong condemnation of pregnancy out of wedlock was averted principally by the use of coitus interruptus.

The effectiveness of coitus interruptus has been traditionally underestimated. Studies have shown the pregnancy rates to be only slightly higher with this method than with other contraceptives in the same populations. Failure is usually attributed to the escape of some semen prior to ejaculation, delayed withdrawal, or deposition of semen within the woman's external genitalia and its seepage into the vagina.

Coitus interruptus no longer appears to be popular in the United States, but it is still the leading form of contraception in many countries of Europe and the Near East. In general, its use appears to be inversely associated with socioeconomic status.

Postcoital douche

A douche is intended to flush the sperms from the vagina; vinegar or some other additive is often included to have a spermicidal effect. It is a very ineffective method and ranks last among the contraceptives currently used. It most frequently fails because it is not used soon enough. Sperms are already entering the uterus within 90 seconds after ejaculation; it is very inconvenient for a female to thoroughly douche so soon after coitus. Lack of thoroughness also contributes to the ineffectiveness, while frequent douching with strong solutions is harmful because it destroys protective bacteria in the vagina and may burn the tissues. Douching became popular in France in the early eighteenth century and was the principal method recommended in the latter part of that century by the earliest American writers on family planning. It was widely used until World War II but has markedly declined since; its current use as a contraceptive is mainly among the uneducated.

106   Prolonged lactation

Prolonged lactation is based on the idea that conception will not occur during the period of breast feeding a child. Breast feeding does postpone the return of normal ovulation, but the extent of postponement is indefinite and unpredictable. Effectiveness is therefore not high, and the disadvantage of a childbirth to initiate the method is overwhelming; this is obviously not an effective population control contraceptive. Breast feeding over long periods is common in many underdeveloped countries; however, it is not known to what extent this is done to intentionally delay conception.

Condom

A sheath of linen, fish skins, or sheep's intestines to cover the penis and prevent the deposition of sperms into the vagina dates back at least to the Middle Ages. The invention of the condom has been attributed to a Dr. Condom who was allegedly the physician of Charles II of England (1660-1685). The king supposedly became alarmed at the number of illegitimate children he was producing, and Dr. Condom recommended the use of a penile sheath that solved the king's problem and won knighthood for the doctor. It is certain, however, that the sheath was well known before the reign of Charles II and, as interesting as the story is, it is doubtful that a Dr. Condom ever existed. More likely, the name first appeared as "condum" derived from the Latin (*condus*, a receptacle) as a euphemism for a long-known, common sense device. Since the early nineteenth century the so-called skin condoms (animal intestines) have gradually been replaced by the cheaper and more convenient rubber sheaths. The condom is highly effective when properly used; a sac must be left at the tip to serve as a reservoir for the semen, and care must be taken to avoid spillage upon withdrawal. It also has the advantage of providing protection against venereal diseases. Defective condoms are seldom encountered; the main disadvantage seems to be the claim by some men and women that they reduce sexual sensations. Prior to the introduction of oral contraceptives, the condom was the most widely used contraceptive in the United States. It is still an important method in the United States, in Europe, and in Japan; its use is increasing in the underdeveloped countries, where it has been incorporated in many national family planning programs.

Vaginal diaphragm

The diaphragm serves as a mechanical barrier to the entry of sperms into the uterus by blocking the entrance (cervix, see Fig. 5-3). The diaphragm is essentially a rubber cup that covers the cervix. It must be professionally fitted and is always used with a spermicidal jelly or cream that also acts as a lubricant for insertion. The diaphragm was invented by a German physician around 1880 and has had only minor modifications since that time. Used consistently, it is highly effective; failure results mainly from improper insertion or displacement during orgasmic expansion of the vagina. Prior to the introduction of oral contraceptives, the diaphragm ranked a close second in use to the condom in the United States; however, it has always been less popular in Europe and elsewhere.

Spermicides

Jellies, creams, foams, and suppositories with strong spermicidal action intended for use without a diaphragm have been developed by the pharmaceutical industry in recent years. However, the insertion of sperm-killing chemicals into the vagina is nearly as ancient as coitus interruptus. Egyptian writers suggested a mixture of honey and acacia tips (a vegetable gum) as a spermicide more than 3,500 years ago. Spermicides used alone have a high failure rate, but when used with a diaphragm or combined with the condom or even the rhythm method, their effectiveness increases. Inadequate quantity or quality or improper distribution within the vagina are the common reasons for failure. Their chief disadvantages are "messiness" and excessive lubrication. Spermicides are not widely used in the United States, but they are more popular in Europe. They are still a major method in the family planning programs of some underdeveloped countries, in spite of their high failure rate.

Rhythm

The basic idea of rhythm, or periodic continence, is to abstain from sexual intercourse during the time that an egg might be fertilized. It is the only method of birth control presently sanctioned by the Roman Catholic Church. Coitus must be avoided for at least 48 hours (sperm life span) before and 15 hours (egg life span) after ovulation. Unfortunately, the exact time of ovulation cannot be determined or predicted with great accuracy. *Calendar rhythm* was developed in the 1920s. According to this method, ovulation is assumed to occur 12 to 16 days before a woman's next menstruation. Adding 2 days before and 1 day after this 5-day span to account for sperm and egg survival makes the total period of abstinence 8 days of each 28-day cycle.

More recently, accuracy has been increased somewhat by *temperature rhythm.* A woman's temperature rises about half a degree right after ovulation and remains elevated throughout the luteal phase of the ovarian cycle. Thus, the length of the luteal phase and the follicular phase (from menstrual flow until the rise in temperature) must be estimated by keeping a daily record of basal body temperature (temperature before getting out of bed) for several months. If the hormonal cycle is fairly regular, the average time of ovulation and the range of variation can be estimated. Adding 2 days before and 1 after the estimated range of ovulation should then give the "unsafe period" for the particular individual. The length of this period depends upon the range of ovulation or, in other words, the regularity of the cycle. Irregularity produces a broad range, which in turn gives an unsafe period that is simply too long to be faithfully practiced. Approximately one out of every six women has a cycle so irregular that the method simply will not work for her. The rhythm method is generally not effective. If this were used as our only means of population control, the irregular women, on whom it works least of all, would produce most of the children. Their irregularity is physiological and undoubtedly has an hereditary component that they would pass on to their daughters each generation. By natural selection irregularity would increase, and the rhythm method would become more and more ineffective. As Garrett Hardin, professor of biology at the University of California at Santa Barbara, put it: "Blessed are the women that are irregular for their daughters shall inherit the earth. . ."

Intrauterine devices

The insertion of an object within the uterus has long been known to prevent conception. Over 2,000 years ago Hippocrates described the insertion of a device into the uterus of a woman to prevent pregnancy, and for centuries Arabian and Turkish camel drivers prevented their animals from becoming pregnant during long trips by inserting small round pebbles into the uterus of each female. In 1928 a Berlin physician, Dr. Gräfenberg, reported on his experience with an intrauterine ring, which he had first made of silkworm gut and later of silver wire. Gräfenberg's device was universally opposed by gynecologists, and not until 1959 were modifications of the original Gräfenberg ring again used successfully. This time the concept was accepted by a significant segment of the medical profession. The modern era of intrauterine devices, generally known as IUDs, began in the 1960s with the availability of inert chemical materials, such as plastics and stainless steel, which could remain in the uterus indefinitely without reacting with, or being rejected by, the body. Most IUDs in use today are made of flexible plastic with a core of metal salts that help reveal the position of the device by X-ray. The most commonly used IUD is the Lippes loop. It is preferred because of the low rates of expulsion, pregnancy, and side effects associated with its use. The newer Saf-T-coil is about as effective as the loop, while the most recent Dalkon shield has prospects of being the most effective IUD of all (Fig. 5-10).

The precise mode of action of the IUD is uncertain in spite of intensive research in recent years. The various possibilities include interference with sperm movement through the uterus and oviduct, or with fertilization of the egg, or with transport of the zygote through the oviduct, or with implantation. There is no evidence to indicate that the action of the IUD involves interference with the embryo at any stage after implantation.

The great advantage of the IUD is that once it is in place, it can be forgotten. It is also very cheap; some devices can be produced at a cost of a few cents each. This method is therefore particularly suitable for large-scale programs where the educational background and motivation of the individuals

may vary considerably. Furthermore, the fact that its use is disassociated from the sexual act itself enhances its acceptance by many couples. Expulsion of the device is its major disadvantage. About 10% of the women using the IUD spontaneously expel it during the first year; the expulsion rate drops sharply thereafter. However, about two out of five women who experience an expulsion are eventually able to retain the device. Insertion must be by a physician or paramedical person and removal, which restores fertility immediately, should be done professionally. Side effects such as bleeding, spotting, cramps, or backache occasionally occur shortly after insertion then tend to disappear. In some cases their persistence requires removal of the device. The more severe side effect associated with IUD is pelvic inflammatory disease that occurs in 2% to 3% of the users during the first year. This incidence is interpreted as reactivation of pre-existing conditions brought on by insertion. Whether or not the insertion of an IUD into a woman with healthy pelvic organs can produce the disease is not known.

IUDs have been the mainstay of several national family planning programs, especially in Asian countries such as India, Pakistan, South Korea, and Taiwan. Use in the United States is estimated to be around 2 million.

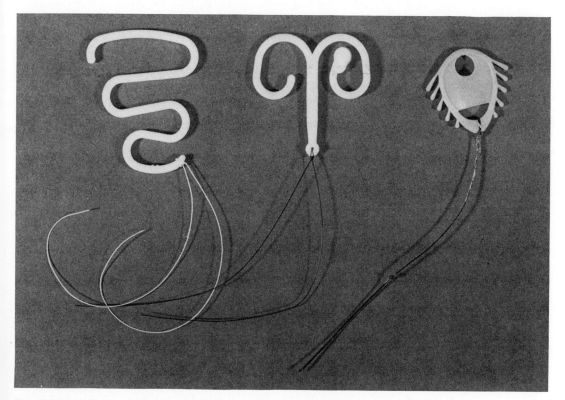

Fig. 5-10. The most commonly used IUDs from left to right: Lippes Loop, Saf-T-Coil, and Dalkon Shield; shown actual size.

110   Oral contraceptives

Although oral contraception has long been an attractive idea (ancient manuscripts contain prescriptions for potions that never worked), "the pill" is certainly the most recent innovation in birth control methods. The modern steroid oral contraceptives, developed in the 1950s by Doctors Gregory Pincus and John Rock, rank high among the scientific achievements having important and immediate social impact. This contraceptive method works directly on the hormonal cycle of the female to suppress ovulation, thereby making pregnancy impossible. The pill utilizes synthetic forms of the two female hormones, estrogen and progesterone, that are prescribed according to two different dosage regimens—combined and sequential. Under the combined regimen, 20 (or 21) identical pills containing both hormones are taken from the fifth to the twenty-fourth (or twenty-fifth) day of the typical 28-day cycle of the reproductive system. Under the sequential regiment, 15 (or 16) pills containing estrogen only are followed by 5 pills containing both hormones. In each case the pills create a hormonal pseudopregnancy by essentially providing enough synthetic estrogen and progestrone to inhibit FSH and LH secretion by the pituitary, thereby preventing follicle development and ovulation, just as a true pregnancy does. Suspension of the pills after 20 or 21 days usually results in menstrual flow, which is often noticeably reduced. This rigid sequence of pills regularizes the menstrual cycle to exactly 28 days, even in women who were never regular before.

The outstanding advantage of oral contraceptives is their effectiveness. According to the available clinical studies, when taken without fail as instructed they are virtually 100% effective. Another advantage, like that of the IUD, is that their application is not directly related to the time of sexual activity. The only important reason for failure is the omission of one or more pills in the sequence; the chances of pregnancy increase with each forgotten pill. Thus, a fairly high degree of motivation is required for the woman to ensure adherence to the regimen.

Oral contraceptives, like any other potent medication, carry the risk of undesirable or detrimental side effects to certain individuals. The most frequent symptoms are similar to those occurring in true pregnancy such as nausea, vomiting, breast swelling and tenderness, breakthrough bleeding, weight gain, headache, dizziness, and a brownish discoloration of the facial skin known as chloasma. Most of these wear off within a few months or can be controlled by adjusting the dosage or changing brands. In addition, a wide range of adverse experiences from vaginitis (vaginal infestation by yeast or fungus) to cancer have been attributed to the pill. Available data have failed, as yet, to establish or to exclude a statistical association with oral contraception for most of the adverse experiences reported. The one important condition for which an association has been established is thromboembolic disease ("wandering" or floating blood clots). The incidence of this disease appears to be several times higher among women using the pill compared to nonusers. The excess mortality from pulmonary embolism attributable to oral contraceptives is estimated at 3 deaths per year per 100,000 users. This risk, however, must be weighed against a death rate of about 25 per 100,000 pregnancies, exclusive of death from illegal abortions. The fact is that many human activities, such as automobile travel, and many other drugs commonly used to combat pain and infection are potentially much more dangerous than the pill. The advantages must be weighed against the disadvantages in every case on an individual level and, as with any medication, a physician familiar with the medical history of the patient should prescribe and monitor the use of the pill.

The first oral contraceptive was approved by the Food and Drug Administration in 1960, and within 5 years the users numbered 3.8 million in the United States, accounting for nearly one-fourth of all contraceptive practices. By 1968 more than 8 million women were using the pill in the United States, and currently it is estimated that 30 million women throughout the world are "on the pill."

Surgical sterilization

A London physician, Dr. Blundell, is credited with first suggesting surgical sterilization to protect women whose life or health was threatened by pregnancy or child delivery as early as 1823. But it wasn't until the latter part of the nineteenth century when aseptic surgery and anesthesia became available that effective techniques were developed for women and men. Surgical sterilization consists of the cutting or tying (ligation) of the oviduct in the female or the sperm duct in the male.

The operation in a female, called a *salpingectomy* or *tubal ligation*, is less common today than the comparable operation in a male, known as a *vasectomy*. In women the abdomen must be opened under anesthesia and a section of the oviducts cut and removed so that eggs cannot pass to the uterus nor sperms reach the egg. The operation is most easily done right after the birth of a baby but can be performed at any time. A new procedure called Band-Aid surgery may make female sterilization much more common in the future. The doctor uses a long slender tube, called a laparoscope, which is inserted into the abdomen through a tiny incision in the navel. A cold light passes through the tube and permits the doctor to see what he is doing internally. A second small incision is made just below the navel through which forceps that carry an electrical current are inserted. Then, watching through the laparoscope, the doctor cauterizes the oviducts. A Band-Aid over the minute incisions is all the further care needed. The whole procedure takes only about 20 minutes and can be done on a 1-day out-patient basis. Thus, the hospitalization and expense of the former method are greatly reduced.

A vasectomy takes only 15 to 20 minutes and can be performed in the doctor's office. The procedure involves a small incision in the upper region of the scrotum made under local anesthesia. The sperm duct is located; two ligatures are applied a small distance apart, and the portion between them is cut out. This prevents sperm from traveling from the testes through the urethra. However, the seminal fluids are not affected, and ejaculation still occurs normally, except that no sperms are found in the ejaculate.

In both females and males, sterilization has no effect on the normal hormonal function of the reproductive system and no effect on normal sexual performance, except in some cases to enhance it by removing all fears of pregnancy. Surgical sterilization is considered to be a permanent means of contraception. Nevertheless, successful reversal operations can now be achieved 50% to 80% of the time in men and 52% to 66% of the time in women. New methods, such as plastic inserts to act as removable plugs, show promise of complete reversibility.

Surgical sterilization is virtually 100% effective but, until reversal can be guaranteed, it has the disadvantage of being essentially permanent. In the late 1950s it was estimated that 110,000 sterilizations were performed annually in the United States, of which 65,000 were on women and 45,000 on men. Since that time tubal ligations have increased slowly, while vasectomies have increased dramatically to 275,000 during the first 6 months of 1970. Many underdeveloped and overpopulated countries encourage vasectomy as a method of population control. In India over 5.5 million men have been sterilized and Pakistan has a program to sterilize 50,000 men a month.

Abortion

Birth can also be controlled after fertilization and implantation occur by aborting the developing embryo at some point prior to normal termination. Abortion is not considered to be a desirable means of birth control, but in view of the fact that our contraceptive methods and practices are not 100% effective, most people in favor of population control consider abortion to be a necessary, and hopefully temporary, backup to contraceptive failures. Some women, on the other hand, consider the termination of an unwanted pregnancy a woman's individual right. In any case, the termination of unwanted pregnancies is becoming more common as it is gradually legalized around the world. Where it is practiced, it has contributed significantly to controlling population growth. For example, 1.5 million legal abortions per year are performed in Japan, and Hungary's abortion rate is now higher than its birth rate.

There are four medically safe methods of abortion currently in use. The most common operation is known as *dilation and curettage* (D and C). This involves the dilation of the cervical opening to allow the manipulation of an instrument with a spoon-like tip, called a curette, to be inserted into the uterus to scrape out the embryo and placenta. A newer method that is quickly replacing the D and C as the standard procedure is the *vacuum curettage* or *uterus aspiration* in which a hollow tube, called a vacurette, connected to a vacuum collection bottle by a plastic tube is inserted into the uterus to remove the embryo and placenta by vacuum pressure. This method is quicker, involves less blood loss and less risk of uterine perforation, and requires less anesthesia than the D and C. Both the D and C and the vacuum curettage are used up to the twelfth or thirteenth week of pregnancy. After this time other methods are needed that require surgical techniques and hospital stays. The common method of abortion beyond the fourteenth week has been the *hysterotomy*. This is essentially a miniature cesarean section in which the abdominal wall and uterine wall are incised and the fetus and placenta removed. This method is rapidly being replaced by the newer method of *intra-amniotic hypertonic saline*, commonly known as salting out. By this method a small area of skin below the navel is locally anesthetized, a long needle is inserted through the abdominal and uterine walls into the amniotic cavity, and several ounces of amniotic fluid are replaced by a strong saline solution (or sometimes a glucose solution). The solution kills the fetus and inhibits the release of placental hormones. Within 24 to 48 hours the endometrium breaks down, and the woman miscarries.

Abortions are virtually 100% effective, although there are rare cases where extremely late abortions by hysterotomy have resulted in surviving infants. Abortions are also quite safe when performed under appropriate conditions by a qualified physician, especially before the twelfth week of pregnancy when the risk of death to the mother is only about one sixth the risk of death through a full term pregnancy. Beyond the twelfth week, the risk of complications or death rises considerably.

In many countries where abortion is not legal, illegal abortions, most with high risks to the patient, are prevalent. It has been estimated that at least 30 million abortions are performed in the world each year, most of them illegal. In the United States the most often quoted estimate of illegal abortions is 1 million or more. These figures indicate that many women are making the critical decision about unwanted pregnancies, even at the risk of death or penalty by law.

Contraceptive effectiveness

The effectiveness of a contraceptive method is defined in terms of its capacity to prevent unwanted pregnancies. This is usually expressed as a *failure rate* (pregnancy rate) per hundred woman-years of use. This calculation can be thought of as the number of unwanted pregnancies that would occur among 100 women using a particular method for 1 year. In the absence of any contraceptive measures, the pregnancy rate (failure rate) is about 80%. This figure is relatively constant from society to society and among different socioeconomic groups. This is, no doubt, a fundamental biological reproductive potential that evolved in our ancestors many millenia ago. Our ancestors had a much shorter life expectancy than modern man; Neanderthal man had an expectancy under 30 years, and our earlier ancestors probably averaged about 20 years. Early man, like all other animals, did not have a life span that exceeded the reproductive period. This short life expectancy coupled with a high death rate meant that a high reproductive capacity was necessary to ensure survival of the species. Today, through cultural evolution, man's life span has been greatly extended, and in many parts of the world the female's total expectancy exceeds her reproductive life span. The fundamental biology of the reproductive system has not been changed by our cultural advancements, however (except perhaps made more efficient by medical science). The reproductive potential in primitive man with a 20-year life expectancy resulted in about four children per female, barely enough to keep ahead of the death rate. But in modern man with the expectancy often exceeding 50 years, the potential is about fifteen children per female. Considering the current rapid growth of the world's population, it would seem that birth control may be a necessary adjunct to our cultural advancement to compensate for a biological system that evolved under much different circumstances than man has created today.

Absolute failure rates of contraceptive methods are difficult, if not impossible, to ascertain. Data come from divergent sources, for example, different countries, and are collected by varied methods. Some data come from clinical studies, whereas some come from demographic surveys. In addition, method-failure cannot always be separated from use-failure. Thus, many sets of data

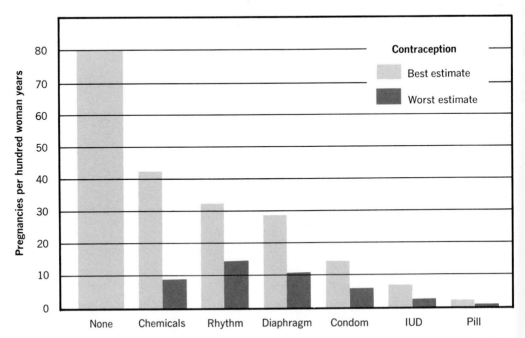

Fig. 5-11. Failure rates of the most common contraceptive methods. (Data from Peel, J. and Potts, M.: Textbook of contraceptive practice, New York, 1970, Cambridge University Press, p. 47.)

are not always comparable and cannot be simply averaged to get a single failure rate. Fig. 5-11 indicates the failure rates of the most commonly used contraceptive methods. For each method, the best and worst estimates derived
from broadly analogous studies are given. The resulting rank order of the
different methods is nearly always the same regardless of the source of the data.

Research actively progresses on the many possible ways of interfering with
the reproductive process. Table 5-3 lists some of the more promising developments that appear to be feasible in the near future. Each method is self-explan-
atory in light of the previous discussions in this chapter; therefore no further
discussion of these methods will be undertaken. Most developments relate to
women because the many steps in the female cycle make it more susceptible
to regulation. Furthermore, the sites of fertilization, implantation, and development make regulation at these stages necessarily female. The male has not
been neglected, however, and there are feasible prospects for new methods of
male contraception in the near future. It should also be noted that any measure
of birth control, to be practical as a means of population control, must be in-
expensive, simple to use, highly efficient, and easily marketed and distributed.

Table 5-3. Prospective methods of birth control

For females	For males	For either sex
Low-dosage daily pill (synthetic progesterone only)	Oral contraception (male pill)	Long-lasting injection
Once-a-month pill	Reversible surgical steriliza-tion (plastic plugs or clips)	Long-lasting implant
Morning-after pill (massive dose of estrogen)		Immunization (vaccine against sperm maturation or motility)
Contact absorption (IUD or vaginal insert impregnated with contra-ceptive chemical)		
Improved detection of ovulation		
Improved IUDs		
Improved method of abortion		

Suggestions for further reading on reproductive processes

Austin, C. R., and R. V. Short, editors. 1972.
Reproduction in mammals.* Cambridge
University Press, New York. A series of
five short books consisting of (1) Germ
cells and fertilization; (2) Embryonic
and fetal development; (3) Hormones
in reproduction; (4) Reproductive patterns; and (5) Artificial control of reproduction. Each book contains several
essays presenting up-to-date and highly
readable accounts of mammalian reproduction written by experts in the field.

*Denotes paperback.

Berrill, N. J. 1968. The person in the
womb.* Dodd, Mead and Company, New
York. Written in a popular vein, this
book addresses such vital questions as
birth control, abortion, and genetic defects in relation to the events that take
place in the womb from conception to
birth.
Francoeur, Robert T. 1973. Utopian mother-
hood: new trends in human reproduction.* A. S. Barnes & Company Inc., Cran-
bury, New Jersey. The emphasis of this
book is on the history and development
of the new technologies of reproduction

with analyses of the present state of the art and projections for the future. The final chapter deals with the social impact of this technology on marriage.

Kerr, Norman S. 1967. Principles of development.* William C. Brown Company, Publishers, Dubuque, Iowa. Designed for the student with little or no background in biology, this book provides a capsule introduction to the field of developmental biology.

Nilsson, Lennart, Ingelman-Sundberg A., and C. Wirsén. 1966. A child is born. Delacorte Press, New York. An outstanding book describing human development and containing dozens of the unprecedented and remarkable photographs of the human embryo and fetus, taken by Mr. Nilsson (two of these appear in Chapter 5). Everyone should make a point of seeing this pictorial masterpiece.

Pierson, Elaine C. 1971. Sex is never an emergency.* Second edition. J. B. Lippincott Company, Philadelphia. This is a birth control guide originally published through the Dean of Students Office at the University of Pennsylvania. The author, a gynecologist, presents authoritative material and asks, then answers, a host of questions that are, or should be, on the minds of everyone participating in adult sexual interactions.

Saunders, John W., Jr. 1968. Animal morphogenesis.* The Macmillan Company, New York. This book, like that of Norman S. Kerr, provides a clear and concise introduction to the study of developmental biology.

Swanson, Carl P. 1969. The cell.* Third edition. Prentice-Hall, Inc., Englewood Cliffs, New Jersey. This book concentrates on an introduction to the study of cellular structure and function. It provides good discussions of mitosis and meiosis.

Hereditary processes

6

The gene and its function

To understand the more specific processes of human heredity, it is important to know what a gene is, what it does, and how it does it. Historically we understood many of the genetic mechanisms before we really knew much about the chemical nature and function of the gene. However, a chronological approach to the understanding of science is *not* always the most logical way. This is the case in genetics where the fundamental principles can be more clearly understood and appreciated in light of our current knowledge of the gene and its function. In this chapter we will endeavor to become acquainted with the modern concept of the gene so that we can apply this knowledge to the discussions of the various modes of genetic transmission in subsequent chapters.

THE GENE

The Austrian monk Gregor Mendel, known as the father of genetics, first described the transmission of the hereditary material in 1865. In his extensive experiments with the garden pea, he referred to genes as "differentiating characters" without having any idea of *how* they produced the various traits he investigated. Some years later, in 1909, the Danish botanist Wilhelm Johannsen coined the term "gene" to describe Mendel's characters. For many more years the transmission and effects of genes were studied, but the gene itself remained a nebulous undefinable entity. Then, in 1953 a major breakthrough occurred. Just as Darwinism represented a milestone in the steady progression of knowledge concerning evolution, the facts and information concerning the genetic material were reaching a critical point, and the time was ripe for someone to put it all together. Again, as with many breakthroughs, several groups of investigators in different parts of the world were simultaneously putting the pieces together, but it was James D. Watson, an American geneticist, and Francis H. C. Crick, an English biochemist, working at the Cavendish Laboratory of Cambridge University who together merged data from chemistry, physics, and biology to build a hypothetical model of *deoxyribonucleic acid*

(DNA) as the genetic material. The now famous Watson-Crick "double helix" was designed to conform to all the specifications of X-ray pictures provided by Maurice H. F. Wilkins, an English physicist. In 1962 Watson, Crick, and Wilkins were awarded the Nobel Prize in Medicine and Physiology for their work on the DNA molecule—an advance in genetics considered by many scientists to be the most significant since Mendel first elucidated the basic principles of hereditary transmission. Within the last decade, research related to DNA and gene function has produced no less than nine more Nobel laureates.

Before we discuss the specific structure of the DNA molecule, it would be well to note what properties would have to be possessed by any substance that proved to be the genetic material. We can then see how the Watson-Crick model of DNA satisfied these prerequisites and thereby more fully appreciate its significance.

The first property required by the genetic material is *variability*. If genes are responsible for determining a multitude of different traits in man and other living organisms, the material of which they are composed must somehow be able to vary in structure sufficiently to be able to account for the tremendous number of different kinds of genes that must exist.

Secondly, since organisms pass their traits on to future generations in a predictable manner, the genetic material must possess a means of *exact replication* in order to account for its filial continuity. This property is also necessary to ensure that the genetic instructions that constitute the hereditary complement of a fertilized egg are passed on to all the cells of the body during embryonic development so that ultimately each appropriate gene will perform its function in some specific cell in a particular tissue or organ. Most genes must be replicated many, many times and passed on to a great number of cells by mitosis before they perform their particular bodily function. Without a means of very exact duplication of the genetic material, embryonic development would be in utter chaos.

Finally, although the stability of exact replication is essential, the genetic material must also have the potential for change. It must possess the property of *mutability,* so that changes or mutations can occasionally occur and then, somehow, the changed form can duplicate itself exactly. The DNA molecule is the only chemical substance known to man that possesses the combination of these properties.

The *nucleic acids* were chemically defined in 1869 as a category of organic compounds characteristic of living organisms. Their name is derived from the fact that they were first discovered in the nuclei of pus cells and were found to have acid properties. Chemically, nucleic acids consist of three subunits, a *pentose sugar* (5-carbon sugar), a *nitrogen-containing base*, and *phosphoric acid*. There are two types of nucleic acids named for the kind of pentose sugar present. One type contains the sugar ribose and is called *ribonucleic acid (RNA)*, and the other contains deoxyribose, which, as the name implies, is ribose minus an oxygen atom and is called *deoxyribonucleic acid (DNA)*. Other differences between RNA and DNA will be discussed later.

The actual units that serve as building blocks of nucleic acids are the *nucleotides*. A nucleotide consists of a sugar, a base, and a phosphate chemically bonded so that the sugar has the phosphate attached to one side and the base attached to the other side. There are four different kinds of nitrogenous bases found in DNA. Two of the bases are *adenine* (A) and *guanine* (G), which chemically consist of two carbon rings and are categorized as *purines*. The other two bases are *cytosine* (C) and *thymine* (T), which chemically consist of only one carbon ring and are categorized as *pyrimidines*. Thus, there are four different kinds of nucleotides identifiable, depending upon which base is present (Fig. 6-1).

The double helix configuration of the DNA macromolecule is produced by two long chains of nucleotides twisted around like a spiral staircase. Each polynucleotide chain is formed by the phosphate of one nucleotide chemically bonding to the sugar of another, producing long sugar-phosphate sequences. This leaves the bases projecting from the side of the chain (Fig. 6-2). The linear order of the bases can occur in all possible combinations and, as we shall see later in the chapter, a particular gene actually consists of some set number and sequence of these bases. The two polynucleotide chains are held together by chemically weak hydrogen bonds that form between the projecting bases. A purine always bonds with a pyrimidine and, more specifically, adenine always bonds with thymine and cytosine with guanine. This specificity is attained by the fact that two hydrogen bonding sites are available on the adenine and thymine molecules, while cytosine and guanine have three sites to be satisfied (Fig. 6-3). Any bonding combinations other than A-T and C-G would leave unbonded active sites and be chemically unstable. The double-stranded structure of the DNA molecule is therefore always a sequence of these specifically matched nucleotide pairs with the bases forming the stairs of the spiral staircase. This *base complementarity* is a very important structural feature of the DNA molecule. Fig. 6-4 diagrammatically illustrates the structure of a segment of DNA.

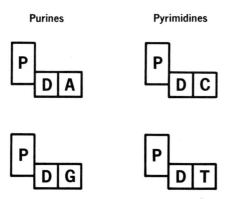

Purines   **Pyrimidines**

Fig. 6-1. Schematic diagram of the four types of nucleotides found in DNA. *P* represents phosphoric acid; *D*, deoxyribose sugar; *A*, adenine; *G*, guanine; *C*, cytosine; *T*, thymine.

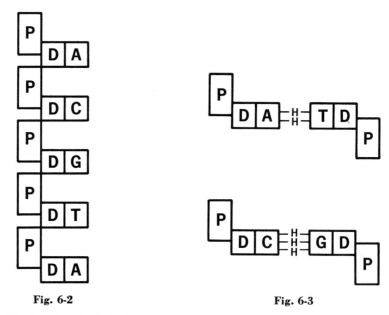

Fig. 6-2

Fig. 6-3

Fig. 6-2. A nucleotide chain. The base sequence may be in any order; here it is shown alphabetically.

Fig. 6-3. Nucleotide base pairing specificity. Note that the phosphates are oriented in opposite directions.

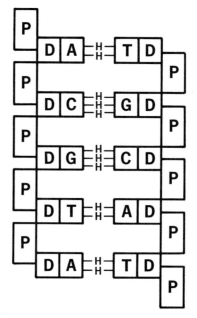

Fig. 6-4. The structure of the DNA molecule.

The helical model proposed by Watson and Crick is shown in Fig. 6-5. The ribbon or backbone of each chain consists of the phosphates and sugars, while the base pairs project across, holding the two strands of the double helix together. The coiling is right-handed, with a complete turn occurring over intervals of ten nucleotide pairs. Since the genetically significant portion of the macromolecule is the base pair sequence, a functional diagram of DNA can be utilized in which the constant sugar-phosphate backbone is simply represented as a straight line with the bases (A, C, G, and T) projecting across like the rungs of a ladder. Fig. 6-6 represents a functional diagram of the DNA sequence of Fig. 6-4. Functional diagrams will be used from this point on in the discussions of gene function.

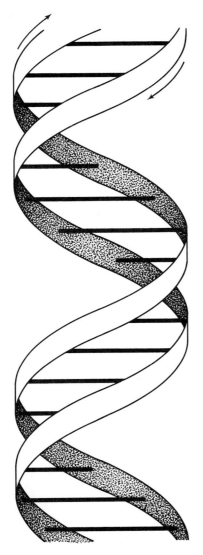

Fig. 6-5. The helical configuration of the DNA molecule as proposed by Watson and Crick.

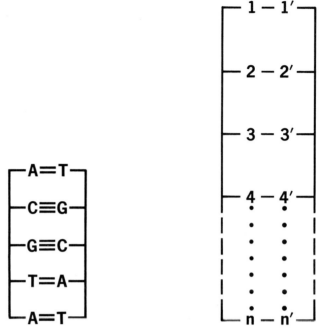

Fig. 6-6. Functional diagram of a segment of DNA.

Fig. 6-7. A hypothetical DNA molecule.

The base pairing specificity means that one side of the DNA molecule dictates the structure of the opposite side, making the molecule redundant in this respect. However, the linear sequence has no predetermined base relationship. Hence, the linear arrangement of bases along one side of the molecule can exist as any possible sequence of the four bases; the opposite side then complements the arrangement according to the pairing specificities. This provides the molecule with a tremendous potential of different linear base sequences limited only by the number of bases making up the sequence. For example, taking a hypothetical short segment of a DNA molecule such as that shown in Fig. 6-7, position number 1 can be occupied by any one of the four bases (A, C, G, or T). If it is occupied by A, then position 1′ must be T (complementary base pairing). If it is occupied by C, then position 1′ must be G, and so forth. Likewise, position number 2 can be occupied by any one of the four bases. Taking positions 1 and 2 together, there are 16 different sequences of the four bases possible (A-A, A-C, A-G, A-T, C-A, C-C, C-G, C-T, G-A, G-C, G-G, G-T, T-A, T-C, T-G, and T-T). The expression 4^n where 4 represents the number of different bases possible and n equals the number of bases in the linear sequence gives the total number of different base sequences possible down one side of the molecule. Considering positions 1, 2, and 3, there would be 4^3 or 64 different sequences possible; adding position 4 to the total length gives 4^4 or 256 possible sequences. These base sequences form the genetic code, and a specific gene consists of a particular sequence in lengths numbering in the hundreds of bases. Thus, conservatively assuming that an average gene consists of a linear sequence of 300 bases, the number of different kinds of genes (different base sequences) possible would be 4^{300} or approximately 4×10^{180}, which is the totally incomprehensible number of 4 trillion, trillion, trillion, trillion, trillion, trillion, trillion, trillion, trillion, trillion, trillion, trillion, trillion, trillion, trillion! DNA clearly possesses the essential property of variability to serve as the genetic material for all the different kinds of genes that constitute the blueprints for life forms from viruses to man.

The double-stranded, redundant structure of DNA provides the molecule with its second essential property—exact replication. The weak hydrogen bonds between the base pairs are easily broken by specific chemical enzymes, allowing the two polynucleotide chains to unwind and separate. In this uncoiled condition each chain can serve as a template to reconstruct its partner by the complementary base pairing of free nucleotides present in the cell (supplied to the cells through nutrition). In this manner each "old" strand forms a complementary "new" strand, and two identical double helices are produced from one (Fig. 6-8). It is the specificity of the base pairs that ensures exact replication, and it is the redundancy of the double-stranded structure that provides the mechanism for replication. Although Watson and Crick only postulated this simple mechanism of replication from their model, it has since been experimentally proved to occur in this way.

The DNA molecule also has the capacity for structural changes—mutations—to occur and be replicated. Mutations will be discussed in more detail later in this chapter. It shall suffice now to point out that under certain circumstances, or perhaps strictly by accident, the wrong base can be incorporated into a position along the polynucleotide chain. Once incorporated it will replicate faithfully by complementary base pairing in a normal fashion. The mutational effect lies in the fact that the specific linear base sequence is changed and, since this represents the genetic code, an altered code may cause a change in gene function resulting in some overt expression recognized as a mutant trait in the organism. The property of mutability is therefore also inherent in the structure of the DNA molecule.

All three properties required by the genetic material, variability, exact replication, and mutability, have been convincingly demonstrated in various ways by many different scientists, and the hypothetical model of Watson and Crick has been validated and universally accepted by the scientific community as the chemical substance of the gene.

THE GENE PRODUCT

The product of gene function is a *protein* (from the Greek *proteios*, meaning primary) or a protein component. Proteins, as their name implies, are of

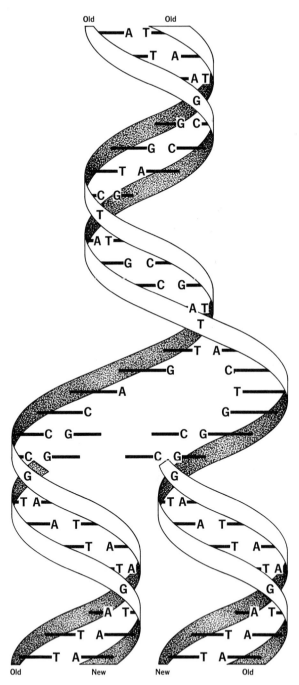

Fig. 6-8. DNA replication. Note that each new strand is produced by complementary base pairing using the old stand as a template.

primary importance to the whole phenomenon of life. They not only serve as the major building material of living organisms but also, in forms we call *enzymes*, govern all the biochemical activities that constitute the very essence of the living state. It is safe to say that any part of our body that is not made of proteins per se is the product of biosynthesis, which is dependent upon proteins (enzymes). Proteins are large, chemically complex molecules that are highly variable in structure and specific in chemical activity. The specificity and variability of proteins enables these molecules to regulate the multitude of vital biochemical processes and also to constitute the molecular basis of structural and functional differences between individuals and between species. To understand how genes function, we must understand the structure of their product and the important relationship between structure and function in the protein molecule.

The building blocks of the protein molecule are *amino acids*. These small nitrogen-containing compounds consist of a central carbon atom, the alpha carbon, to which is attached an acidic carboxyl group (COOH) on one side and a basic amino group (NH_2) to the other side. These components give the amino acids the property of an amphoteric electrolyte, which means they can act either as acids or bases in chemical reactions. Fig. 6-9 illustrates the general chemical structure of an amino acid. The R in the figure stands for chemical side-chains that are different for each kind of amino acid. There are twenty kinds of amino acids commonly found in proteins. R may be as simple as a hydrogen atom (H) or a methyl group (CH_3) producing the amino acids, glycine and alanine, respectively, or it may be a complex organic entity such as a carbon ring, characteristic of phenylalanine, or a carbon chain, which characterizes arginine. Table 6-1 lists the twenty amino acids commonly found in proteins; this list is simply given as a reference to the names and abbreviations that will be used throughout the remainder of the text as we discuss gene products and abnormal hereditary traits.

Because of their amphoteric nature, amino acids are able to form acid-base bonds between themselves by the simple chemical process of dehydration synthesis, that is, chemically bonding by breaking out a water molecule from their structures. As Fig. 6-10 illustrates, this produces a chemical bond between the carbon atom of the carboxyl group of one amino acid and the nitrogen atom of the amino group of the other amino acid with water formed as a

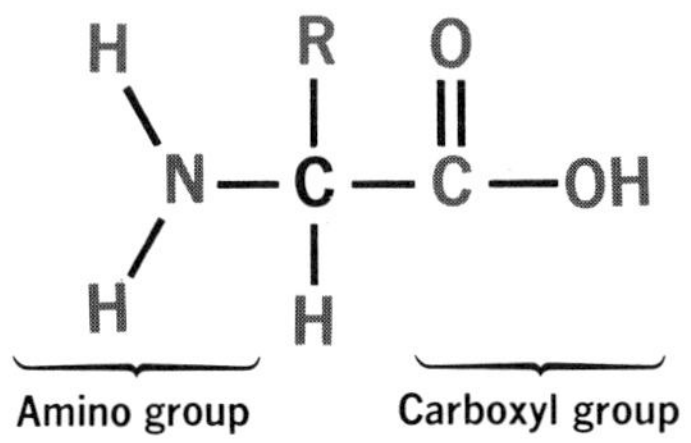

Fig. 6-9. The general chemical structure of amino acids. The alpha carbon is in bold type.

Table 6-1. The twenty amino acids commonly found in proteins

Amino acid	Abbreviation
Alanine	Ala
Arginine	Arg
Asparagine	AspN
Aspartic acid	Asp
Cysteine	Cys
Glutamic acid	Glu
Glutamine	GluN
Glycine	Gly
Histidine	His
Isoleucine	Ileu
Leucine	Leu
Lysine	Lys
Methionine	Met
Phenylalanine	Phe
Proline	Pro
Serine	Ser
Threonine	Thr
Tryptophan	Tryp
Tryosine	Tyr
Valine	Val

by-product. This kind of chemical linkage is known as a *peptide bond*, and long chains of amino acids held together by these bonds are called *polypeptides*.

The *primary structure* of a protein is the sequence of amino acids making up the molecule. The specificity of proteins depends on the particular order of the amino acids, whereas the tremendous variety of proteins stems from the many different sequences possible. 20^n, where n is the number of amino acids in the polypeptide chain, describes the number of possible sequences; estimating conservatively, n would be at least 100 in the "average" protein. The protein molecule is neither a long, two-dimensional polypeptide chain nor a randomly twisted and coiled polypeptide mass. Rather, the molecule is a neatly coiled and folded, three-dimensional entity. Thus, other levels of structure can be recognized in the determination of the geometric configuration of a protein.

The *secondary structure* of a protein molecule is the helical coiling of the polypeptide chain, stabilized by intrahelical hydrogen bonds. The three-dimensional configuration of the molecule results from the folding of the polypeptide into a complex pattern. This folding, which represents the *tertiary structure*, is very precise and important for proper protein (or enzyme) function. For example, enzymatic activity depends on the proper chemically active sites being exposed in the proper position on the molecule. Alteration of the tertiary structure changes or inhibits proper protein function.

One of the strong stabilizing forces in maintaining tertiary structure is the disulfide bond. This is a chemical bond that forms between two sulfur atoms in the oxidation of sulfhydryl groups (SH). The amino acid cysteine is the

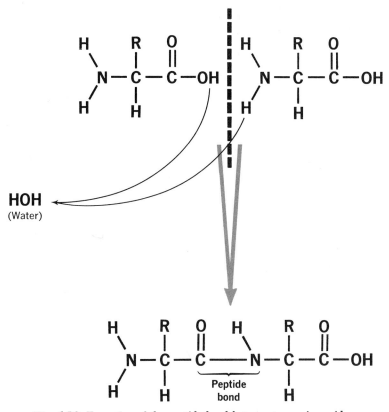

Fig. 6-10. Formation of the peptide bond between two amino acids.

only one of the common twenty that contains an exposed sulfhydryl group as part of its R group; therefore, the particular location of cysteine molecules in the polypeptide sequence (primary structure) greatly determines the folding configurations that can be stabilized by disulfide bonds. Both the secondary and tertiary levels of structure are, in large measure, a consequence of the primary structure.

Finally, many proteins consist of two or more, similar or different, interacting polypeptide chains. This association of polypeptides to produce one functioning protein molecule is referred to as the *quaternary structure*. These polypeptide chains are stabilized by hydrogen bonds, disulfide bonds, and electrostatic bonds.

When we consider that the coiling, folding, and possible associations of polypeptides are largely dictated by the interactions and bonding of critically positioned side chains of particular amino acids composing the chain, it becomes clear how important the primary structure is to normal protein function. Altering the primary structure may drastically change or inhibit a protein's function. We will now discuss how this primary structure is determined and what role DNA plays in its determination.

FROM GENE TO PROTEIN

If proteins (polypeptides) are the product of gene function, we must now ask how the correct amino acids are positioned into the exact sequence characteristic of a particular polypeptide. The fact is that the base sequence of DNA serves as a code to be translated into an amino acid sequence that is assembled in the cytoplasm of the cell. This is accomplished through a very orderly and precise mechanism that we will discuss presently, but first it is important to describe the principle of the *genetic code*.

Recall that the variability inherent in the structure of the DNA molecule is dependent upon the four nitrogenous bases (A, C, G, and T), whereas proteins utilize 20 different amino acids in their structure. Therefore, a coding system must exist whereby the 4 DNA bases can somehow be translated to represent 20 amino acids; this is analogous to having a 4-letter alphabet that must describe 20 different objects. It is obvious that 1 base cannot stand for 1 amino acid; there would be 16 amino acids left over. Likewise, combinations of 2 bases (2-letter words) will not suffice, for this would still leave 4 amino acids unaccounted for (4^2 or 16 different 2-base combinations are possible). Thus, 3-base combinations (3-letter words) are the minimum that can code for 20 different objects. Such a triplet code was originally hypothesized by Watson and Crick and has since been substantiated, although modification was necessary. For instance, a triplet code gives 64 (4^3) potential combinations of which only 20 were at first thought to "make sense" (code for an amino acid), while the other 44 were believed to be "nonsense" (combinations that did not specify any amino acid). Two or more triplets coding for the same amino acid seemed redundant and inefficient, which is not characteristic of nature, so it was said that the genetic code was not *degenerate* (a code in which more than one group of symbols stands for the same word). As researchers began the highly tedious and technically sophisticated task of unraveling the genetic code, it was discovered that, in fact, the code is redundant and that most of the 64 possible triplets, now called *codons*, do code for an amino acid. The genetic code is now deciphered, and we believe that all but three triplets represent amino acids. The three "nonsense" codons (ATT, ATC, and ACC) are called *terminators* because they seem to signal the end of an amino acid sequence (polypeptide). At least two other codons (TAC and CAC) appear to act as *initiators* of polypeptides, in addition to coding for an amino acid. Now that the genetic code is deciphered and it appears to be the universal code of life, we are stuck with the awkward situation of calling it degenerate! The 64 DNA base triplet combinations and the 20 amino acids are arranged in Table 6-2 to represent the genetic dictionary as it is presently known.

The mechanism through which the DNA code is translated into a polypeptide and finally a functioning protein involves three important intermediaries that are all forms of nucleic acids closely related to DNA. They are the *ribonucleic acids* (RNA), which structurally differ from DNA in that the sugar in the sugar-phosphate backbone is *ribose* rather than deoxyribose; the molecule is single-stranded rather than double-stranded; the nitrogenous base uracil (U) replaces thymine (T) as one of the four bases in its structure; and it is usually not self-replicating but is *produced by* DNA through the same kind of complementary base pairing mechanism involved in DNA replication. According to recent research, it appears that the DNA molecule partially unwinds, and under specific enzymatic control one side, and only one side, of the double helix serves as a template for RNA synthesis.

Three kinds of RNA with specific functions in the translation of the DNA code to a polypeptide are produced. The most common type is *ribosomal RNA* (rRNA). It constitutes 85% to 90% of the total cellular RNA and is found associated with proteins in the form of spherical bodies called *ribosomes* located in the cytoplasm. The ribosomes are the sites at which polypeptide synthesis occurs.

Between 5% and 10% of the RNA is actually involved as the coding intermediary. This type is called *messenger RNA* (mRNA) and, as its name implies, it carries the genetic message (code) from the DNA, which remains in

Table 6-2. The genetic dictionary of DNA base codons and the amino acids for which they code*

First base	Second base				Third base
	A	C	G	T	
A	AAA AAG } Phenylalanine	ACA ACG } Cysteine	AGA AGG	ATA ATG } Tyrosine	A G } Purines
	AAC AAT } Leucine	ACC Terminator ACT Tryptophan	AGC AGT } Serine	ATC ATT } Terminators	C T } Pyrimidines
C	CAA CAG CAC CAT } Valine (Initiator)	CCA CCG CCC CCT } Glycine	CGA CGG CGC CGT } Alanine	CTA CTG } Aspartic acid CTC CTT } Glutamic acid	A G C T
G	GAA GAG GAC GAT } Leucine	GCA GCG GCC GCT } Arginine	GGA GGG GGC GGT } Proline	GTA GTG } Histidine GTC GTT } Glutamine	A G C T
T	TAA TAG } Isoleucine TAC Methionine (Initiator) TAT Isoleucine	TCA TCG } Serine TCC TCT } Arginine	TGA TGG } Threonine TGC TGT	TTA TTG } Asparagine TTC TTT } Lysine	A G C T

*Note also the initiators and terminators of polypeptide synthesis.

the nucleus of the cell, to the ribosomes in the cytoplasm. Hence, the base sequence of mRNA is complementary to the template strand of DNA from which it was produced, with the exception that uracil replaces thymine. This means that the DNA codons are *transcribed* into complementary RNA codons before the actual *translation* into an amino acid sequence. For example, the DNA codon CGT would be transcribed to GCA in mRNA; the DNA codon ATA would be UAU in mRNA, and so forth. Based on this relationship, the genetic dictionary is often referred to in terms of RNA codons that can be readily derived from Table 6-2.

The remainder of the cellular RNA, some 5%, is a type called *transfer RNA* (tRNA). It transfers specific amino acids from their random assortment in the cytoplasm to the mRNA for their proper incorporation into a polypeptide. (The amino acids are supplied to the cells by the bloodstream, which picks them up from the intestines as the digested product of the proteins we eat.) The structure of the tRNA molecule determines its specificity for performing its two vital functions in polypeptide synthesis—pick-up of a particular amino acid and recognition of the proper mRNA codon for alignment of the amino acids into the correct sequence. Hence, there must be a different tRNA molecule for each of the 61 amino acid-determining codons.

A transfer RNA molecule is about 80 nucleotides long and, although the RNA is single-strand, it folds back upon itself in some regions, forming a double helix by complementary base pairing, while other regions remain unpaired and form protruding loops. There are usually three loops with the overall form of the molecule resembling a cloverleaf. Functionally there are three important regions in the molecule. One of the loops serves as an *anticodon*, which is a sequence of three unpaired bases complementary to a codon in mRNA; this typical base complementarity provides the mechanism for tRNA-mRNA recognition. In addition to the helical regions and loops, all tRNAs have a "tail" that consists of an unpaired base sequence attached to the end of the molecule. The last three bases of the tail are the same for all tRNA molecules, CCA, with the adenine serving as the point to which the amino acids are chemically bonded to their specific tRNA. Such amino acid–carrying tRNA molecules are known as "activated" tRNAs. Finally, the tRNA specificity for one, and only one, particular kind of amino acid must lie somewhere in its internal structure, since the tail attachment is identical in all tRNAs. Fig. 6-11 represents a stylized version of tRNA illustrating its essential genetic features.

For the anticodon of tRNA to pair properly with its complementary mRNA codon, the attachment must be stabilized by the ribosome. Without the ribosome, the attachment would probably not be strong enough or last long enough to allow the amino acids of adjacent tRNAs to be linked by peptide bonds. The ribosome seems to move along the mRNA by some, as yet unknown, mechanism in such a manner that two tRNA molecules are attached to one ribosome at any one time. Attachment at the forward portion of the ribosome

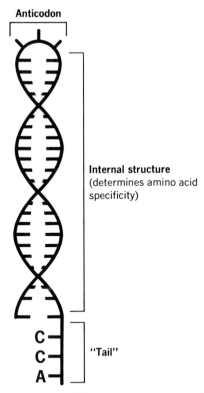

Fig. 6-11. Stylized version of tRNA, illustrating its essential genetic features.

is known as the *growing site;* it is here that new tRNAs with the proper anticodon become paired with the appropriate mRNA codon. As the ribosome moves foreward, the tRNA at the back portion is said to be on the *binding site.* The amino acid attached to the tRNA on the binding site forms a peptide bond with the amino acid of the tRNA on the growing site. Then, as the ribosome moves on, the tRNA on the binding site is released, the tRNA on the growing site becomes positioned on the binding site, and a new tRNA moves in on the next codon that now is positioned on the growing site. The tRNA released from the ribosome is "deactivated" (released from its amino acid) and becomes free to pick up another amino acid molecule for insertion into another polypeptide at the appropriate place. In this way the amino acid chain grows until a terminator codon is reached in the message. At this point the ribosome detaches from the mRNA, and the polypeptide synthesis is completed. The chain then coils (secondary structure), folds (tertiary structure), and, in many cases, associates with other polypeptides (quaternary structure) to produce an active protein molecule. The entire process of gene function is schematically illustrated in Fig. 6-12. We can now define a gene as a sequence of nucleotides in DNA that serves as the code for the assembly of a specific polypeptide.

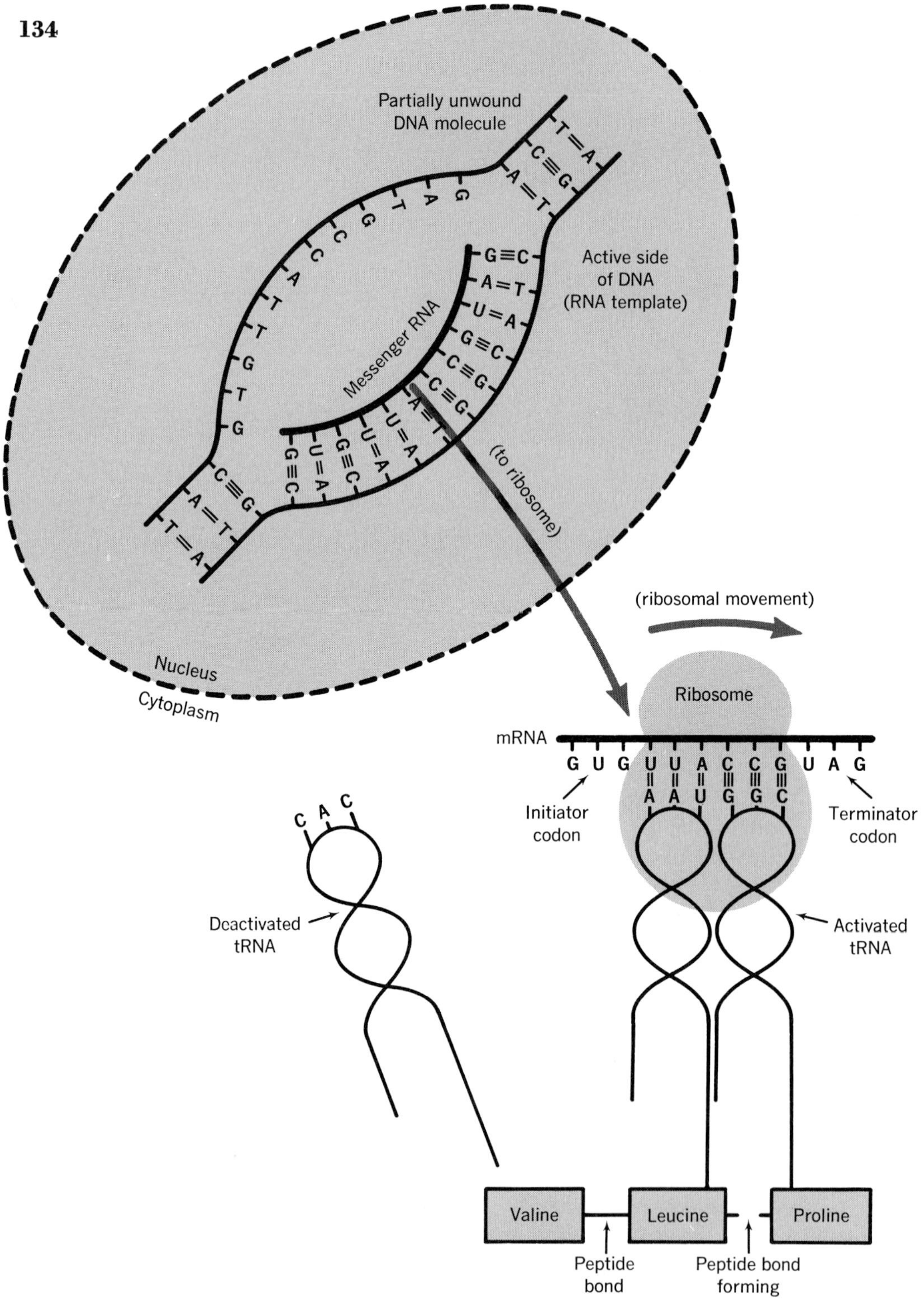

Fig. 6-12. Schematic summary of gene function.

GENE REGULATION

Polypeptide production proceeds sequentially from one particular end of the chain (at an initiator codon) to the other end (at a terminator codon). Usually, a number of polypeptides in various stages of completion are being manufactured at any one time as one ribosome follows another along the mRNA molecule; such a grouping of ribosomes is known collectively as a *polysome.* Just how much polypeptide synthesis occurs is dependent upon such things as the rate at which ribosomes attach to and move along the mRNA, the availability of ribonucleotides for mRNA synthesis, the stability of the mRNA molecule, the availability of amino acids, and the cellular conditions under which the various chemical reactions proceed.

However, more direct control of gene function must exist in order to regulate the highly organized activities of a living organism. Even though every cell in our body receives an identical genetic complement through mitosis, it is obvious that all genes do not function in all cells at all times. Digestive enzymes (gene products) are not produced in the cells of the eyes, the protein of hair is not produced by the pancreas, and so forth. Likewise, the phenomenon of epigenetic development is dependent upon particular gene functions at precisely specified times. It is therefore apparent that some kind of regulatory mechanism turns genes on and off. Locked up in this mechanism must lie the key to such extremely complex problems as regeneration and the development of a completely new individual from the "unlocked" genetic complement of a typical body cell (asexual reproduction). The implications of bringing this mechanism under human control will be discussed in Chapter 14.

In 1961 two French scientists, F. Jacob and J. Monod, formulated the *operon hypothesis* to explain the control of gene function in the colon bacteria *Escherichia coli.* Subsequent research around the world has substantiated this mechanism as the foundation of gene control, and current work is proceeding toward an understanding of the complex gene regulatory systems in higher animals, including man. Jacob, Monod, and their collaborators at the Pasteur Institute in Paris established the existence of two classes of genes—*structural genes* and *control genes.* Structural genes are what we might consider to be the "typical" genes responsible for the actual synthesis of polypeptides to be incorporated into the specific proteins and enzymes of the body as described previously. The control genes are responsible for the regulation of this production. Two categories of control genes have been recognized, *operator genes* and *regulator genes.* The operator gene is located immediately adjacent to the structural gene, or genes, it controls; it may even constitute the initial portion of a structural gene. The operator together with the structural genes it controls constitutes an *operon.* The function of the operator gene is the initiation of synthesis by the structural genes (mRNA production). This synthesis-initiating function is in turn controlled by the regulator gene; regulators may or may not be located close to the genes they regulate. The regulator gene synthesizes a product, called the *repressor,* that can chemically combine with the operator gene and inhibit its initiating function, thereby blocking polypeptide synthesis by the structural genes of the operon.

Two systems of gene regulation are known in bacteria, which very likely operate in a similar fashion in all organisms to meet the needs of normal bodily functions. One is the *inducible system,* which involves the production of structural gene enzymes, known as *inducible enzymes,* only when they are needed. These are enzymes normally involved in catabolic activities, that is, the chemical breakdown of some substance derived from the environment and used as a source of energy or as molecular building blocks for some synthetic process. In these cases the genes for enzyme production have to be "turned on" only when the substrate to be broken down is present in the body. In this system the repressor interacts with the operator to block the structural genes until the substrate to be acted upon is present. When the substrate is present, it acts as an *inducer* to turn on the structural genes by interacting with the repressor in such a way that the repressor no longer combines with the operator, and, thus, the operator is freed to initiate mRNA production by the structural genes. When the substrate is completely broken down, the repressor again interacts with the operator and the structural genes are turned off (Fig. 6-13).

The other gene regulation system is called the *repressible system* and controls the production of *repressible enzymes.* These are enzymes involved in anabolic activities, that is, the synthesis of a biochemical product for use in the body. These enzymes are produced only when the product they form is lacking or being utilized by the body as fast as it is produced. In this system the repressor produced by the regulator gene does not interact with the operator, and transcription of the structural genes takes place. However, when an excess of a particular product accumulates, it acts as a *corepressor* and combines with the repressor so that the repressor-corepressor complex blocks the operator and turns off the operon. Once the accumulated product is used up by the body and no excess is available to act as a corepressor, the system is turned on again (Fig. 6-14).

The operon model of induction and repression forms the foundation of gene regulation. In complex organisms, including man, it is thought that integrated operon systems may form complex relationships where the product of one operon may act as an inducer or corepressor for another operon. A complex interrelated system of such controls could quite easily be formed where the normal functioning of some operons keep their precursors permanently turned off. This could explain why regeneration and asexual reproduction cannot occur in complex organisms. However, if man learns how to bypass the normal control mechanisms and turn such "permanently" off genes back on, he may gain control over both of these phenomena.

MUTATION OF GENE ACTION

Mutations are sudden heritable changes in the genetic material. These changes are usually divided into two broad categories. One type consists of

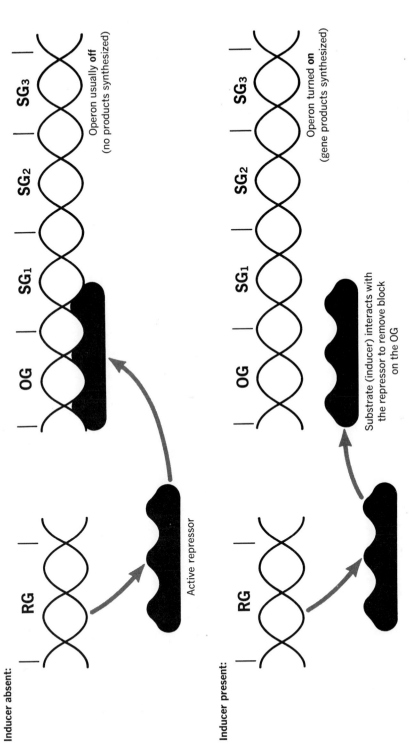

Fig. 6-13. Model of the inducible system of gene regulation according to the operon concept. *RG* represents regulator gene; *OG*, operator gene; and *SG*, structural genes.

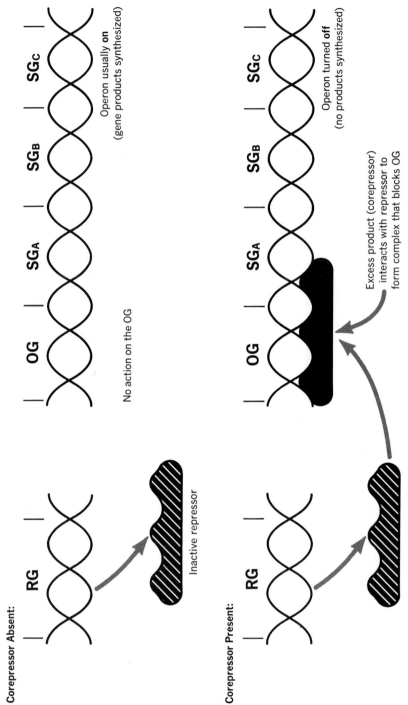

Fig. 6-14. Model of the repressible system of gene regulation according to the operon concept. *RG* represents regulator gene; *OG*, operator gene, and *SG*, structural genes.

gross chromosome changes, usually visible under the light microscope. Such chromosomal abnormalities, referred to as *chromosome aberrations*, will be discussed in Chapter 12. The other type of mutation is much more subtle and consists of chemical or physical changes within the genes. These tiny alterations, far below the resolution of any microscope, are called *point mutations*. Point mutations involve a change in one or a few nucleotide bases in DNA and, although subtle, they affect the integrity of the genetic code and often have profound detrimental effects upon the organism concerned. Nearly all alternative inherited traits, including most genetic diseases, are the result of point mutation and the abnormal polypeptide production that ensues from an altered code.

The human protein *hemoglobin* serves as an outstanding example of the effect of point mutations. Hemoglobin is the oxygen-carrying pigment of the red blood cells. The molecule is composed of four polypeptide chains (quaternary structure) called globins, each possessing an iron-containing moiety called a heme. Two different kinds of polypeptide chains form the normal adult hemoglobin molecule (HbA). There are two identical alpha (α) chains and two identical beta (β) chains, each the product of a separate gene. In the developing fetus the hemoglobin, called *fetal hemoglobin* (HbF), is composed of two alpha chains and two gamma (γ) chains. Thus, a third gene that produces the γ chains also exists, but it is normally turned off at birth, and the gene for β chain production is turned on.

The hemoglobin molecule has been extensively studied and analyzed, and its complete chemical structure is known. The α polypeptide consists of 141 amino acids, and the β polypeptide consists of 146, with the exact sequence in each chain known. More than thirty different abnormal hemoglobins have been described; we shall consider two types that illustrate the effect of point mutations on gene action.

Sickle cell hemoglobin (HbS) produces a severe genetic disease called *sickle cell anemia,* which is usually fatal in early childhood because of drastically lowered oxygen-carrying capacity of the hemoglobin and vascular obstructions in various tissues, such as the spleen, lungs, and bones. The abnormality is characterized by a tendency of the red blood cells to become distorted and sickle-shaped when deprived of oxygen; normal cells remain disk-shaped. The disease occurs most frequently in equatorial Africa, less commonly in the Mediterranean area and India; it is found universally in black populations. Approximately 1 in 500 black babies is born suffering from the disease, and some 8% to 10% of the black population carry the abnormality in one of their hemoglobin genes and can be so identified as having what is known as *sickle cell trait* (a mild anemia that is not fatal).

Another relatively common abnormal hemoglobin is *hemoglobin C* (HbC) that produces a milder anemia than sickle cell disease and displays no distortion of the red blood cells under oxygen stress. Hemoglobin C disease occurs mainly in equatorial Africa.

Thorough analyses of the amino acid sequences of the α and β polypeptides of HbS and HbC have revealed that in each, the α chain is normal but that the β chain is abnormal; that is, the primary structure of the β chain has been altered. It so happens that both HbS and HbC have *one* changed amino acid in the 146 making up the β chain, and the alteration occurs at the same site, position number 6, in the polypeptide sequence. In HbA, position 6 is occupied by glutamic acid (glu), whereas HbS has valine (val) in that position, and HbC has lysine (lys) there. Table 6-3 lists the amino acid sequence in positions 1 through 8 of the β polypeptides of HbA, HbS, and HbC. The remaining 138 amino acids are also known and are identical in all three hemoglobins. (Note that valine, the product of one of the initiator codons, begins the polypeptide.) Such subtle changes in polypeptide structure can bring great suffering and death to the individuals who bear them. Let us now see how this is related to gene function.

Now that the genetic code has been deciphered, we can relate the amino acid changes in polypeptide structure to changes in DNA structure that would account for the alteration and, in this manner, we can gain insights into how point mutations occur. Referring to Table 6-2, we find that the two DNA codons that code for the glutamic acid found in HbA are CTT and CTC. From this we find that if the C in the first position of either codon should be changed to a T, the new triplets (TTT and TTC) are the codons for lysine, the amino acid substitution in HbC. If, instead, the second position in either glutamic acid codon should be changed from T to A, the new triplets (CAT and CAC) are two that code for valine, the amino acid substitution in HbS. Here we can see that just one base change in the DNA segment that constitutes the β polypeptide gene can produce either of these two genetic diseases.

The occurrence of mutations such as those described above is supported by experimental evidence. The change from C to T in the first position of the glutamic acid codons represents the substitution of one pyrimidine for another. This type of mutation is called a *transition*. (Substitution of one purine for another also fits this category.) Three types of mechanisms have been shown to cause transitions in DNA: (1) rearrangements of the hydrogen atoms of the base molecules, called tautomeric shifts, (2) chemical mutagens, called base analogues, that substitute for the normal bases, and (3) chemical mutagens that alter the normal bases.

Table 6-3. The amino acid sequence of positions 1 through 8 of the β polypeptide of HbA, HbS, and HbC

Hemoglobin type	Position number									146
	1	2	3	4	5	6	7	8	....	
HbA	val –	his –	leu –	thr –	pro –	glu –	glu –	lys	....	his
HbS	val –	his –	leu –	thr –	pro –	val –	glu –	lys	....	his
HbC	val –	his –	leu –	thr –	pro –	lys –	glu –	lys	....	his

On the other hand, the change from T to A in the second position of the glutamic acid codons represents the substitution of a purine for a pyrimidine. This kind of change (or vice versa) is called a *transversion*. Agents that depurinate (remove purines) DNA, such as the combination of acid and heat or chemicals known as alkylating agents, leave a gap in the molecule during replication into which any of the four bases may be inserted. Some of these result in transversions. Nonionizing radiation, such as ultraviolet light, can also alter the bonding characteristics of the bases and lead to transitions or transversions.

Point mutations are difficult to detect and analyze in the human population. However, the mode of chemical action that causes transitions and transversions has led many geneticists, physicians, chemists, and increasing numbers of the public to become very concerned over the potential mutagenic hazards of numerous drugs, industrial chemicals, pesticides, food additives, and other substances that may be causing increased harm to the individual through somatic mutations, and damage to the human species in general through germinal mutations (gametes), the effect of which may not be fully felt for several generations.

7

Hereditary transmission

Having discussed the processes of reproduction, especially the meiotic mechanism, and the structure and function of the gene, we are now prepared to embark on discussions of inherited traits and the patterns of hereditary transmission. It is important to recognize at the onset that characteristics, abnormalities, and diseases that are hereditary in nature are not all expressions of identical genetic mechanisms and, more importantly, the pattern of transmission is not identical in all cases. Therefore, to understand human heredity and to appreciate its implications for the future of mankind, it is necessary to know *how* the genes react and interact to produce their final phenotypic expression. Most of the remainder of the book will deal with the various modes of hereditary transmission. In this chapter we shall discuss the fundamental mechanisms involved in the transmission and expression of qualitative inherited traits, that is, attributes of individuals that can be described as sharply contrasting alternative characteristics. Chapters 8 and 9 will discuss the circumstances that produce an array of different patterns in the expression of these traits; in Chapter 11 more complex traits will be discussed.

SINGLE-FACTOR INHERITANCE

A rather severe genetic disease, *phenylketonuria* (PKU), is caused by a defective enzyme (phenylalanine hydroxylase) that, in turn, reflects a mutation in the code of the gene responsible for the production of the enzyme. The absence of this enzymatic action blocks the biochemical breakdown of the amino acid, phenylalanine, which is present in much of our food, and concentrations then build up in the blood serum, sweat, urine, and cerebrospinal fluid. Thus, the body accumulates abnormal amounts of a substance that it normally metabolizes. Children born with PKU are normal at birth, but after 6 months of age the symptoms of the disease, including mental retardation, become apparent and rapidly grow more severe. Many untreated children with PKU have IQs below 20; in rare cases the IQ may reach the normal range.

The life expectancy of phenylketonurics is considerably shortened; only about 25% live beyond age 30, and the vast majority are institutionalized at an early age. As a result, afflicted individuals seldom reproduce, and children with PKU are usually born to "normal" parents. The incidence of PKU is estimated at 1 in 10,000 births; it accounts for 1% to 2% of all mentally retarded patients. But how do "normal" parents transmit a genetic disease, which they do not have, to their offspring?

The principles that Gregor Mendel elucidated in 1865 with his experiments on garden peas universally apply to the inheritance of qualitative traits in all diploid, sexually reproducing organisms, including man. From his extensive data on seven different qualitative characteristics (Table 7-1), he deduced the following principles. First, whenever two "pure breeding" plants with alternative traits of the same characteristic were cross-fertilized, the F_1 offspring (first filial generation) always displayed *one*, and only one, of the two traits and did so repeatedly. Mendel called those traits that were transmitted unchanged to the offspring *dominant* and those that became latent he termed *recessive* because they withdrew in the F_1 progeny. However, when F_1 individuals were intercrossed, the recessive trait reappeared unchanged among some members of the F_2 generation. Mendel also noted that the number of individuals with the recessive trait was only about one third the number that possessed the dominant trait. Considering the entire number of individuals composing the F_2 generation, this relationship of three dominant types to every one recessive type actually means that one fourth of the F_2 progeny was recessive in character and three fourths were dominant. Table 7-2 summarizes the actual numbers and ratios that Mendel obtained in his experimental crosses. By combining the principle of dominance and recessiveness with the observation of the three to one F_2 ratio, Mendel concluded that the alternative traits were determined by "differentiating characters" (genes) that existed as distinct entities and were present in pairs (two doses) in each individual. This second principle of *paired differentiating characters* then formed the basis for his third principle—*character segregation*. This principle, now often called the first Mendelian Law, states that during reproduction the paired hereditary

Table 7-1. Characteristics of the garden pea studied by Gregor Mendel, with the dominant and recessive traits indicated

Characteristic	Dominant trait	Recessive trait
Form of ripe seeds	Round	Angular
Color of seed albumen	Yellow	Green
Color of seed coat associated with flower color	Grey-brown Violet-red	White White
Form of ripe pods	Inflated	Wrinkled
Color of unripe pods	Green	Yellow
Position of flowers	Axial (distributed along the stem)	Terminal (bunched at top of stem)
Height of plant	Tall (6 to 7 feet)	Short ($\frac{3}{4}$ to $1\frac{1}{2}$ feet)

Table 7-2. Actual data and ratios obtained by Mendel in his experimental crosses with garden peas

Characteristic	Results		Ratio
Form of ripe seed	5,474 round	1,850 angular	2.96:1
Color of seed albumen	6,022 yellow	2,001 green	3.01:1
Seed coat/flower color	705 grey-brown/red	224 white/white	3.15:1
Form of ripe pods	882 inflated	299 wrinkled	2.95:1
Color of unripe pods	428 green	152 yellow	2.82:1
Position of flowers	651 axial	207 terminal	3.14:1
Height of plant	787 tall	277 short	2.84:1

characters (genes) segregate from each other during gamete formation (meiosis) in such a way that each egg or sperm receives only one member of each pair. At fertilization the zygote reconstitutes the paired condition from which the appearance (phenotype) of the individual is determined by its hereditary complement (genotype).

Through these three principles Mendel was able to explain the results he obtained in the F_1 and F_2 generations of his experiments in the following manner. Using letters of the alphabet to represent the various hereditary traits, he assigned a letter to represent a particular characteristic and designated the dominant trait by a capital and the recessive trait by the lower-case form of the same letter. This system is still the conventional method of genetic nomenclature although modifications and elaborations have occurred. For example, Mendel used the letter *A* for seed form with *A* (read "big A") representing round seeds and *a* (read "little A") representing angular seeds. Such alternate forms of the same basic gene are called *alleles*. Allelic genes represent differing genetic codes within a DNA segment that serves as the gene determining some particular characteristic. The "pure breeding" plants Mendel used in his initial crosses or parental generation (P_1) bred true because they possessed two identical doses of a particular allele. Thus, the round-seeded parent may be symbolized as *AA* and the angular-seeded parent as *aa*. Each case represents a *homozygous* genotype; that is, both members of a given gene pair are identical. When these homozygotes are crossfertilized, a new zygote is formed by the union of one gamete from each parent. According to the law of segregation, each gamete has one member of each gene pair, but since each parent was a homozygote, each can produce only one kind of gamete. If the round-seeded plant were the male parent, it would produce all *A* sperms, and if the angular-seeded plant were the female parent, all eggs would be *a*. The zygote formed by this cross, or vice-versa, would have a genotype of *Aa*; this is called a *heterozygous* genotype (different alleles make up the gene pair). These heterozygotes constitute the F_1 generation that, Mendel observed, appear as all round-seeded plants similar to the round-seeded parent. Thus, round dominates angular, and both the *AA* and *Aa* genotypes express the round phenotype; the recessive phenotype is only expressed when no *A* allele is present, and the

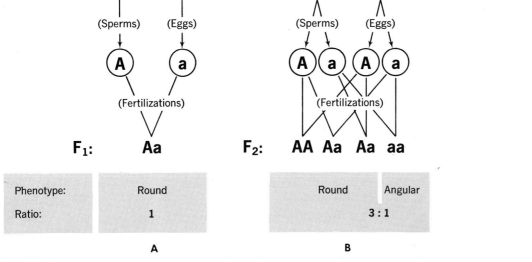

Phenotype: Round   Round Angular

Ratio: 1   3 : 1

A   B

Fig. 7-1. Single-factor genetic cross illustrating Mendel's hypothesis of character segregation and recombination. **A,** Pure-breeding parental cross; **B,** intercrossing of F_1 individuals.

genotype is *aa*. Fig. 7-1, *A* diagrammatically illustrates the parental cross and F_1 generation.

Following the same logic that explained the production of F_1 progeny, we can see how Mendel explained the 3:1 ratio of dominant to recessive traits in the F_2 generation. Fig. 7-1, *B* represents the intercrossing of male and female F_1 plants. Since the F_1 organisms are heterozygotes, they will produce two kinds of gametes, half with *A* and half with *a* alleles. And, since fertilization occurs at random, either kind of egg is likely to be fertilized by either kind of sperm. Thus, a *A* egg may be fertilized by a *A* sperm to give a *AA* homozygote, or it may be fertilized by a *a* sperm, producing a *Aa* heterozygote. On the other hand a *a* egg may be fertilized by a *A* sperm, producing another *Aa* heterozygote, or it may be fertilized by a *a* sperm to give a *aa* homozygote. Each of these four possibilities is equally likely to occur and, on the average, in any large random sample each combination would be expected ¼ or 25% of the time. The ratio of genotypes expected would then be 1*AA*: 2*Aa*: 1*aa*, and the ratio of phenotypes (observed by Mendel) would be 3 round-seeded:1 angular-seeded. Among the dominant types, ⅓ would be homozygous and ⅔ heterozygous.

This was also proved by Mendel to substantiate his interpretations by utilizing what is now referred to as a *testcross* to reveal the genotype of individuals expressing the dominant trait. By using an individual with the recessive phenotype as a "testor," the genotype of an individual with the dominant phenotype can be ascertained in the manner outlined in Fig. 7-2, *A* and *B*. The basis for the testcross lies in the fact that the homozygous recessive testor can produce only one kind of gamete, that carrying the recessive allele. If it is crossed to a homozygous dominant individual, all the offspring will be hetero-zygotes displaying the dominant phenotype (Fig. 7-2, *A*). If, on the other hand, the testor is crossed with a heterozygous dominant individual, the offspring will be half heterozygous dominants and half homozygous recessives. This 1:1 phenotypic ratio of a testcross reveals the heterozygosity of the tested indi-vidual. This strongly substantiated Mendel's principles concerning single-factor inheritance. But as we shall see shortly, since testcrossing is not feasible in man, other methods must be devised to identify heterozygotes in the human population.

Why should the identification of heterozygotes in man be important or de-sirable anyway? Returning to our original example of PKU, it is now apparent that this is a recessive genetic disease and the "normal" parents who have an afflicted child are both heterozygotes. In situations where both parents are "carriers" of a detrimental gene for the same trait, we can predict a $\frac{1}{4}$ prob-ability of producing an abnormal child. This prediction is based on the simple Mendelian phenotypic ratio of 3:1; the result would occur as the average from a large sample. A probability statement simply expresses the number of times a particular event is expected to occur out of the total number of events oc-curring. Thus, one out of every four children of heterozygous parents is ex-pected to have PKU.

If we assign the letter *P* to symbolize the gene for phenalanine hydroxylase, then *P* would represent normal production and *p* would represent inactive enzyme production. An individual with PKU would have the genotype *pp*, and

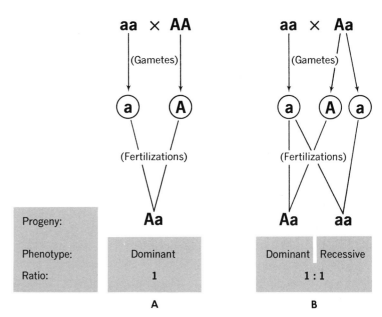

Fig. 7-2. Test crosses illustrating the identification of **A,** homozygous, and **B** heterozygous genotypes for a dominant trait.

"normal" individuals would be either *PP* or *Pp*. Among "normal" people the only parental combination that can produce a PKU child is *Pp* × *Pp*. Note that the parental combinations *PP*×*PP* and *PP*×*Pp* will not produce any children with PKU; however, the latter combination would produce 50% carriers of the disease (heterozygotes). How then can we identify these carriers so that the critical *Pp* × *Pp* parental combination can be determined and these couples counseled about the disease and their probability of having an afflicted child?

HOW DOMINANT IS DOMINANT?

It is obvious that we cannot testcross humans to identify carriers of genetic defects. Even if our morals and ethics permitted it, the long gestation period and the fact that only one offspring per mating is normally produced would make positive results temporally impractical. For example, for PKU it would take six normal siblings in a row to give 99% assurance that the testor was not heterozygous. This is undoubtedly too time-consuming a test to perform on a couple contemplating marriage! In theory, our knowledge of gene function implies that for every genic abnormality, there must be some abnormal gene product that can be detected and assayed. However, in practice it is no simple matter to determine the gene product responsible for every trait, let alone devise feasible means of detection. What do we want to assay for anyway?

Dominance on the phenotypic level does not also occur on the biochemical level. That is to say, two doses of a particular gene usually result in about twice as much gene product as does one dose. Since most enzymes are only needed in small quantities, the phenomenon of dominance occurs when the amount of enzyme produced by a single dose of a gene is sufficient to promote normal, or very near normal, development and/or function in the heterozygous individual. Although homozygotes and heterozygotes are generally similar, close investigation usually reveals subtle differences between the two types. For example, in the case of PKU, "normal" individuals are able to produce enough of the enzyme phenylalanine hydroxylase to metabolize phenylanine under normal conditions and therefore avoid the problems of phenylalanine accumulation. Under stress conditions, however, the reduced enzyme quantity in heterozygotes may be detected. Carriers of PKU can now be distinguished from normal homozygotes with 90% accuracy through the so-called phenylalanine tolerance test. After a standard amount of the amino acid is ingested, the level of phenylalanine in the plasma of carriers usually falls about midway between the values found in normal homozygotes and afflicted persons. This biochemical analysis can thus be applied to help counsel prospective parents who have histories of PKU in their family backgrounds and are particularly concerned about the possibility of having afflicted children.

The relationship of dominance to recessiveness is further illustrated by *galactosemia*. This is another recessive trait inherited in the same manner as

PKU. It is characterized by an inability to convert galactose (a common sugar prevalent in milk and many other foods) to glucose for normal use by the body. When untreated, the accumulation of galactose and its by-products often leads to early infant death or severe mental retardation and partial blindness in surviving children. Estimates of the incidence of the disease range as high as 1 in 18,000 births; most are produced from "normal" parents.

The primary genetic defect in galactosemia is the lack of an active form of an enzyme known as *galactose-1-phosphate uridyl transferase,* which is one of the enzymes necessary to convert galactose to glucose. In this case, direct determinations of enzyme activity can be made. Affected individuals lack any activity, and their "normal" parents (heterozygotes) display enzymatic activity that is intermediate as compared to known normal homozygotes. Heterozygotes are phenotypically normal because the amount of enzyme supplied by one dose of the functional allele gives enough enzymatic activity to metabolize galactose sufficiently. However, such carriers may pass on the abnormal allele of their gene pair for this enzyme to their children; in fact, half the gametes of heterozygotes are expected to possess the abnormal allele.

Mendel's principle of dominance and recessiveness still applies on the gross phenotypic level for most characteristics, while the biochemical nature of dominance in relation to normalcy adds greatly to our understanding of genic functions, especially in cases of hereditary diseases such as those described. Nevertheless, even at the phenotypic level a dominant-recessive relationship does not always exist between a pair of alleles. *Sickle cell anemia,* which served as one of our examples of the effect of point mutation in the previous chapter, is also an example of *dominance lacking.* Recall that the mutation that produces sickling occurs in the β polypeptide (a gene product) of the hemoglobin molecule, whereas the α polypeptide (product of a different gene) is normal. Therefore, considering only the gene for the β polypeptide, we can symbolize the normal form as β^A and the allele that produces sickling as β^S. A normal person (HbA) with respect to this condition has the genotype $\beta^A\beta^A$ for his double dose of the gene, and a person with sickle cell anemia (HbS) has the genotype $\beta^S\beta^S$. Heterozygotes ($\beta^A\beta^S$) are the individuals with mild anemia and sickling of only some red blood cells. This produces a third phenotype, *sickle cell trait,* intermediate to either homozygote. In cases such as this, where each genotype produces a distinguishable phenotype, the alleles are said to be *codominant,* and the phenotypic ratio is the same as the genotypic ratio. The marriage of two individuals with sickle cell trait produces a $\frac{1}{4}$ probability of having a child with sickle cell anemia and the same probability of having a normal child, while the chance is $\frac{1}{2}$ that any child they conceive will have sickle cell trait just like its parents (Fig. 7-3).

GENE BEHAVIOR

Dominant, recessive, or codominant relationships are not the only factors that affect the behavior of genes with respect to the observable phenotype pro-

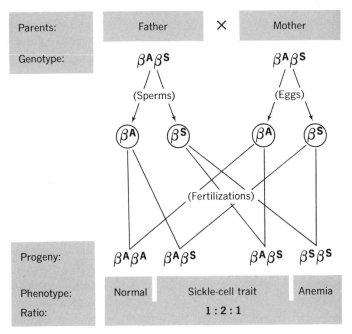

Fig. 7-3. Genetic cross of parents with sickle cell trait, illustrating a phenotypic ratio of 1:2:1 characteristic of codominant alleles.

duced by their allelic forms. In some instances individuals with the appropriate genotype fail to express the expected phenotype. Alleles that behave in this manner are said to be *incompletely penetrant*. The phenomenon of incomplete penetrance is most easily traced for a dominant trait, although recessively conditioned traits may also exhibit less than 100% penetrance. *Camptodactyly* is a dominantly inherited abnormality characterized by a permanently bent and stiff little finger. Since this is a dominant characteristic, heterozygotes are expected to have the abnormal trait or, to put it another way, anyone receiving the allele for camptodactyly should have the stiff little finger. Cases are known, however, where "normal" sons or daughters of a parent with camptodactyly have married a normal spouse and have produced children having stiff little fingers. How can a "normal" parent pass a dominant abnormality on to his children without exhibiting the trait himself? It appears that about 25% of the time the camptodactyly allele is present, it fails to express itself. Penetrance is therefore measured on the populational level as the percent of genetically susceptible individuals that actually display the trait. Accordingly, camptodactyly can be described as a dominantly conditioned trait with 75% penetrance. Most commonly discussed hereditary traits are 100% penetrant.

At the individual level, penetrance is an all-or-none phenomenon, that is, a person with the appropriate genotype either develops the phenotype or lacks it completely. On the other hand, some traits that are demonstrably inherited in a simple single-factor pattern exhibit great latitude in their expression. This phenomenon, called *variable expressivity,* often encompasses an array of phenotypic expressions that ranges from extremeness to a mildness that resembles "nonpenetrance." Another dominantly inherited abnormality, *juvenile cataract* of the eyes, exemplifies this phenomenon. This type of cataract develops in children and progressively impairs vision as the lens of the eye becomes packed more and more densely by the deposition of flakes and granules. The pattern produced in the lens by the flakes and granules is quite variable. In some individuals the flakes and granules are concentrated toward the front of the lens; in others they compact into a dense central mass, and in some cases the cataract consists of fibrous rather than flakelike particles. These diverse patterns can all be traced through single family lineages by pedigree analysis (Chapter 10) and shown to follow a transmission pattern for a single-factor dominant allele. Yet a parent with the trait expressed as a dense central mass may pass the allele to a portion of his children (we would expect half on the average to receive his dominant allele if he is heterozygous himself), some of whom express the trait in a fashion similar to the parent, whereas others express it concentrated toward the front of the lens or as fibrous particles. A child in either of the latter two categories may then produce children who express the trait as their grandparent did. Thus, in cases such as this the transmission pattern is clear-cut, but the phenotype is variably expressed.

Although the phenomena of penetrance and expressivity are recognized behavioral attributes of some genes, they are not fundamental properties of the gene per se. It must be kept in mind that no gene works in isolation. Each gene is a segment of DNA among other segments, located in a chromosome among other chromosomes, and this whole genetic complement exists in the environment of the cell. Incomplete penetrance and variable expressivity are the result of interactions between the primary gene involved and the surrounding genetic and environmental factors. It is well established that the genetic background within which any particular gene operates may enhance, dilute, or even prevent its phenotypic expression. This is accomplished through systems of *modifying genes* that interact with the *major gene.* We have already discussed extreme cases of environmental effects on gene expression in the production of phenocopies. It is not unexpected, therefore, that milder environmental insults or variations can produce variable expressions of the phenotype.

Genes may behave in still another manner related to their influence on phenotypic expression. It is not unusual for a particular gene to produce multiple phenotypic effects. Such genes are said to be *pleiotropic* (which literally means several changes), and the phenomenon is called *pleiotropy.* In contrast to penetrance and expressivity, pleiotropy *is* a fundamental gene property in that the multiple effects are produced from the basic gene product, its polypeptide. Once the polypeptide, for example an enzyme, is produced, it may enter into several different biochemical pathways or it may affect the production of a substrate for several different biochemical processes. In any case, when a gene product is involved in more than one bodily function, any alteration of the product may affect all the functions. It is probable that most genes, including those with a clear-cut major effect, are pleiotropic, even though the secondary effects might be insignificant.

An interesting example of pleiotropy concerns a recently described recessively inherited syndrome (set or group of abnormalities) known as *pyknodysostosis,* the name literally means thick, dense bone formation. The several manifestations of this disease that appear in individuals homozygous for the abnormal recessive allele are:

1. Densification of the skeleton, resulting in fragility of the bones
2. Shortened stature, resulting in heights that range from 4 feet 5 inches to 5 feet (This range is above the true dwarfs produced by other genetic abnormalities.)
3. A rather large skull characterized by failure of the fontanel ("soft spot") to knit in adulthood
4. A receding chin and abnormally formed lower jaw bone
5. Abnormal development of the extremities of the fingers and toes, causing a shortened effect

Here we have a situation where the gene product apparently affects bone development, resulting in an array of abnormalities. It now appears that the famous artist Toulouse-Lautrec was afflicted with this genetic disease (Fig. 7-4). It was previously thought that he was an *achondroplastic dwarf,* which is a *dominantly* inherited trait. Since neither of his parents was abnormal, and no history of dwarfism was evident in the family background of either, Lautrec was thought to represent the product of a newly arisen mutation in either the egg or the sperm that combined to constitute his genetic complement. This is not unreasonable, since a large portion of dominant abnormalities appear de novo as the direct, unmasked product of recurrent mutation. However, Lautrec was about 5 feet tall, beyond the range of true dwarfs, and suffered broken thighs twice (fragile bones). His other characteristics also fit the description of pyknodysostosis. He had a disproportionately large skull, and no doubt he grew a beard to hide his receding chin, but this abnormality was still prominent enough to be stressed in caricatures of him. One acquaintance commonly referred to "his comical little hands," and his biographer, Perruchot, confirms the distinct possibility that his fontanel lacked strength. Not only do all these traits conform to the pyknodysostosis syndrome, but the circumstances of Lautrec's birth add credence to this more recent interpretation. Toulouse-Lautrec was the child of a marriage between first cousins (both normal), and such inbreeding (mating between closely related individuals) is known to greatly increase the chances of making a recessive allele homozygous because the allele may exist in both parents by virtue of its descent through their common family lineage. This situation will be more thoroughly discussed in Chapter 13. All available evidence, phenotypic and genetic, strongly supports Toulouse-Lautrec's condition as the pleiotropic effects of the recessive, pyknodysostosis allele.

Fig. 7-4. Toulouse-Lautrec. Was he the result of a newly arisen dominant mutation or the product of a genetically hazardous first cousin marriage? (Culver Pictures)

The behavioral aspects of genes described here must be recognized as important contributing factors, beyond simple genetic transmission, in the development of certain phenotypes. The degree of penetrance and expressivity is especially important in individual cases because, for the particular individual, the phenotypic expression in him or her is what that person will have to live with. On the other hand, the recognition of pleiotropy helps us to understand the genetic transmission of groups of abnormalities that are inherited together. In the next chapter we shall discuss other mechanisms that produce more complex genetic patterns.

THE MECHANISM OF SEGREGATION

Regardless of the dominant-recessive relationship or the behavior of genes in respect to phenotypic expression, the basic mechanism of hereditary transmission remains the segregation of paired differentiating characters, the genes, as Mendel first theorized. In the first few years of the twentieth century, scientists studying the chromosomes in gamete formation, fertilization, and development discovered the striking resemblances between the maneuvers of the chromosomes and the segregation and recombination of Mendel's characters. Thus, the connection was made between gene and chromosome, and meiosis was established as the mechanism of gene segregation. It is appropriate to refer to Fig. 4-7 at this time to review the meiotic sequences. The pairing and separation of the homologous chromosomes that result in haploid gametes follow the same pattern as the paired genes, segregating in such a way that only one member of the pair is present in any particular egg or sperm.

In the early part of the century, the relationship between gene and chromosome was assumed through the correlations of the behavior of each to adequately explain and predict hereditary patterns. Today, our knowledge of gene structure proves beyond doubt that the chromosomes are the vehicles of the genes. Each gene (genetic code sequence) exists at a certain place on a particular chromosome; this position is known as its *locus* (Latin for location, plural *loci*), and chromosome replication is in fact gene replication (DNA synthesis).

After reviewing the details of meiosis in Fig. 4-7, we can simplify the process to its essential features and see how it serves as the mechanism of segregation. Using "stick" diagrams to represent chromosomes, Fig. 7-5 depicts the segregation of one gene pair whose locus is indicated as a specified segment of the chromosome. It should also be noted that the specificity of homologous chromosomes is caused by their genetic similarities; each gene has a corresponding locus on each of the homologous chromosomes. When meiosis is completed, the haploid condition represents not only one member of each chromosome pair but genetically one member of each gene pair. This pattern is also related to the segregation of maternal and paternal chromosomes without the effects of crossing over, which will be discussed in the next chapter.

The end result from heterozygous parents represents an equal ratio (1:1) of *A*-bearing and *a*-bearing products. Hence, if sperms are being formed, half would carry *A*, and half would carry *a*. On the other hand, if oogenesis is occurring, the single egg cell produced would correspond to one of the end products of meiosis so that any particular egg from a heterozygous woman has a 50-50 chance of bearing *A* or *a*. On the average, half of all eggs produced would carry *A*, and half would carry *a*.

These "chances" of occurring are probabilities that can be quite easily determined. A probability indicates a degree of uncertainty about the occurrence of an event. We use the concept informally when we make statements like "she'll probably be late," or "nine out of ten times I don't hear the alarm."

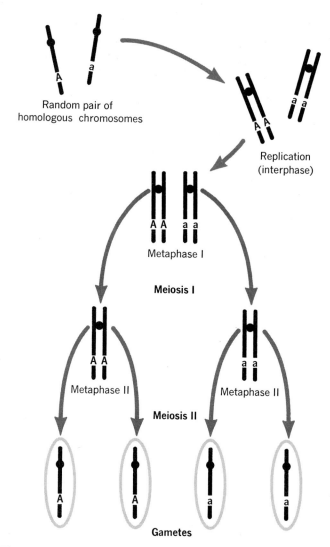

Fig. 7-5. The essential features of meiosis that illustrate how the maneuvers of the chromosomes serve as the basis for gene-pair segregation.

More precisely, as previously stated, a probability is defined as the number of times a particular event is expected to occur in relation to the total number of possible events. Thus, the probability of a man of genotype *Aa* producing a *A*-bearing sperm is 2 (the expected meiotic products with a *A* gene) out of 4 (the total number of products). This probability is written as a fraction $\frac{2}{4}$ equals $\frac{1}{2}$ or as a percentage derived by dividing out the fraction, in this case, .50 or 50%.

The random combinations of eggs and sperms from two individuals, representing the possible zygotes that can be produced, can be generated by placing the potential gametes of each sex on lines at right angle to each other and

A

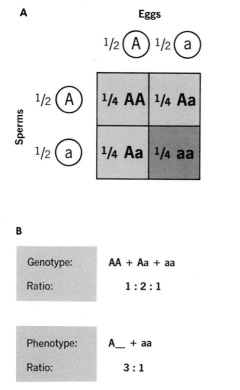

B

Genotype:	AA + Aa + aa
Ratio:	1 : 2 : 1

Phenotype:	A__ + aa
Ratio:	3 : 1

Fig. 7-6. A, Possible combinations and frequencies (probabilities) of eggs and sperms from parents heterozygous at a single locus. **B,** The genotypic and phenotypic ratios of the zygotes, considering *A* dominant over *a*.

forming all possible combinations between eggs and sperms, as illustrated in Fig. 7-6, *A*. Note also that probabilities of occurrence are also assigned to each zygotic combination. These are calculated by applying a simple rule for combining probabilities, which states that the probability that *both* of two independent events (the formation of eggs in the female and sperms in the male) will occur together (form a particular zygotic combination) is the product of the individual probabilities. In this case multiplying the probability of a certain kind of egg by the probability of a certain kind of sperm gives the probability of their occurrence together as a particular zygotic combination. These probabilities are also given in Fig. 7-6, *A*. Summarizing these zygotic combinations by genotypes or phenotypes (assuming *A* dominant over *a*) indicates the classic Mendelian ratios (Fig. 7-6, *B*). Note that in summarizing phenotypes when one allele is dominant over the other, the notation *A__* (read "*A* blank") serves to identify the phenotype as that expressing the dominant trait. The "blank" stands for either *A* or *a* and in this case includes the homozygous *AA* genotype and both heterozygous genotypes. The correlation of these theoretical predictions with actual hereditary patterns supports meiosis as the mechanism of genetic segregation, for it is the process responsible for establishing the gametic probabilities in the first place.

MULTI-FACTOR INHERITANCE

Mendel's experiments with the garden pea were not limited to tracing one hereditary factor at a time; he also studied the transmission of two or more characteristics at one time. From these studies he formulated the principle of *independent assortment*, often called the second Mendelian Law. To paraphrase Mendel's own words, the offspring of individuals heterozygous at several loci exhibit the combined traits as a series of combinations of the separate traits; that is, each gene pair in the series segregates independent of the other gene pairs of the parents. In other words, two or more sets of genes being considered together segregate in gamete formation and recombine in zygote formation randomly and independent of each other. This principle is true as long as the sets of genes in question are located on different sets of homologous chromosomes; the special case of different genes located on the same pair of chromosomes will be considered in the next chapter.

To predict the kinds and frequencies of zygotes in a cross involving two or more sets of genes, the eggs and sperms must be randomly combined according to their expected frequencies (probabilities) to generate the zygotic array. The first step then is to determine the kinds of gametes each parent can produce and the probabilities of their occurrence. Considering two or more sets of genes on different chromosomes is analogous to considering the shuffling of maternal and paternal chromosomes during meiosis, disregarding crossing over. Another review of Fig. 4-7 will reveal that the maternal and paternal chromosomes (gray and black chromosomes) of different homologous pairs are shuffled at random in the end products; that is, they assort independently. Thus, any genes located on these chromosomes would likewise be randomly assorted. The kinds of gametes possible from individuals heterozygous at two loci *(AaBb)* located on different chromosome pairs are illustrated in Fig. 7-7. Alternatives 1 and 2 are equally likely orientations of the chromosome pairs at metaphase I of meiosis; therefore, each event has a $\frac{1}{2}$ probability of occurrence. Considering both of these alternatives as representing the total set of events possible, we find that 4 different kinds of haploid products are generated: *AB, ab, Ab,* and *aB* gametes. Each different haploid genotype occurs in 2 of the 8 possible products, thus, each kind of gamete has a $\frac{2}{8}$ or $\frac{1}{4}$ probability of occurrence. If both parents are heterozygous at these loci, the random combination of their potential gametes will generate the zygotic combinations possible. Putting the possible egg genotypes on lines at right angles to the possible sperm genotypes produces a 4×4 checkerboard with 16 zygotic combinations. Fig. 7-8, *A* illustrates these combinations with the genotypic and phenotypic ratios (considering *A* and *B* dominant over their respective alleles) summarized in part *B* of the figure.

As the number of gene sets involved in any particular genetic analysis increases, this system of diagramming the crosses becomes complex and unmanageable. With 3 sets of genes assorting independently, heterozygous individuals would produce 8 kinds of gametes each, resulting in 64 zygotic combinations to summarize for genotypic or phenotypic ratios, and these numbers increase exponentially as more gene pairs are considered. The relationship of gene pairs to gamete types, zygotic combinations, and distinguishable genotypes and phenotypes is given in Table 7-3. Clearly, a method simpler than plotting a checkerboard and summarizing each cross is needed.

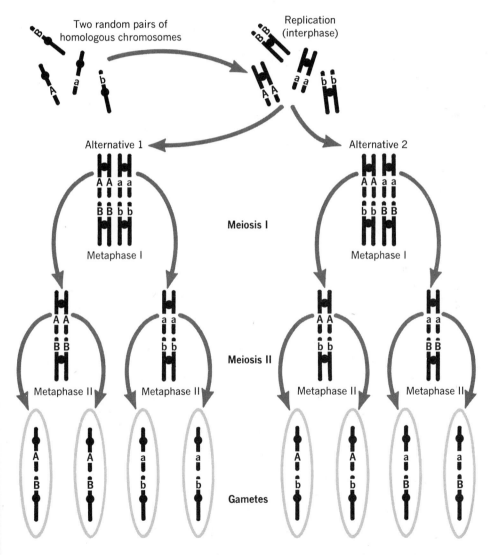

Fig. 7-7. Independent assortment of two sets of genes in meiosis of a heterozygous individual. Alternatives 1 and 2 are equally likely orientations of the chromosome pairs at Metaphase I.

A

Eggs

1/4 (AB) 1/4 (Ab) 1/4 (aB) 1/4 (ab)

	1/4 (AB)	1/4 (Ab)	1/4 (aB)	1/4 (ab)
1/4 (AB)	AABB 1/16	AABb 1/16	AaBB 1/16	AaBb 1/16
1/4 (Ab)	AABb 1/16	AAbb 1/16	AaBb 1/16	Aabb 1/16
1/4 (aB)	AaBB 1/16	AaBb 1/16	aaBB 1/16	aaBb 1/16
1/4 (ab)	AaBb 1/16	Aabb 1/16	aaBb 1/16	aabb 1/16

Sperms

B

Genotype:	AABB	+	AABb	+	AAbb	+	AaBB	+	AaBb	+	Aabb	+	aaBB	+	aaBb	+	aabb
Ratio:	1	:	2	:	1	:	2	:	4	:	2	:	1	:	2	:	1

Phenotype:	A_B_	+	A_bb	+	aaB_	+	aabb
Ratio:	9	:	3	:	3	:	1

Fig. 7-8. A, Possible combinations and frequencies (probabilities) of eggs and sperms from parents heterozygous at two independent loci. **B,** The genotypic and phenotypic ratios of the zygotes, considering A and B dominant over their respective alleles.

A technique known as a *tree diagram* can be used to generate combinations and assign probabilities of occurrence for independently assorting gene pairs. The advantage of this technique over the checkerboard approach is that it eliminates the need to summarize separate zygotic combinations, and it gives genotypic or phenotypic probabilities directly.

Since the phenotypic array is simpler to generate, we shall take it first. To generate the phenotypic ratio in a multi-factor cross, each gene pair is considered separately with respect to the phenotypes that may be produced and the probabilities of their occurrence. In the cross of $AaBb \times AaBb$, the phenotypes of the A gene pair would be $A\_\_$ and aa with probabilities of $\frac{3}{4}$ and $\frac{1}{4}$ respectively. At the B locus the phenotypes and probabilities would be $\frac{3}{4}B\_\_$ and $\frac{1}{4}bb$. A tree diagram starts with the phenotypes of one gene pair and branches each to the possible combinations that may be formed with the next gene pair, and so on until all gene pairs under consideration have been ac-

Table 7-3. The relationship of gene pairs to gamete types, zygotic combinations, distinct genotypes and distinguishable phenotypes produced from independent assortment in crosses between individuals heterozygous at all loci concerned

Number of gene pairs	Number of different gamete types	Number of random zygotic combinations	Number of different genotypes	Number of different phenotypes*
1	2	4	3	2
2	4	16	9	4
3	8	64	27	8
4	16	256	81	16
.	.	.	.	.
.	.	.	.	.
.	.	.	.	.
.			.	
n	2^n	4^n	3^n	2^n

Assuming complete dominance of one allele over the other at each locus.

counted for. Since the probabilities associated with each gene pair are independent of the other gene pairs, the probability of any combination along a particular pathway of the diagram is obtained by multiplying the separate probabilities (rule for combining probabilities of independent events).

Fig. 7-9 represents the phenotypic tree diagram of a typical dihybrid (double heterozygous) cross. Note that it automatically summarizes the zygotic combinations generated in Fig. 7-8. A tree diagram of the genotypic array from the same cross is given in Fig. 7-10. In this case $A\_\_$ is replaced by the two specific genotypes incorporated in that phenotype, AA and Aa, and their probabilities of $\frac{1}{4}$ and $\frac{2}{4}$ respectively are assigned to their pathways. Three branches representing the B gene combinations are then coupled to each A gene combination to generate the genotypic pathways.

As more sets of genes are considered, the tree diagram grows in complexity also. However, in many instances of genetic predictions the entire phenotypic or genotypic array is not needed. Rather, it is often the case that the probability of some specific phenotype (or genotype) is desired to indicate the likelihood of occurrence of some particularly good or bad or otherwise important combination. The situation may be as mundane as desiring to know the chances of having a child with blue eyes, blond hair, and freckles or as serious as professional genetic counseling to determine whether or not the odds are favorable for the production of normal children in parents with particular genetic diseases in their backgrounds.

Taking the first situation described, the genetics of the traits are as follows. *Eye color* is determined mainly by the deposition of a brownish pigment, called *melanin*, in the iris. This pigment is involved in several predominant traits of the human species including the other two presently under consideration. The further involvement of melanin in skin color and albinism will be discussed in later chapters. In the eye, coloration depends mainly on where and how the

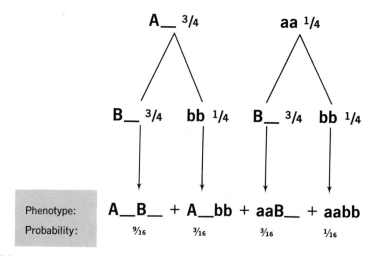

Fig. 7-9. Phenotypic tree diagram of a dihybrid cross with *A* and *B* dominant over their respective alleles.

melanin is deposited in the iris (the pigmented diaphragm in front of the lens that regulates the amount of light passing through by adjusting the size of its central opening, the pupil). Dark eyes contain abundant amounts of the pigment, and lighter eyes possess lesser amounts. The particular color of the eye depends on the concentration, distribution, and reflective qualities of the melanin; different colored eyes do not possess different colored pigments. In blue eyes the pigment is minimal in amount and is deposited in the rear of the iris; the blue color results from the way the light is reflected to the viewer through the colorless front portion of the iris (this is the same optical effect that makes the sky appear blue). The other eye colors are produced by additional amounts of melanin deposited in the front portion of the iris. In gray eyes, scattered melanin in the front "grays down" the blue reflection. Scattered yellowish pigment (possibly dilute melanin) in the front portion of the iris blends with the blue reflection to produce green eyes. A uniform distribution of melanin in the front of the iris simply reflects its brown color, and this deepens to darker intensities, approaching black eyes as the amount increases.

These various arrangements of melanin are genetically controlled by a series of alleles at a single locus in such a way that an allele for deeper intensity dominates one of less strength. Thus, all other alleles dominate that for blue eyes, while the rest of the series is: green dominated by gray dominated by brown dominated by black. The genetics of such a multiple allelic system will be discussed in more detail in the next chapter. For our present purpose we can lump all the alleles that dominate blue into one category called "dark" (most commonly brown) and treat the group collectively as *B* (for dark eyes) dominant over *b* (the allele for blue eyes). This simplifies our considerations to only two phenotypes, blue or dark eyes.

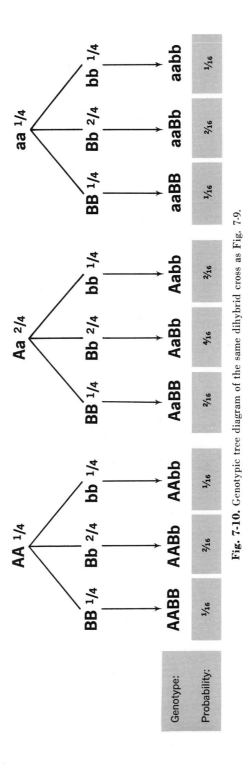

Fig. 7-10. Genotypic tree diagram of the same dihybrid cross as Fig. 7-9.

The second trait under consideration, *hair color,* is also determined by a set of multiple alleles that determines the amount of melanin that will be deposited in the hair cells. Blond hair, which consists of dilute yellowish melanin, is dominated by more active alleles that govern hair intensity from light brown to black. Since all alleles dominate blond, we can again consider two alternatives, blond or "dark" (brown or black), to simplify our discussion. We shall designate the gene pair as Y (for dark) dominant over y (blond or yellowish). Red and reddish hair colors are governed by genes at a different locus; this will be discussed in conjunction with the brown hair genes in the next chapter.

Freckling is associated with melanin in respect to its distribution, not its amount, in the skin. The gene for producing freckles acts dominant over its allele for no freckles. The freckling gene apparently causes the pigment to clump rather than to be evenly distributed throughout the skin. The intensity of freckles increases upon exposure of the skin to sunlight because the sun's rays stimulate added pigment production as a protective mechanism (tanning), and this is reflected as larger clumps under the influence of the freckling gene. This is why freckles tend to fade during the winter and blossom in summer. We shall designate the dominant freckling gene as F and its recessive, inactive allele as f.

Hence, if a dark-eyed, dark-haired, freckled woman married a man with similar features and *both* were heterozygous at all three loci, what is the likelihood of their producing a child with blue eyes, blond hair, and freckles? Assuming independent assortment, a phenotypic tree diagram can be constructed, as in Fig. 7-11, to generate the entire phenotypic array. Having done this, we find that the particular phenotype we are interested in corresponds to the pathway leading to the combination *bb yy F__* (outlined in the figure) and that its probability of occurrence is $3/64$ or about a 5% chance.

Once the tree diagram concept of generating random combinations and the biological basis (meiosis) for assigning probabilities to particular allelic combinations are understood, shortcuts can easily be taken. Considering the cross in question and the phenotype of the child in question, we can simply ask the probability of getting each of the independent events involved then multiply them together to obtain the probability of their combined occurrence. Thus, from the parental cross *Bb Yy Ff* × *Bb Yy Ff,* the probability of blue eyes *(bb)* is $1/4$, the probability of blond hair *(yy)* is $1/4$, and the probability of freckles *(F__)* is $3/4$. Their product ($1/4 \times 1/4 \times 3/4$) equals $3/64$ which gives the probability of all occurring together *(bb yy F__).*

Obviously not all crosses among individuals for various traits concern heterozygosity at each locus. However, considering each gene pair independently to derive the probabilities involved requires only a slight and easily managed modification. For example, if the child discussed above were a female with the exact genotype of *bb yy Ff (F__* could have been *FF* also) and she married a man with the phenotype and genotype of her father *(Bb Yy Ff),*

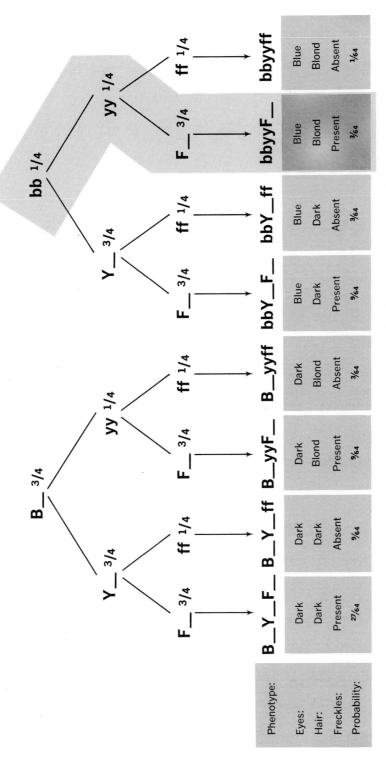

Fig. 7-11. Phenotypic tree diagram of the cross *Bb Yy Ff* × *Bb Yy Ff* (see text for description of traits and the particular pathway outlined).

what would be the likelihood of this union producing a child with the pheno-type of the mother; that is, the same phenotype *(bb yy F__)* as desired in the first situation? Here the cross is *bb yy Ff* ♀ × *Bb Yy Ff* ♂, and we must determine the separate probabilities of getting the desired combination in a child when two of the three gene sets are not the "ideal" heterozygous × hetero-zygous crosses. In each case the female can produce only *b*-bearing and *y*-bearing eggs. Taking each case separately, a *b*-bearing egg may be fertilized by either a *B-* or *b*-bearing sperm, each sperm with a probability of $\frac{1}{2}$. This pairing will give two kinds of zygotic combinations, *Bb* and *bb*, with *bb* (the desired combination in their child) occurring as one of the two possible events; in other words the probability of the *bb* combination in this case is $\frac{1}{2}$. The same situation pertains to the *Y* locus, which indicates that the *yy* com-bination has a $\frac{1}{2}$ probability also. The *F__* combination is still the product of *Ff* × *Ff*, as in the first situation, and its probability remains $\frac{3}{4}$. However, the probability of occurrence of the *bb yy F__* phenotype in this example is $\frac{1}{2} \times \frac{1}{2} \times \frac{3}{4} = \frac{3}{16}$ or about 19%. This is much more likely than the 5% probability in our first example. The difference is caused by the different hereditary make-up of the parental cross in each case. Nevertheless, the mech-anism of transmission (meiosis) and the procedure of combining probabilities remain the same; only the actual probabilities change. This illustrates that genetic predictions, such as the probability of a particular phenotype (*bb yy F__*), are not fixed probabilities but the product of a consistent biological system (meiosis) and a specific procedure (tree diagram concept) that enable us to make accurate determinations, providing sufficient genetic data are available.

8

Hereditary patterns

Although Mendel's principles discussed in the previous chapter correctly explained the genetic patterns observed in his experiments with garden peas, other, more complex patterns of heredity are also quite commonplace in the human species and all other organisms. In this chapter we shall discuss several non-Mendelian hereditary patterns that involve a host of traits important in human affairs. The various modes of action that produce modifications of the classical Mendelian phenotypic ratios must be recognized if a reasonable understanding of human heredity is to be achieved.

Before embarking on any discussion of hereditary patterns, it should be noted that meiosis still provides the basic mechanism of transmission in every case even though the patterns (phenotypic ratios) produced are altered in various ways, depending on the particular manner in which the alleles or gene sets in question govern their respective phenotypes. Relating these patterns to the fundamental meiotic process places the action in the proper context and makes interpretation of the various patterns logical and easily comprehensible.

MORE THAN TWO ALLELES

In the discussion of eye color and hair color in the previous chapter, it was pointed out that the several phenotypes for each characteristic are genetically controlled by a series of alleles at a single locus. A series of this type forms a *multiple allelic system* that in turn produces a modified hereditary pattern when the entire system is considered (rather than everything being lumped into only two alternatives such as blue and "dark" or blond and "brown"). The genetic control of eye color has not been thoroughly explained, and it may well be more complex than a multiple allelic system. Some investigations indicate a polygenic basis (a mode of inheritance to be considered in Chapter 11), and others indicate some involvement of sex-linked genes (Chapter 9). Therefore, we shall not consider eye color any further, but it should be clearly understood that blue-"dark" is a gross oversimplification of a complex human characteristic.

Hair color is also a complex characteristic, but it is amenable to building a genetic model that exemplifies a multiple allelic system for the inheritance of brown colors. Also, it can be expanded in the next section of this chapter to elucidate the phenomenon of gene interaction by adding the locus for red hair to our consideration. The simplest multiple allelic model is one with three alleles. However, in any case of multiple alleles, it must be recognized that any particular individual normally possesses only *two* doses of any particular gene, one on each of the two homologous chromosomes carrying the locus in question. Considering a three-allele model for hair color, we see listed in Fig. 8-1 a system of nomenclature, the nine diploid genotypes possible (all combinations of the three alleles taken two at a time), and phenotypic interpretations of the genotypes. The system of nomenclature for multiple alleles uses a base letter for the gene (*M* for melanin here) and superscripts to describe the various alleles. Note that the phenotypic interpretations conform to the observed phenomenon of darker hair dominating lighter colors, but biochemically the phenotypes simply result from the cumulative effects of heavier and heavier deposits of pigment. The two genotypes interpreted as medium browns may not be identical in actuality but are theoretically so if light to medium to

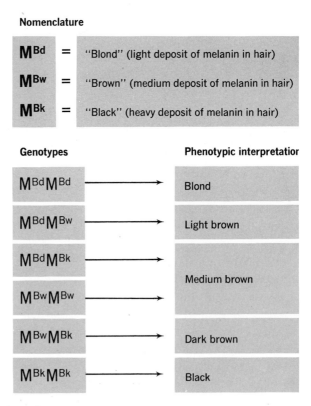

Nomenclature

M^{Bd} = "Blond" (light deposit of melanin in hair)

M^{Bw} = "Brown" (medium deposit of melanin in hair)

M^{Bk} = "Black" (heavy deposit of melanin in hair)

Genotypes		Phenotypic interpretation
$M^{Bd}M^{Bd}$	→	Blond
$M^{Bd}M^{Bw}$	→	Light brown
$M^{Bd}M^{Bk}$	→	Medium brown
$M^{Bw}M^{Bw}$	→	Medium brown
$M^{Bw}M^{Bk}$	→	Dark brown
$M^{Bk}M^{Bk}$	→	Black

Fig. 8-1. A three-allele model for hair color indicating the system of nomenclature for multiple alleles, the diploid genotypes, and a phenotypic interpretation.

heavy is considered as a linear progression of equal increments. The hereditary pattern in a multiple allelic system is modified only with respect to the increased number of genotypes and phenotypes possible. In any particular cross, segregation of the alleles occurs normally, and random recombination produces zygotes in the usual manner. However, as illustrated in Fig. 8-2, particular crosses can yield phenotypes in the offspring that would be inexplicable by simple Mendelian principles.

Hair form is also genetically controlled by a multiple allelic system that determines the shape of the hair in cross section. Straight hair is circular in cross section; the curlier hair forms are attributed to progressively more oval and flattened hair shapes. The order of dominance among the hair form phenotypes is: wooly dominates kinky that dominates curly that dominates wavy that dominates straight. A three-allele model can also be used to explain this situation, as illustrated in Fig. 8-3. In the Negro race the allele for wooly hair is in very high frequency, producing a preponderance of wooly and kinky hair types. All alleles appear in the Caucasian race, but the wooly allele is in low frequency. As in the hair color example, the two phenotypes interpreted as curly hair may not actually be identical. An exception to the hair forms described here exists in the Mongolian race, where the thick, straight hair characteristic of these people seems to dominate all other types. In crosses between Mongolians and Caucasians or Mongolians and Negroes, the thick, straight hair form nearly always prevails. This special situation could be caused by a fourth allele in the multiple allele system which dominates all others, or it

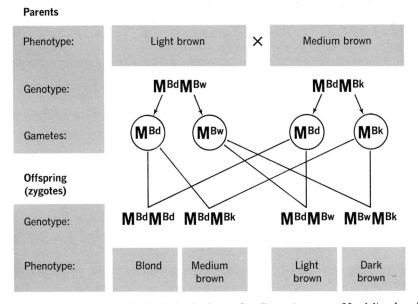

Fig. 8-2. Cross involving multiple alleles for hair color illustrating a non-Mendelian hereditary pattern. Note that the medium brown genotype $M^{Bw} M^{Bw}$ would give results different from those shown here.

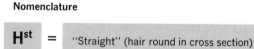

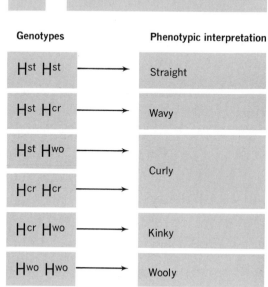

Fig. 8-3. A three-allele model for hair form, indicating nomenclature, diploid genotypes, and a phenotypic interpretation based on codominance among the alleles and linear, cumulative effects of their action on the shape of the hair in cross section.

could be produced by a dominant gene at a different locus that overrides the effect of the multiple allelic system.

Multiple alleles also form the genetic basis for the inheritance of the major *blood* types. Of the fourteen or more blood group systems known in the human species, the ABO system and the Rhesus system are of greatest importance. The ABO system is genetically controlled by a series of three multiple alleles, as illustrated in Fig. 8-4. The gene symbol I stands for a protein called isoagglutinogen that is the precursor of the different agglutinogens or *antigens* produced by the various allelic forms of the gene. In effect, I^A changes the isoagglutinogen precursor to the A-antigen, I^B changes it to the B-antigen, and I^O leaves it unchanged. This precursor is known as *H-substance*. (A rare mutant of the gene responsible for H-substance will be discussed in the next section of this chapter.) The behavior of these three alleles is such that I^A and I^B are codominant (each produces its particular antigen), and both are dominant over I^O (which produces nothing beyond H-substance). Thus, the six genotypic combinations of these alleles result in only *four* phenotypes, as described in Fig. 8-4.

Blood type in man is extremely important in transfusion therapy. The blood

Nomenclature

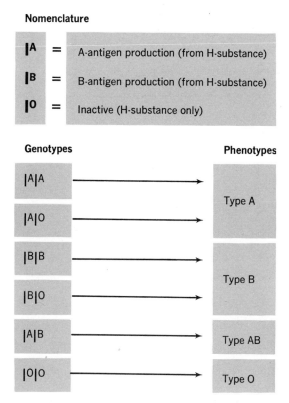

Fig. 8-4. The multiple allelic system governing the hereditary pattern of the ABO blood groups.

of donor and patient must be compatible in order to avoid an adverse reaction that could lead to the death of the patient. The basis of this reaction, which results in the clumping, or agglutination, of the red blood cells, is a result of a normal bodily protective mechanism against "foreign intruders" known as the *immune response*. The body is capable of producing an array of molecules, known as *antibodies*, that attack "foreign" proteins, mainly viruses and bacteria that invade the body. The antibodies chemically deactivate proteins foreign to the body (that is, proteins different in amino acid structure from any produced by the body itself) and remove them from the system. This immune response operates against many different proteins other than viruses and bacteria, however. Specific reactions against pollen, certain food proteins, or virtually any foreign protein source produce an array of allergies in man, and the same processes account for organ transplant incompatibilities and tissue rejection.

It is usually the introduction of the foreign protein that stimulates production of the antibody that deactivates that particular protein; the body does not store all possible antibodies for proteins it has never encountered. The ABO

Table 8-1. Relationship between the antigens on the red blood cells and the antibodies in the blood serum for the ABO blood group system

Blood type	Antigens on red blood cells	Antibodies in blood serum	Frequency in U. S. black population	Frequency in U. S. white population
A	A	B	27%	41%
B	B	A	21%	10%
O	Neither	A and B	48%	45%
AB	A and B	Neither	4%	4%

blood system is an exception to this rule, however. In this system the antibodies that agglutinate the antigens are naturally occurring (present without previous stimulation) whenever the antigen (A or B) is *not* present. For example, a person with type A blood (genotype $I^A I^A$ or $I^A I^O$) would possess the A-antigen and the B-antibody (anti-B). As a result, any blood possessing the B-antigen would be agglutinated by the anti-B of a type A person if it were transfused to him. The general rule in transfusions is never to put an antigen into a person that his or her blood does not naturally possess. Table 8-1 summarizes the antigen-antibody relationships in the ABO blood group system. It also indicates the frequency of these blood types in the white and black populations of the United States.

The other important blood group system in man is the Rhesus, or Rh, system (named for the Rhesus monkey in which it was discovered). Some geneticists believe that this system is controlled by multiple alleles, whereas others favor the theory that a system of three separate loci are linked closely together on the same chromosome. In either case, individuals who are Rh-positive possess an antigen (protein) on their red blood cells that is lacking in Rh-negative individuals. Although three different antigens are involved (either as the products of alleles in a multiple allelic system or the active alleles at three separate loci), most Rh-positive individuals (about 75%) possess a protein called the D-antigen. The other two antigens, called C and E, are milder in antigenic strength and of such low frequency that routine checks of Rh factor for blood transfusions usually deal with the D-antigen only. When only the major D-antigen is considered, the gene for its production dominates its inactive allele. We can therefore consider the Rh situation in its simplest genetic form as *R__* producing Rh-positive, and *rr* producing Rh-negative individuals. The antigen-antibody immune response mechanism also creates incompatibility problems of great importance to man with regard to the Rh factor. The reaction in this case is somewhat different from that in the ABO system, however, because the anti-D (anti-Rh) antibody is *not* naturally occurring in the absence of the D-antigen but must have its production stimulated by exposure to the antigen (sensitization). The action of both systems will be discussed in the following paragraphs.

The immune reactions of the ABO and Rh blood groups can produce incompatibilities between the blood of a mother and her developing fetus. Thus, from a genetic point of view, certain marriages pose the problem of producing a blood phenotype in the fetus that is not compatible with the phenotype of the mother's blood. Incompatible marriages in this respect are any genotypic combinations in which the father may transmit a gene to the fetus for the production of an antigen that the mother does not possess—this defines that antigen in the blood of the fetus as foreign to the mother and thus capable of evoking an immune response. In such situations specific antibodies from the mother may diffuse across the placenta and shorten the life span of the fetus or completely destroy fetal red blood cells. This is called *hemolytic disease* and is characterized by anemia, sometimes severe enough to cause death or physical impairment.

Hemolytic disease involving the Rh factor is the one most often involved in human affairs. In this case an incompatible marriage is one between an Rh-negative female and an Rh-positive male. The combination genotypically would be either $rr \times Rr$ or $rr \times RR$. In the former the probability of producing an Rh-positive child (having an antigen foreign to the mother) is $\frac{1}{2}$, and in the latter it is 1.

Since anti-D (anti-Rh) is not naturally occurring, a female must be sensitized by exposure to the D-antigen before any antibody production takes place. Normally the maternal and fetal blood systems are completely separate, with the exchange of materials limited to molecules, much smaller than red blood cells, that can diffuse across the placental membrane. However, in the latter stages of pregnancy small ruptures in the placental barrier frequently occur, permitting fetal red blood cells to enter the maternal system. Also, at birth, when the placenta pulls loose from the uterus, fetal cells are quite likely to gain entrance to the mother's bloodstream. In the Rh-negative mother, the Rh-positive fetal red blood cells stimulate production of anti-D as a typical immune response. Once sensitized, antibody production continues for significant lengths of time (years). In contrast to red blood cells, anti-D is of small enough molecular size to freely diffuse across the placental membrane. This means that any subsequent Rh-positive fetus will suffer a reaction with the antibody (Fig. 8-5). A blood transfusion of Rh-positive blood to an Rh-negative woman could, of course, also sensitize her and produce a reaction in her *first* Rh-positive child.

Anti-D rapidly removes red blood cells from circulation, causing the fetus to become anemic. The fetus responds to the loss by releasing large numbers of immature red blood cells, called *erythroblasts;* for this reason the condition is commonly called *erythroblastosis fetalis.* Intrauterine death resulting in a stillbirth may occur because of heart failure as a secondary result of severe anemia. If the baby is born alive, serious liver and brain damage may result. The rapid destruction of red blood cells causes an abnormal accumulation of a reddish bile pigment called *bilirubin,* a substance that causes jaundice to develop within 24 hours. If the excess bile persists in the blood, severe brain damage results,

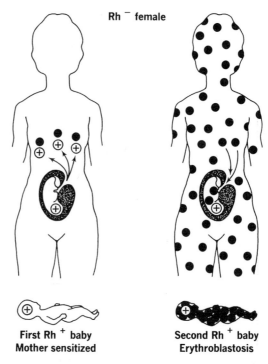

Rh⁻ female

First Rh⁺ baby
Mother sensitized

Second Rh⁺ baby
Erythroblastosis

Fig. 8-5. Rh sensitization and erythroblastosis fetalis in Rh-positive babies of an Rh-negative woman. Solid circles represent mother's antibodies formed against baby's antigens.

leaving the child deaf, mentally retarded, and with cerebral palsy, if it survives.

The treatment for erythroblastic babies is to give them an exchange transfusion of Rh-negative blood. The reason for putting Rh-negative blood into an Rh-positive baby is to avoid further reactions of D-antigen (present in positive blood) with maternal anti-D in the baby's circulation. The Rh-negative blood carries the infant over the period of time required to exhaust the maternal anti-D; normal red blood cell production then gradually replaces the negative red blood cells with the child's own positive ones, and the crisis is over. More recently, a preventative treatment has been developed that is highly effective. In an Rh-negative mother who has not been previously sensitized, sensitization can be prevented by injection of a substance called RhoGAM (Rh-immunoglobulin) immediately after delivery if her child is Rh-positive and fetal cells are demonstrated in her circulation. RhoGAM contains Rh-antibodies that destroy the Rh-positive fetal cells before they stimulate the mother's immune mechanism to produce her own antibodies. RhoGAM is obtained from Rh-negative men or postmenopause women who voluntarily become sensitized by controlled injection of Rh-antigens to produce the Rh-antibodies. The blood serum of these persons then supplies the RhoGAM in the form of a concentrated protein gamma globulin substance.

Hemolytic disease involving the ABO blood group system also occurs, causing natural abortions and jaundice, especially when a mother is type O and her fetus possesses the A- or B-antigen. In this case no sensitization is necessary because the antibodies are naturally occurring. Oddly enough, this ABO incompatibility is beneficial in certain cases because it provides protection against the more serious Rh incompatibility. In the Caucasian race about 15% of the people are Rh-negative, and one marriage in seven is between an Rh-negative woman and an Rh-positive man. Erythroblastosis fetalis occurs in only 10 to 15% of such marriages, however. The ABO combination affords protection in many instances. For example, if a type O, Rh-negative woman is carrying a type A, Rh-positive fetus, fetal red blood cells that accidently enter the maternal circulation are destroyed by the mother's naturally occurring anti-A before they can sensitize her against the Rh-antigen (D-antigen). Thus, the ABO "compatible" marriages are usually the ones in which the Rh incompatibility is expressed. Rh-hemolytic disease is very infrequent in the Negro race, where only about 5% of the people are Rh-negative, and very rare in the Mongolian race, where nearly all are Rh-positive.

Our understanding of the genetic transmission of the ABO, Rh, and a dozen other blood group systems, coupled with the accurate and relatively easy classification of phenotypes by the antigen-antibody reactions, has led to the use of these hereditary traits as valid legal evidence in court cases involving disputed paternity and in cases where hospitals mix up the identification of newborn infants or are accused of doing so. Considering only the ABO system, if a woman of type A blood had a child with type B blood, the multiple allelic pattern of inheritance in this case dictates that the father of the child must have had either type B or AB blood but could not have had type A or O blood. Hence, a putative father can be proved innocent of the accusation, but guilt cannot be made positive, only possible. In our example, for a type A mother to have a type B child, the allele for the B-antigen (I^B) must have been contributed by the father (the mother could not have had a B-antigen and be type A), while the allele the child received from the mother must have been I^O (if the mother gave an I^A allele, the child could not have type B blood), therefore the mother's genotype must be $I^A I^O$. Thus, the father of the child must have possessed at least one I^B allele in his genotype; the only blood types with this allele present are B and AB. Adding more systems to the analysis increases the probability of excluding an innocent male because the genetic impossibility of only one system would clear a wrongly accused man; the hereditary pattern would be explicable in *every* system in the real father. The same kind of analysis can nearly always solve a case of actual or alleged misidentification.

GENES THAT INTERACT

The phenotype expressed for any particular characteristic is not always governed by alleles at a single locus. It is not at all unusual for the product of one gene to influence the product and phenotypic potential of a gene at a different locus. Thus, interactions can occur in which the action of a gene at one locus suppresses, masks, or alters the action of *nonallelic* genes at one or more other loci. (Note that the typical dominant-recessive interaction is between *alleles* at *one* locus.) This phenomenon, known technically as *epistasis,* modifies phenotypic expression and thereby produces modified hereditary patterns. An array of different patterns is produced depending on the actions of the interacting genes—several different types will be discussed.

For our first example of nonallelic gene interaction, we will continue our discussion of blood with the rare, but interesting, "Bombay" trait in the ABO system. Recall that the A- and B-antigens are produced from H-substance that remains unchanged in individuals with type O blood. Thus, type O individuals possess both anti-A and anti-B. In 1952 two men were discovered in a Bombay hospital whose blood serum contained anti-A, anti-B, and anti-H. What was their ABO blood type? Family studies (pedigree analysis—Chapter 10) led to the conclusion that individuals with the Bombay trait were homozygous for a gene that inhibited the production of H-substance. This recessive gene, called *h,* is not allelic to the alleles of the ABO system. Most people are *HH,* or rarely *Hh* (*h* has an extremely low frequency in the population), and normally produce H-substance that then permits the ABO alleles to determine their phenotypes. Without H-substance (*hh* individuals), the ABO alleles have no precursor to work on and are therefore suppressed. A typical blood test for the ABO system would reveal the presence of anti-A and anti-B only. This would indicate type O blood. However, a transfusion of type O blood to a Bombay phenotype would result in a severe reaction, because the undetected anti-H would agglutinate the H-substance of typical type O blood. Fig. 8-6 illustrates the relationship between the locus for H-substance and that of the ABO alleles, which explains their interaction. A person with the Bombay trait has his or her "true ABO blood type" masked but will still pass the appropriate alleles to all progeny who will express them if not homozygous for *h* as their parent; Fig. 8-7 illustrates this situation.

A more commonly occurring gene interaction involves inherited *congenital deafness* (deaf-mutism). It appears that any one of two or more recessive alleles at separate loci can produce deafness when homozygous. In this case homozygosity at any of the loci concerned can override "normal" alleles at other loci and produce deafness. As a result, the hereditary pattern is often modified in the transmission of this trait. Sometimes two deaf parents have only deaf children; this is the typical pattern for an autosomal recessive trait. However, in other cases, two deaf parents have all normal-hearing children— a quite unusual pattern for a recessive trait. Furthermore, if the normal offspring of such deaf parents should intermarry, the probability that they will have deaf children is nearly 50%. Assuming two sets of separate, independently assorting loci, *D-d* and *E-e,* that interact in such a way that *either* recessive homozygote causes deafness, the situation described above can be explained. Fig. 8-8 shows the genetic crosses that result in *A* all deaf children from deaf parents, *B* no deaf children from deaf parents, and *C* a high probability ($\frac{7}{16}$)

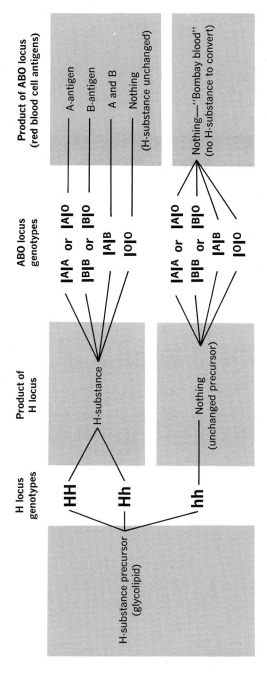

Fig. 8-6. Nonallelic interaction between the locus for H-substance and that of the ABO alleles.

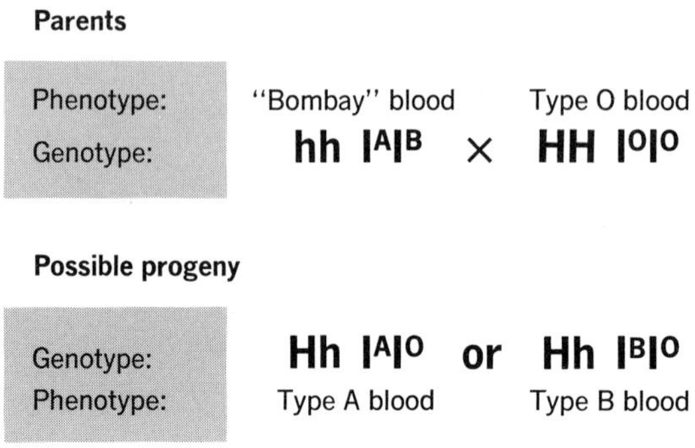

Parents

Phenotype:	"Bombay" blood	Type O blood
Genotype:	**hh I^A^I^B** ×	**HH I^O^I^O**

Phenotype: | "Bombay" blood | Type O blood
Genotype: | $hh\ I^AI^B$ × | $HH\ I^OI^O$

Possible progeny

Genotype: | $Hh\ I^AI^O$ **or** $Hh\ I^BI^O$
Phenotype: | Type A blood | Type B blood

Fig. 8-7. Hereditary transmission of ABO blood types involving nonallelic interaction with the "Bombay" locus.

A All deaf progeny from deaf parents:

Parental genotypes   Progeny genotypes

$$DDee \times DDee \longrightarrow DDee$$
$$ddEE \times ddEE \longrightarrow ddEE$$

B All normal-hearing progeny from deaf parents:

$$DDee \times ddEE \longrightarrow DdEe$$

C A high probability of deaf progeny from the intermarriage of the normal-hearing offspring of deaf parents:

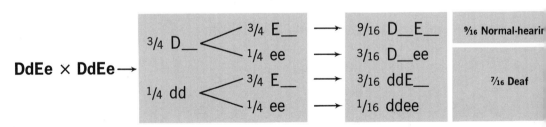

Fig. 8-8. Genetic crosses involving nonallelic interaction for loci governing inherited congenital deafness.

of deaf children from the intermarriage of the normal-hearing offspring of deaf parents. It should be noted that although the frequency of inherited congenital deafness is about 1 in 3,000 births, the combinations described here are much more likely to occur than chance or random mating would dictate because socially, individuals with this trait (or normal individuals from such families) are likely to be drawn together selectively through special schools, organizations, and activities, which, coupled with their "common interests," tend to significantly enhance their chances of intermarriage. This phenomenon of positive assortative mating maintains the frequency of deafness and many other genetic abnormalities at much higher levels in the population than would be expected through purely random breeding.

Another gene interaction with quite obvious phenotypic modifications concerns the action of the gene that produces *albinism*. This trait results from a homozygous recessive condition that blocks the normal production of the pigment melanin. This gene therefore interacts with the loci for brown hair and eye colors, which, you will recall, control the deposition and distribution of melanin to produce their phenotypic effects. Without melanin production, the activities of these loci, along with those controlling skin color (to be discussed in Chapter 11), are suppressed. The typical albino has white hair, white skin, and pink eyes (caused by the reflection of the unmasked blood in the capillaries of the eye). Although an albino person carries genes for hair, eye, and skin color, they cannot be phenotypically expressed when melanin is lacking; however, those genes are transmitted to an albino's offspring in the normal fashion. This interaction is illustrated in Fig. 8-9 by the modified phenotypic ratio expected in the cross between two double heterozygotes at the melanin locus (albinism) and the eye color locus (considering the simplified blue-"dark" alternatives). Positive assortative mating is not as strong for albinos as for the individuals with deafness previously described. (Perhaps this is because albinos do not normally receive specialized schooling and so forth, except to some extent when they attend schools for the blind because of the visual impairment caused by the lack of pigment in the iris.) The incidence of albinism is about 1 in 20,000 births, and most come from "normal" parents who are carriers. It is also noteworthy that 20% to 30% of albinos have consanguineous parents (mainly first cousins—see Chapter 13).

The incidence of albinism runs high in certain isolated populations, however, where albinos seem to have a selective advantage (high fitness). For example, in the Hopi Indians of Arizona, the frequency of albinism is about 1 in 200 (100 times higher than average). Apparently albinos have been accepted in the dark-skinned Hopi society where traditional Hopi comments state: "Albinos are smart, clean, nice, and pretty. There is nothing wrong with them." Although albinos are not often selected as marriage partners, the culture of the Hopis and the "attractiveness" of the albino have given the male albino a reproductive advantage. Since the Hopis are an agricultural society, Hopi men and boys spend most of the day in the field, often quite a distance from their mesa village.

However, because of their susceptibility to sunlight, the albino men are not expected to tend the fields; they remain in the village to perform other tasks such as weaving. In the village they are protected from the sun and, more importantly, they are provided with ample time and opportunity to engage in sexual activity. Whatever bias the females have against marrying albinos does not hold true for sexual activity. As a result, albino men produce more offspring, on the average, than typical Hopi males. This, by definition, is high fitness and provides the mechanism for increasing the frequency of the albino gene. Hopi tradition says of one legendary albino:

> Some say he had about 12 kids; others say about 15. He never married. They say he was always around trading with the ladies. He would make babies with them. He was real funny, and knew a lot of good stories. Everyone liked him. (From Woolf, Charles M., and Dukepoo, Frank C.: Hopi Indians, inbreeding and albinism, Science **164**[3875]:36, 1969.)

Nomenclature

M	=	Normal production of the pigment melanin
m	=	Inability to produce melanin
B	=	"Dark" eye color (melanin deposition in iris)
b	=	Blue eye color (no melanin in front of iris)

Parents
(double heterozygotes)

Phenotype:	"Dark" eyes "Dark" eyes
Genotype:	**MmBb × MmBb**

Possible progeny

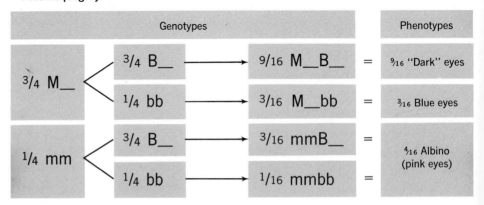

Fig. 8-9. Modification of the hereditary pattern of eye color caused by nonallelic interaction between the albino and eye color loci.

Nomenclature

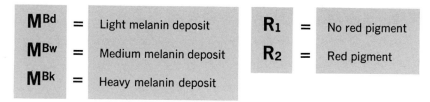

M^{Bd} = Light melanin deposit

M^{Bw} = Medium melanin deposit

M^{Bk} = Heavy melanin deposit

R_1 = No red pigment

R_2 = Red pigment

Dominance relationship among interacting alleles

$$M^{Bk} > M^{Bw} > R_2 > M^{Bd} > R_1$$

Possible diploid genotypes and phenotypic interpretation of interacting alleles

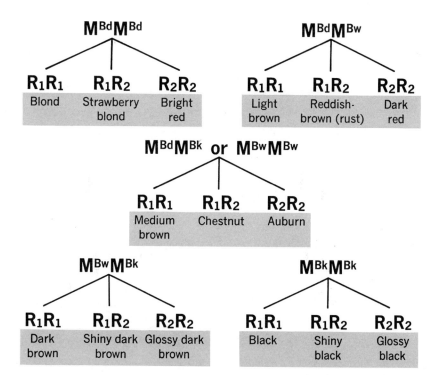

$M^{Bd}M^{Bd}$

R_1R_1	R_1R_2	R_2R_2
Blond	Strawberry blond	Bright red

$M^{Bd}M^{Bw}$

R_1R_1	R_1R_2	R_2R_2
Light brown	Reddish-brown (rust)	Dark red

$M^{Bd}M^{Bk}$ or $M^{Bw}M^{Bw}$

R_1R_1	R_1R_2	R_2R_2
Medium brown	Chestnut	Auburn

$M^{Bw}M^{Bk}$

R_1R_1	R_1R_2	R_2R_2
Dark brown	Shiny dark brown	Glossy dark brown

$M^{Bk}M^{Bk}$

R_1R_1	R_1R_2	R_2R_2
Black	Shiny black	Glossy black

Fig. 8-10. Nonallelic interaction model for the brown and red hair loci in man.

Finally, we can return to hair color at this point and add the locus for red pigment production to our previous model for brown hair colors (Fig. 8-1). An interaction model consisting of 18 genotypes that can be interpreted into at least 15 different phenotypes is constructed in Fig. 8-10. This model assumes that the red hair locus consists of only two alleles—R_1 for no red pigment and R_2 for red pigment production—which are codominant and assorted independent of the brown alleles. The phenotypic descriptions are derived not only from the dominance relationships within each locus but, more importantly,

Parents

Phenotype:	Reddish-brown	Chestnut

Genotype: $\quad M^{Bd}M^{Bw}R_1R_2 \times M^{Bd}M^{Bk}R_1R_2$

Possible progeny

Genotype		Phenotype	Probability
	$1/4\ R_1R_1$	Blond	$1/16$
$1/4\ M^{Bd}M^{Bd}$	$2/4\ R_1R_2$	Strawberry blond	$2/16$
	$1/4\ R_2R_2$	Bright red	$1/16$
	$1/4\ R_1R_1$	Light brown	$1/16$
$1/4\ M^{Bd}M^{Bw}$	$2/4\ R_1R_2$	Reddish-brown	$2/16$
	$1/4\ R_2R_2$	Dark red	$1/16$
	$1/4\ R_1R_1$	Medium brown	$1/16$
$1/4\ M^{Bd}M^{Bk}$	$2/4\ R_1R_2$	Chestnut	$2/16$
	$1/4\ R_2R_2$	Auburn	$1/16$
	$1/4\ R_1R_1$	Dark brown	$1/16$
$1/4\ M^{Bw}M^{Bk}$	$2/4\ R_1R_2$	Shiny dark brown	$2/16$
	$1/4\ R_2R_2$	Glossy dark brown	$1/16$

Fig. 8-11. Genetic transmission of brown and red hair loci illustrating an "atypical" hereditary pattern which transgresses the parental hair color range. (Parental types among progeny are outlined.)

from the dominance relationships among the alleles from both loci considered together. As described in the figure, the overall dominance relationship among the interacting alleles appears to operate in such a way that heavy and medium brown deposits tend to mask red pigment, whereas the red overrides light deposits of brown. It is also thought that in dark brown and black colored hair, the presence of red pigment produces a shiny or glossy effect, depending on the dosage of the red pigment allele. Close inspection of Fig. 8-10 will reveal that each phenotype is a blend of brown and red pigment intensities producing the range of color from blond to glossy black, yet genetically controlled by only two loci and *four* active pigment-producing alleles.

Nonallelic interaction is a mechanism that generates an array of diverse hereditary patterns from still rather simple transmission. Fig. 8-11 provides an example of typical dihybrid genetic transmission involving the brown and red pigments and the production of an "atypical" hereditary pattern that significantly transgresses the hair color range of the parents in both directions. Lack of knowledge concerning the explanation of such phenotypes in the offspring of quite similar parents is often a source of confusion, and sometimes "trouble," for married couples. A clear understanding of human hereditary can explain a great deal about patterns formerly attributed to the "milkman."

GENES THAT STICK TOGETHER

When we consider the pattern of two or more different traits in their transmission from generation to generation, we do not always find a ratio, such as 9:3:3:1, that can be explained by independent assortment of the chromosomes. Quite unusual patterns can arise when the loci of two or more traits are located on the *same* chromosome pair. In such cases all the genes on the chromosome tend to "stick together" as they are carried about through gamete formation (meiosis) as a single unit. The phenomenon of two or more loci located on the same chromosome is known as *linkage,* with the genes so located forming *linkage groups* that correspond in number to the pairs of chromosomes typical of the species—in man, 23.

The reason unusual patterns arise regarding linked genes is that linkage groups do not always remain intact, that is, although these genes tend to stick together, they can also assort from one another. But the assortment of linked genes is not caused by the random shuffling of independent chromosomes; rather, it is *crossing over* in prophase I of meiosis that shuffles the loci within a chromosome through the physical interchange of segments between homologous chromosomes (see Fig. 4-7). Thus, if two genes are on the same chromosome, it takes the occurrence of a crossover between them to break their linkage. And since crossovers occur more or less at random along the length of a chromosome, the farther apart two linked genes are from each other, the greater are the chances that a crossover will occur between them. Linkage is nearly always incomplete; it is complete only when two loci are so close together that the probability of a crossover separating them is virtually nil. The strength or tightness of the linkage between two loci can thus be ascertained by the amount of crossing over that takes place between them. This also serves as a relative measure of distance between linked loci, which can be used to construct a linear diagram of the order and "distance" of loci along the length of the chromosome—such diagrams are known as *chromosome maps*.

Linkage is most easily illustrated in the cross between an individual who is doubly heterozygous (*AaBb*) at two linked loci and one who is homozygous for the recessive allele at each locus (*aabb*)—this is a double testcross. The "testor" in this case always produces *ab* gametes, thereby allowing us to focus our attention solely upon the genetic assortment in the gamete production of the double heterozygote. Fig. 8-12 presents a comparison of independent assortment *A* to complete linkage *B* and incomplete linkage *C*. Both cases of linkage produce patterns different from independent assortment, but incomplete linkage, the most common situation, presents an array of patterns that are dependent on the degree or "strength" of linkage. To understand incomplete linkage, it is important to recognize the effect of crossing over on gamete production. Part *A* of Fig. 8-12 depicts the four haploid combinations of gametes expected from the random assortment of two independent loci (see Fig. 7-7), while part *B* presents the only two combinations possible if both loci are on the same chromosome pair with the two dominant alleles on one chromosome and the two recessive alleles on the homologous chromosome. Note that the patterns (phenotypic ratios) produced in these different situations are quite different even though the basic transmission still occurs through normal meiosis.

The pattern produced by incomplete linkage is a function of the amount of crossing over that occurs between the loci in question. Incomplete linkage is detected in the first place when the expected frequencies of phenotypes do not conform to a "typical" pattern of independent assortment even though *all* phenotypes are observed. For example, in part *C* of Fig. 8-12 the four phenotypes appear but in an unusual pattern (ratio of 1:4:4:1). In this pattern two combinations are higher in frequency than expected from independent assortment, and two combinations are lower. Such a pattern indicates incomplete linkage if the pattern is consistent and the relationship of the genotypes concerned is a reciprocal one.

Reciprocal genotypes (or gametes) are those produced by the meiotic segregation of linked gene pairs. Thus, in the double heterozygote depicted in part *B* of Fig. 8-12, the gametes *AB* and *ab* are reciprocal. On the other hand, *Ab* and *aB* are also reciprocal combinations that are possible from a double heterozygote. Also, if the allele arrangement in part *B* had been with *A* and *b* on one chromosome and *a* and *B* on the other chromosome, then, with complete linkage, the only two kinds of gametes produced would have been *Ab* and *aB*. By tracing the pathway of the two genotypes at high frequencies in part *C* (*AaBb* and *aabb*), we find that they were generated from reciprocal gametes derived from the doubly heterozygous parent when crossing over *did not* alter the original

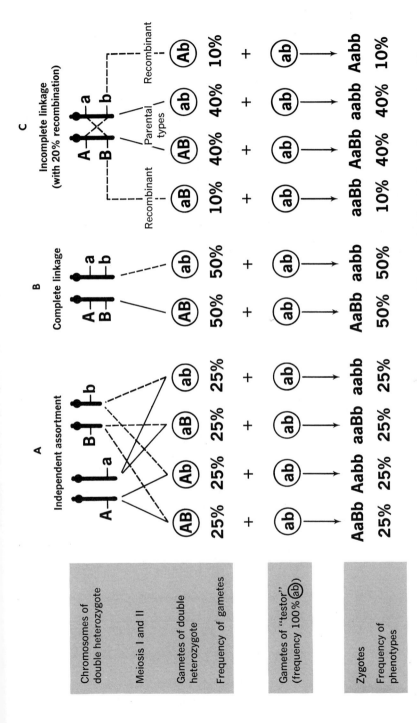

Fig. 8-12. A comparison of the hereditary patterns (phenotypic ratios) produced from two loci by **A,** independent assortment; **B,** complete linkage; and **C,** incomplete linkage (with 20% recombination) in a "double testcross" when the "testee" is doubly heterozygous.

arrangement of the alleles on the chromosomes. Tracing the two genotypes at low frequencies (*aaBb* and *Aabb*), we also find that they were generated from reciprocal gametes derived from the double heterozygote, but in this case a crossover between the *A* and *B* loci (actual exchange of chromosome segments) was necessary to produce these gametes. The reciprocal gametes produced by crossing over are called *recombinants;* this name also applies to the genotypes and phenotypes they determine. Thus, when a phenotypic ratio such as that in part *C* is consistently produced, incomplete linkage explains the pattern, and the strength of the linkage is indicated by the percentage of recombination observed. In our example recombinants constitute 20% of the ratio. In mapping the linear sequence of loci along a chromosome, each recombination percentage is interpreted as a "map unit," which, in our example, would place the *A* locus 20 units away from the *B* locus on the same chromosome.

Once linkage is detected and the recombination percentage determined, it is possible to predict and assign probabilities to particular situations. This knowledge may be quite valuable for genetic counseling, as our next example will illustrate. But before predictions can be made in cases of incomplete linkage, one other important factor must be considered. It is essential to know the exact arrangement of the alleles on the homologous chromosomes in order to determine which reciprocal gamete combination will represent parental, or noncrossover, types and which reciprocal combination will represent recombinant, or crossover types. The allelic arrangement in a double heterozygote (*AaBb*) may have both dominant alleles (*A* and *B*) on the same chromosome and both recessive alleles (*a* and *b*) on the other chromosome (that is, *A* and *B* came from one parent and *a* and *b* from the other parent); this is known as the *coupling* phase. On the other hand, if the *A* and *b* alleles are on one chromosome and the *a* and *B* alleles are on the homologue (one dominant and one recessive allele on each chromosome), the linkage is said to be in the *repulsion* phase.

Therefore, simply indicating that an individual is heterozygous (*AaBb*) is not sufficient information when the loci are linked; we must know whether the alleles are in coupling or in repulsion before any predictions can be made. Fig. 8-13 illustrates the difference between the coupling and repulsion arrangements of the alleles and the influence of these arrangements on gametic frequencies. Note that, since the recombination percentage is equal to the sum of the reciprocal recombinant gametes (Fig. 8-12, *C*), in making predictions each recombinant gamete is expected to constitute one half of the recombination value. (This expectation is derived from the segregation of chromosome pairs in meiosis.) Coupling and repulsion depend on how the alleles come into the individual concerned, that is, which alleles were on the chromosome in the egg and which were on the homologous chromosome in the sperm that combined to form the zygote in question.

Although linkage has been extensively studied in many organisms—the chromosomes of the tiny fruit fly, *Drosophila melanogaster*, have been "mapped" in fine detail—studies in man have only scratched the surface. Without pre-

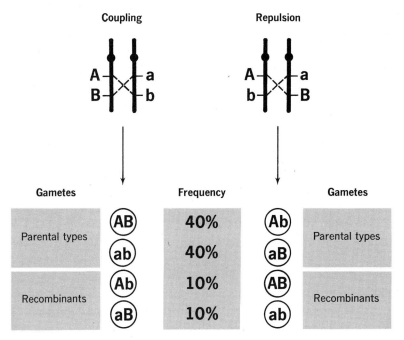

Fig. 8-13. The difference between the coupling and repulsion arrangements of linked alleles and the influence of these arrangements of gametic frequencies when 20% recombination occurs. The gametes occur as reciprocal pairs in each case; however, the parental types (at high frequency) and the recombinant types (at low frequency) are determined by the initial allelic arrangement.

scribed experimental crosses and large numbers of progeny to detect patterns, linkage data in man are difficult to obtain. However, one case in which recombination data are available will serve to review the phenomenon of linkage and illustrate its potential for genetic counseling.

One of the few linkage groups known in man consists of the locus for the Rh blood factor (*R___* producing the Rh-positive phenotype and *rr* producing Rh-negative) and a locus at which a dominant allele *(E)* produces oval-shaped red blood cells causing a form of anemia called *elliptocytosis*. The recessive allele *(e)* at the locus allows normal red blood cell formation. These loci exhibit a recombination value of approximately 4%.

Suppose that the following situation were presented to a genetic counselor. A married couple, each of whom is Rh-positive and suffers from elliptocytosis, wants to know the chances of producing a child that is Rh-negative and free of elliptocytosis. (Here we are simply asking for the double homozygous recessive genotype, *rree*, because it is the easiest to work with in our illustration. When more linkage groups are known, the potential importance of counseling will be in those cases where a particular combination or interaction is detrimental or undesirable, or perhaps advantageous or highly desirable.) Knowing that these are linked loci with 4% recombination, the exact genotype and coupling-repulsion configuration for each individual must be determined before any prediction can be made. This requires information on the family history and the application of the basic principles of hereditary transmission to ascertain the genotypes of the concerned couple. This technique is known as pedigree analysis; it is discussed more extensively in Chapter 10.

Tracing the family backgrounds of the couple in question, the counselor finds that each had one parent with elliptocytosis (*E___*, the other parent therefore being *ee*) and each had one Rh-negative parent *(rr)*. Since both the husband and the wife concerned are Rh-positive *(R___)* and suffer from elliptocytosis *(E___)*, we can establish their exact genotype as *RrEe* (double heterozygotes). Because each had one parent homozygous for the *r* and *e* alleles, therefore, each had to receive at least one *r* and one *e*; their phenotypes tell us that each had a *R* and a *E* also.

We must still establish the coupling-repulsion situation in these double heterozygotes before we can proceed with a prediction, however. To accomplish this, further refinement of the family history is required. Specifically, we need the genotypes of the parents of the couple in question in order to determine how the genes went into the crosses that produced each—this will tell us the coupling-repulsion situation. Now, if we find that on the husband's side his father was Rh-negative and free of elliptocytosis (genotype rree) and his mother was Rh-positive and had elliptocytosis *(R___E___)*, then he had to receive his *R* and *E* alleles from his mother (on one chromosome in the egg) and his *r* and *e* alleles from his father (on the homologous chromosome of the pair in the sperm). The husband, therefore, possesses the alleles in the coupling phase. If on the wife's side we find that her father was both Rh-negative and elliptocytotic *(rrE___)* and her mother was Rh-positive and nonanemic *(R___ee)*, it becomes evident that she received her *R* allele from her mother and her *E* allele from her father and possesses the alleles in the repulsion phase. A summary of this situation is given in Fig. 8-14, *A*.

When the genotype, coupling-repulsion phases, and recombination value are known, it is quite easy to make a probability statement on the likelihood of producing the progeny genotype in question (or any other one for that matter). Fig. 8-14, *B* lists the genotypes and expected frequencies (noncrossovers and crossovers) of the gametes of the husband and the wife. The double recessive genotype, *rree*, can be produced in only one way—a *re* sperm fertilizing a *re* egg. The probability of this combination equals the product of the probabilities of each independent event (sperm and egg formation), which is .48 × .02 = .0096 or approximately 1%. To determine the probability of any other phenotypes, it would be necessary to summarize the probabilities of the several zygotic combinations (genotypes) that constitute a particular phenotype. Fig. 8-15 illustrates the entire pattern (phenotypic ratio) for the cross in question utilizing a checkerboard. Gamete frequencies from Fig. 8-14, *B* have been used to calculate the expected proportions of the various genotypes. These proportions are then summarized to give the phenotypic ratio (see Fig. 7-8 for review of procedure).

The effect of linkage and the coupling-repulsion configuration on the final hereditary pattern can be seen by comparing the possible coupling-repulsion combinations among doubly heterozygous parents with each other and with independent assortment (no linkage). These comparisons are given in Table 8-2 with respect to the double homozygous recessive genotype; concomitant changes would attend the whole phenotypic ratio and could be calculated for each case, as in Fig. 8-15. Thus, the three linkage combinations described produce three different hereditary patterns, and *all* are distinct from the independent assort-

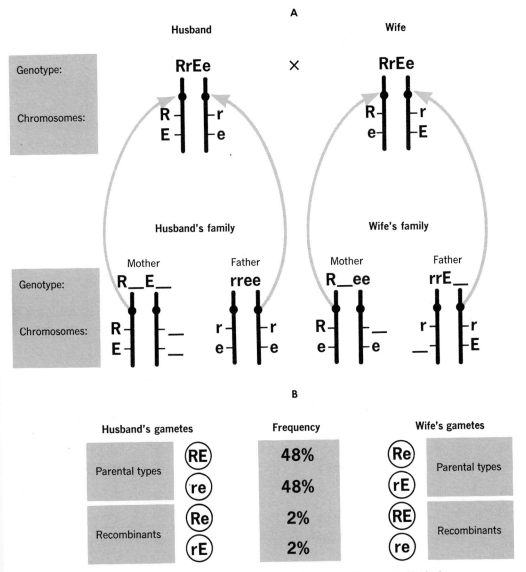

Fig. 8-14. A, Pedigree information for determining coupling and repulsion in individuals heterozygous at two linked loci; **B,** expected frequencies of the gametes of each individual when recombination is 4%.

Eggs

.02 (RE) .48 (Re) .48 (rE) .02 (re)

		.02 (RE)	.48 (Re)	.48 (rE)	.02 (re)
.48 (RE)		RREE .0096	RREe .2304	RrEE .2304	RrEe .0096
.02 (Re)		RREe .0004	RRee .0096	RrEe .0096	Rree .0004
.02 (rE)		RrEE .0004	RrEe .0096	rrEE .0096	rrEe .0004
.48 (re)		RrEe .0096	Rree .2304	rrEe .2304	rree .0096

Sperms (vertical label on left)

Phenotypic ratio
(summarized from left to right
and top to bottom of checkerboard)

R__E__   .0096 + .2304 + .2304 + .0096 + .0004 + .0096
           + .0004 + .0096 + .0096 = .5096 $\approx$ 51%

R__ee   .0096 + .0004 + .2304 = .2404 $\approx$ 24%

rrE__   .0096 + .0004 + .2304 = .2404 $\approx$ 24%

rree   .0096   = .0096 $\approx$ 1%

Fig. 8-15. Hereditary pattern (phenotypic ratio) from a cross between two double hetero-zygotes having linked loci displaying 4% recombination with the husband in the coupling phase and the wife in repulsion.

ment pattern of 9:3:3:1. Note again, however, that no alteration of the trans-mission mechanism (meiosis) occurs.

As more linkage groups are determined in man, linkage will inevitably become a more important factor in genetic counseling. At present only four other linkage groups on the autosomes of man are known. (Loci linked on the sex chromosomes will be discussed in the next chapter.) These are (1) a blood group locus, known as Lutheran (L^+ and L^-), linked to the secretor locus (S = secretor, and s = nonsecretor), which causes the A, B, and H blood antigens to be secreted in the saliva—with a recombination value of 13%; (2) the ABO blood group locus linked to the nail-patella syndrome locus (N = abnormally reduced fingernails and kneecap [patella], and n = normal de-velopment)—with 10% recombinations; (3) the locus for a blood serum pro-tein called transferrin linked to the locus for a serum enzyme called choles-

Table 8-2. Comparison of the expected frequencies of the homozygous double recessive genotype produced from the mating of two double heterozygotes when the loci are independent and when they are linked in all possible coupling-repulsion combinations with a recombination value of 4%

Parents Arrangement of loci	I × II* Independent assortment	I × II Coupling-coupling	I × II Coupling-repulsion	I × II Repulsion-repulsion
Expected frequency of re gamete from parent I	.25	.48	.48	.02
Expected frequency of re gamete from parent II	.25	.48	.02	.02
Fertilization	.25 × .25	.48 × .48	.48 × .02	.02 × .02
Expected frequency of $rree$ zygotic combination	.0625 = 6.25%	.2304 ≈ 23%	.0096 ≈ 1%	.0004 = 0.04%

*Designation of parents as I and II simply indicates that the cross represents both husband-wife and wife-husband combinations.

terase—with 16% recombination; and (4) the locus of a component of the serum protein, globulin, linked to the locus of another serum protein, albumin—with a recombination value of approximately 2%.

When we consider the fact that the thousands of genes that control our development and function form only 23 linkage groups, we can recognize the important role linkage plays in producing hereditary patterns. Computers are now being used to analyze quantities of information about phenotypes within families, marriage combinations, progeny produced, and other data. Geneticists are using this information in attempts to compare the probability of obtaining certain family data for two loci if they are linked (in varying degrees) with the probability of obtaining the same data if the loci assort independently. Such computer analysis holds promise for learning much more about the linkage groups of man.

Hereditary patterns are influenced by two other major factors, namely, the presence of the loci on the sex chromosomes and polygenic inheritance (two or more loci governing the degree of development of a particular trait), which will be discussed separately. This chapter and Chapters 9 and 11 will constitute the fundamental principles of explaining and predicting most hereditary patterns known in man at the familial level.

9

The implications of sex

Aside from the differences societies have fostered between the sexes and the roles that have been traditionally assigned to, or excluded from, one sex or the other, there is a basic biological difference between the sexes that is undeniable. The differences in reproductive systems and their function (Chapter 5) are inherited attributes. Even so-called "sex reversals" do not produce one *functional* sex from another *functional* sex. They either correct an abnormal development that caused a misclassification of sex in early childhood or change the sex from a functional state to a nonfunctional (incapable of reproducing) mimic of the opposite sex through superficial surgical procedures, and perhaps hormone treatments, in order to satisfy a psychological desire.

In this chapter we shall discuss the primary genetic control of sex differentiation and expand our repertoire of mechanisms that produce hereditary patterns different from the simple Mendelian patterns.

SEX DETERMINATION AND SEX RATIO

Various mechanisms have evolved in different groups of animals and plants for determining sex and controlling sexual development. In man sex differentiation is genetically controlled by the sex chromosomes (X and Y). As previously noted in the karyotypes of the two sexes (Figs. 4-3 and 4-4), a normal female possesses two X chromosomes, and a normal male has an X and one Y as his sex chromosome constitution. Having two Xs making up the sex chromosome pair means that a female normally produces only X-bearing gametes, while the XY male is expected to produce half X-bearing and half Y-bearing sperms from normal spermatogenesis. Thus, as Fig. 9-1 illustrates, the sex ratio would be expected to be 1 male:1 female. Note that the Y chromosome of a male always comes from his father; if the egg is fertilized by an X-bearing sperm, a female will develop. The combined action of genes on both X chromosomes is apparently necessary to produce a fertile female. (An individual with only a single X chromosome [XO] will develop toward femaleness but will be sterile, see Chapter 12.) On the other hand, in the human species, genes on the Y chromosome are required for the development of maleness. (In some or-

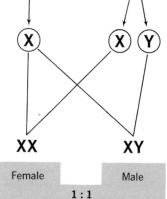

Parents

Phenotype: Female   ×   Male

Sex chromosomes: **XX**   **XY**

Gametes: Ⓧ   Ⓧ Ⓨ

Offspring
(zygotes)

Sex chromosomes: **XX**   **XY**

Phenotype: Female   Male

Ratio: 1 : 1

Fig. 9-1. Sex determination according to typical meiotic segregation involving the sex chromosomes and producing a theoretical ratio of 1:1.

ganisms, such as the fruit fly commonly studied in genetics, sex determination is based on the ratio of X chromosomes to autosomes, and the Y chromosome ensures fertility but is not necessary for development toward maleness. In these cases the possession of only an X chromosome [XO] produces a sterile *male.*) Thus, in our species, individuals possessing a Y chromosome are phenotypically male, and those without a Y are phenotypically female. Even in cases of abnormal numbers of the sex chromosomes (Chapter 12), individuals with XO, XXX, XXXX, or XXXXX develop toward femaleness and individuals with XXY, XXXY, or XXXXY develop toward maleness (YO does not develop at all). The gonads do not usually develop normally in individuals with abnormal sex chromosome complements, and nearly all are sterile. Although sex is determined at conception, sexual differentiation is not recognizable until the seventh week of development. By this time the primordial gonadal tissue is stimulated under genetic control to develop into either ovaries or testes.

Although a 1:1 ratio of males to females would be expected at conception (Fig. 9-1), in actuality the sex ratio is not equal at conception or at birth; furthermore, the ratio changes chronologically. Thus, when speaking of the "sex ratio," a time factor should be used as a qualifier. Three categories of sex ratio are recognized. The *primary sex ratio* is the ratio of the sexes at the time of conception; the *secondary sex ratio* is the ratio at birth; and the *tertiary sex ratio* refers to the proportion of males to females at any specified time after birth. These ratios are most conveniently expressed as the number of males

Table 9-1. The approximate sex ratios for the human species

Time	Male/Female
Primary ratio (conception)	120 to 150 : 100
Secondary ratio (birth)	106 : 100
Tertiary ratio (postnatal)	
Second to fourth decades	100 : 100
Fifth decade	90 : 100
Sixth decade	70 : 100
Eighth decade	50 : 100
Tenth decade	20 : 100

compared to every 100 females conceived, born, or alive at the particular time (they *do not* imply a constant survival rate of females). Table 9-1 lists the generally accepted sex ratios for the human species.

It seems quite definite that the female of our species, and for that matter most animal species, tends to outlive the male and be biologically "stronger." There are, for instance, very few diseases that do not affect men more than women. Of the 64 specific causes of death listed by the U. S. Census Bureau, the rate in 57 is higher for men at all ages than for women at comparable ages. Five of the seven not higher in men pertain to female-specific causes, such as uterine cancer or child-bearing. Even in the embryonic stage, the death rate of males is disproportionately high. Natural abortions (miscarriages) and still-births (born dead), which account for 20% of all recognized pregnancies, consist of 2 to 4 times more males than females, and very early intrauterine losses (often not even recognized as pregnancies) may be even more disproportionate. Interestingly enough, whatever the mechanism is that produces such unequal primary and secondary sex ratios, it appears to be an evolutionary adaptation that ensures a nearly perfect ratio during the prime reproductive years. This is very advantageous for a species that tends to form long-lasting pair bonds as a significant feature of its social structure. Apparently, when man's life span was shorter and the death rate higher, the ratio became equal at an earlier age, which again corresponded to the prime reproductive period of our predecessors.

Intersexes: boy or girl?

It is often asked whether or not *hermaphrodites* (individuals having both male and female reproductive organs) exist in the human species. Technically the answer is no, because animals such as earthworms that are hermaphroditic possess *functional* male and female reproductive systems. In man there occasionally arise individuals called hermaphrodites who possess *both* ovarian and testicular tissues, but they do not possess functional organs and systems of both sexes. In fact these individuals are usually sterile and function as neither sex. For this reason, it is more appropriate to refer to these as *intersexes*, rather than hermaphrodites, since the latter term connotes reproductive ability. Less than 100 cases of hermaphrodites have ever been reported and confirmed. Far more sexual anomalies are categorized as *pseudohermaphrodites*, but even this term is misleading, for these individuals are not really even good mimics of true (functional) hermaphroditism. These are individuals who have physical abnormalities that cause a misclassification of sex or an apparent mixture of both sexes. In pseudohermaphrodites only one kind of gonadal tissue is present, but some characteristics of the opposite sex also occur. Such abnormal development is quite often the result of an inherited condition.

An example of inherited pseudohermaphroditism in each direction will illustrate this point. An abnormality inherited as a typical recessive trait (on one of the 22 pairs of autosomes) that often leads to misclassification in females is the *adrenogenital syndrome*, which occurs about once in every 67,000 live births. In genetic males (XY chromosome constitution) there is usually no difficulty in sex identification, although this syndrome often causes sexual precocity. In genetic females (XX), however, the clitoris may be enlarged and the labia majora pronounced and wrinkled or even fused, leading to misclassification. The condition results from overactivity of the adrenal cortex, which, you will recall, normally produces small amounts of androgens in the female to promote normal growth (Chapter 5). Although this excessive production of male hormone leads to masculinization, it does not result in a functional male. In females with adrenogenital syndrome who have not been misclassified sexually, the abnormality may not be diagnosed until puberty when they undergo a change in voice, begin to grow a beard, and develop other masculine traits that lead them to seek medical help.

An inherited type of pseudohermaphroditism that affects males is *testicular feminization*. It appears to be inherited as a recessive sex-limited trait; this mode of inheritance will be discussed later in the chapter—suffice it now to point out that only males are affected by the gene action. A person afflicted with this anomaly is "female" in external appearance but has an XY chromosome constitution and possesses testes that may be located in the abdominal cavity, inguinal canal, or labia. Although no ovarian tissue is formed, estrogens are secreted from the testes instead of androgens; this characteristic gives the condition its name. Internally, except for a rudimentary vagina, no female structures accompany the female external secondary sex characteristics developed in response to estrogen stimulation. Such individuals often appear to be particularly well-developed "females" and often marry male partners. Their condition is not often recognized until the late teens, or later, when they may seek medical help because of amenorrhea (absence or suppression of menstruation), inguinal hernia (when surgery often reveals the testes), or sterility, which always occurs.

Surgical correction of the physical abnormalities of pseudohermaphrodites, often accompanied by hormonal treatments to correct their imbalance, accounts for most cases of "sex reversals" dramatized from time to time by the popular press. This most often involves restoration of the true genetic sex

(XX or XY) following initial misclassification because of the abnormal physical development. More rare are "sex reversals" that change a functional male or female to the opposite sex. These consist of essentially nothing more than plastic surgery to make *transvestism* (cross-dressing) or *transsexualism* (the desire to be the opposite sex) as complete as possible—even anatomically! There is no convincing evidence that this behavior or *homosexuality* has any genetic basis. However, some genetic predisposition to abnormal sexual behavior (even overaggressiveness) cannot be entirely discounted.

GENES ON THE SEX CHROMOSOMES

At this point all traits discussed in the explanation of hereditary transmission (Chapter 7) and the various hereditary patterns (Chapter 8) have been governed by loci located on one or more of the 22 pairs of autosomes. This is known as *autosomal inheritance* and is distinguished from the patterns produced by loci on the sex chromosomes. The so-called *sex-linked traits* are governed by genes on the X or Y chromosome. Because females and males differ with respect to sex chromosome constitution, these traits produce patterns (phenotypic ratios) different from autosomal transmission and different between the two sexes.

More information is available concerning genetic loci on the X chromosome than on any other single member of the chromosomal complement. The reason for this is that in the male XY condition, *all* loci on the X chromosome, whether dominant or recessive with respect to their allelic forms, are phenotypically expressed. This is because the small Y chromosome has very few, if any, loci in common with those of the X chromosome, thereby giving the male only one dose of the genes on the X chromosome. (It should be noted, however, that there is enough similarity between the X and Y chromosomes to make them homologous pairing partners in meiosis.) Males cannot be homozygous or heterozygous at these loci; instead they are *hemizygous* (which literally means half zygotes). Loci found only on the X chromosome are specifically referred to as *X-linked* and are always expressed in the hemizygous males. As we shall see shortly, this alters the hereditary pattern and exposes recessive mutants in such a way that makes X-linked traits much easier to analyze and trace than those whose loci are on the autosomes. Over 50 X-borne mutants have been identified in man, causing such well-known conditions as colorblindness, hemophilia (bleeder's disease), and progressive muscular dystrophy.

The most common sex-linked abnormality is partial colorblindness of the red-green type. Actually there are two different varieties of red-green colorblindness now recognized. About 75% of the cases are known as *deutan types* and the other 25% as *protan types*. The former is basically green insensitivity, whereas the latter is red insensitivity; both cause a confusion in red-green distinction, however. It has been demonstrated that each type is produced by an X-linked recessive allele at two different loci. Considering both types together, about 8% of the males in the United States are red-green colorblind, although in females the frequency is only 0.4%. This great disparity in the incidence of red-green colorblindness between the sexes is a common phenomenon of sex-linked traits of all kinds.

X-linked recessive traits always appear in higher frequency among males because, as hemizygotes, a single dose of the gene has no possibility of being masked by a dominant allele. Thus, males cannot be normal carriers (heterozygotes) of an X-linked recessive trait.

Since two different loci are involved in producing essentially the "same" trait (red-green colorblindness), an additional form of nomenclature must be introduced at this point to fully explain the implications of sex-linked inheritance on hereditary patterns. When multiple loci govern the "same" phenotype, a general symbol for the phenotype may be used with subscripts to designate the separate loci. (Do not confuse this with the nomenclature for multiple alleles that all occur at *one locus* and that are distinguished by *superscripts*, see Chapter 8). Thus, we can use the letter C as the gene symbol for red-green colorblindness, with C as dominant normal vision and c as colorblindness. The two varieties can then be described as C_d and c_d for the deutan alleles and C_p and c_p for the protan alleles.

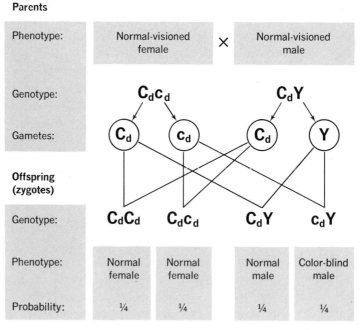

Fig. 9-2. The typical mode of transmission of an X-linked recessive trait, using the deutan type of red-green colorblindness as the example. Note that half the males are expected to display the trait, whereas in females the probability of colorblindness is zero. The symbol Y is used to denote presence of the Y chromosome in males to account for both sex chromosomes; the Y does not possess the locus concerned. The symbol for the X-linked locus also represents the X chromosome.

The typical mode of transmission for X-linked recessive traits (especially the severe abnormalities to be discussed later) is illustrated for the deutan variety only in Fig. 9-2. (The protan type considered alone would be identical to the deutan examples used here.) In this cross both parents are normal, but ¼ of the children are expected to inherit the defect. Note that although this expectation in general is the same as that for an autosomal recessive trait from two heterozygous parents, only males display the abnormal phenotype in this case. To be more specific, no females are expected to have colorblindness (although ½ will be carriers like their mothers), whereas ½ of the males are expected to be colorblind and the other ½ normal. Only if the father *has* the abnormality (and thus the mutant allele on his single X chromosome) is there any possibility of producing a female with the trait. Fig. 9-3 illustrates the most common cross in the production of colorblind females. It should also be noted that a colorblind mother (c_dc_d) and a normal-visioned father (C_dY) would produce *all* colorblind sons (c_dY) and *no* colorblind daughters (C_dc_d), although all the daughters would be carriers. Recessive X-linked traits appear in patterns modified mainly by the hemizygosity of the male, and therefore, these and all sex-linked traits must be interpreted in relation to the sex of the individuals in question in addition to the phenotype of the characteristic under

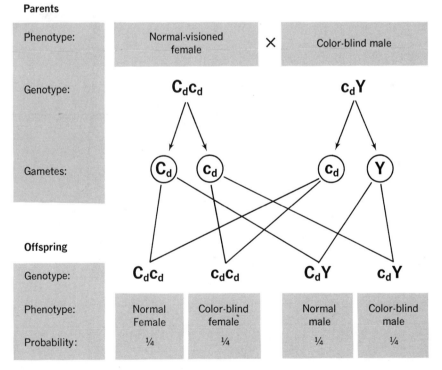

Fig. 9-3. The most common cross in the production of colorblind females. Note that half the females are colorblind and the other half are carriers, whereas half the males are color-blind and the other half are completely normal.

consideration. This explains the main reason for the high incidence of such traits in males as compared to females.

With red-green colorblindness the proportion of males to females with the trait is further enlarged by the fact that two different loci are involved. In some cases where *both* parents are colorblind it is sometimes observed that within certain families all daughters have normal vision, but all sons are color-blind. Even with an X-linked recessive, when the mother has the trait (homozygous recessive) and the father has it too (hemizygous recessive), we would expect *all* children to have the trait. However, if the mother has the deutan type $(c_d c_d)$ and the father has the protan type, $(c_p Y)$ or vice versa, all daughters would be double heterozygotes $(C_d c_d C_p c_p)$ with normal vision, but all sons

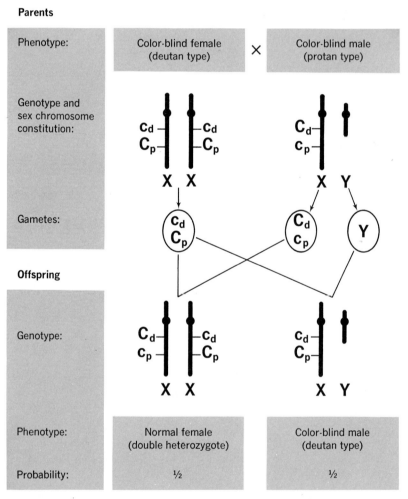

Fig. 9-4. A cross involving both types of red-green colorblindness in which two colorblind parents produce normal-visioned daughters but colorblind sons. Note that the sons inherit the mother's type.

would inherit the mothers' type of colorblindness $(c_d Y)$ (Fig. 9-4). In addition, as Fig. 9-5 illustrates, such daughters would themselves produce *all* sons with colorblindness (½ deutan and ½ protan) even if married to a normal visioned male (compare this to Fig. 9-2). Only an occasional normal son would appear in such crosses as a result of crossing over. Thus, when both red-green color-blind alleles are segregating within the same family, the probability of color-blindness is usually increased for males but decreased for females, thereby magnifying the disparity in phenotypic expression between the sexes.

For the more severe X-linked recessive abnormalities, such as *hemophilia*, the incidence remains disproportionately high in males for yet another reason. There are several different kinds of hemophilia, each affecting a particular

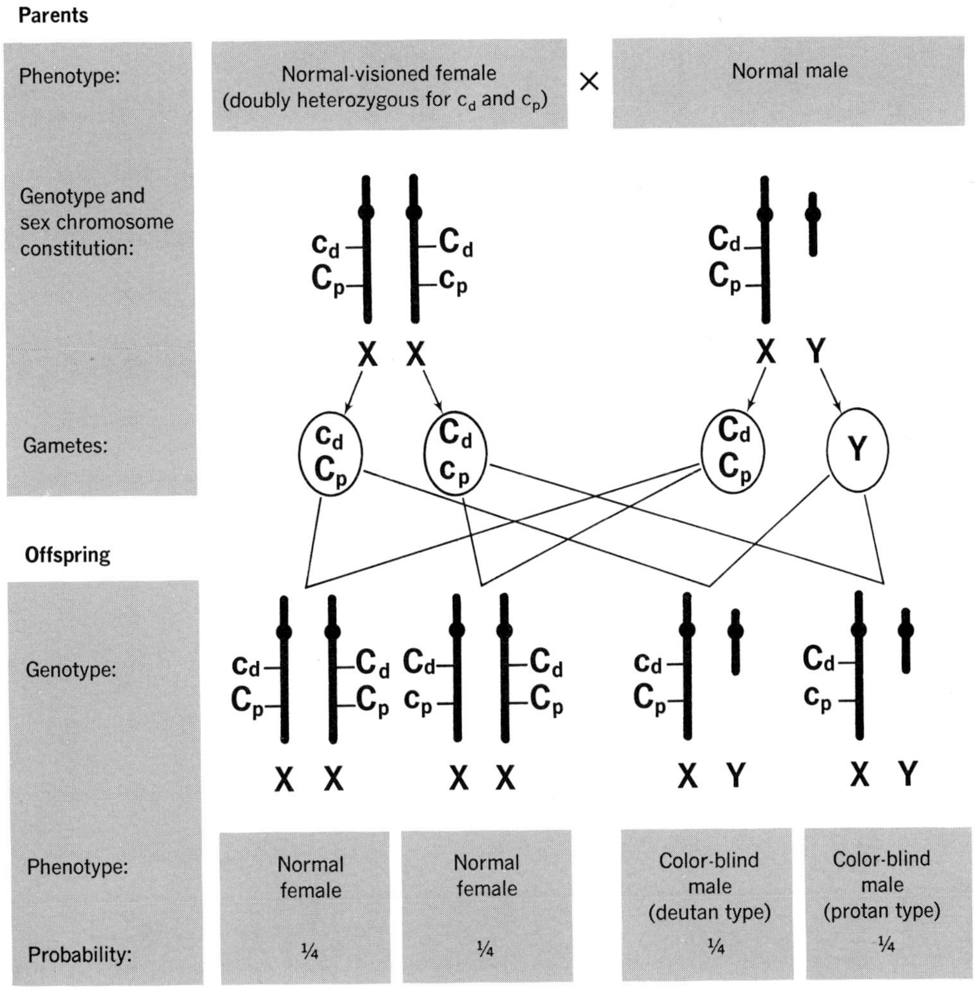

Fig. 9-5. The production of *all* colorblind sons from two normal-visioned parents when the female is heterozygous for both the deutan and protan types. Note that all daughters are normal but carry one or the other defective allele.

chemical constituent of the complex blood clotting process. Nearly 75% of the cases are *hemophilia A*, while another 20% or so represent the milder *hemophilia B* (also known as *Christmas disease*). Both of these are inherited as X-linked recessives at separate loci (similar to deutan and protan colorblindness). A few extremely rare forms are autosomally inherited, but our discussion will center on the classical hemophilia A.

The occurrence of a female with hemophilia is extremely rare because a female must be homozygous to express it, which means that her father, as well as her mother, would have to possess the allele (see Fig. 9-3). In the case of a severe disease such as hemophilia, males seldom survive through reproductive age or they reproduce at a very low rate, thus making the possibility of a homozygous recessive female quite low. The first substantiated case of a female hemophiliac was described in 1951, and since then only a few have been recorded.

On the other hand about 1 male in every 10,000 is born with hemophilia. According to the calculations of population genetics (Chapter 13), this frequency in males would lead to a prediction of 1 hemophiliac female for every 200 million children born. When we also consider the fact that male hemophiliacs have a reproductive disadvantage perhaps 4 or 5 times greater than normal males, the expectation of a female hemophiliac decreases to barely 1 in a billion births. This would place the number expected in the entire world at perhaps four or five!

Great strides have been made in the treatment of hemophilia in recent years. Individuals with hemophilia A lack activity of a blood clotting protein (gene product) known as antihemophilic factor (AHF), also called Factor VIII. In the past, for many victims, the lack of AHF meant long and frequent hospital stays, a highly restricted life spent in constant fear of injury, crippling of the body as a result of internal bleeding into the joints and, of course, greatly reduced reproductive activity. Now this vital clotting factor can be precipitated from normal blood plasma in a highly concentrated and purified form that can be injected intravenously to allow hemophiliacs for the first time to live near-normal lives. In the near future hemophiliacs may be able to treat themselves with AHF in much the same way as insulin is injected to control diabetes. The long-range implications for society of medical treatment of genetic diseases are often completely overlooked. However, with a disease as severe as hemophilia, which, because it is X-linked, is practically absent in females although males are highly "susceptible," it is imperative that consideration be given to future prospects.

The general situation is that when *any* genetic defect that produces a reproductive disadvantage (through early death, incapacitation, sterility, or any other reason) is treated through medical technology to produce near-normal life, including reproductive potential, the frequency of that defect will rise in the population over time. A compounding of this effect through the treatment of more and more genetic diseases of greater and greater severity is what leads to concern over the so-called deterioration of the human gene pool. The increased reliance of more and more individuals on medical technology also places increased stress on the resources of society. For example, it is estimated that treatment of the present number of hemophiliacs (estimated at over 10,000 in the U. S.) would require the AHF precipitated from a million pints of blood each year at an average cost of $150 per month to each patient. A critical question for society is what to do about the genetic implications of such medical treatment. Should treated hemophiliacs be sterilized to prevent the transmission of their defective gene, which is unaffected by the treatment, to the next generation? Initially the most common cross by far involving treated hemophiliacs would be *hY* × *HH*. Therefore, should prenatal sex determinations be made to abort all female fetuses *(Hh)* who, although normal, would be the transmitters of the defective gene to future generations, but to allow male fetuses *(HY)* who would be normal to develop? Or, when the control of sex determination is fully achieved, should such individuals be *required* to have only sons? These are obviously difficult questions to answer. But sooner or later we will have to reconcile the present with the future in genetic terms as well as in the areas of population size, pollution control, and resource management. Fig. 9-6 presents the possible progression in the rise of treatment-dependent hemophiliacs in the population over time when hemophiliacs no longer suffer a reproductive disadvantage and uninhibited, random mating occurs. The figure is intended only as a general progression indicating how an increased probability of certain marriage combinations would tend to foster an increased production of hemophiliacs in the population. In the numerical sequence indicated, each step simply signifies the new marriage combination whose likelihood has been increased by the previous combinations. At any particular step all previous combinations would also be occurring, and each step would require several generations to merge into the next.

Several types of sex-linked recessive muscular dystrophy, referred to in general as progressive muscular dystrophy because of the relentlessly constant degeneration of the muscles, are also extremely rare among females because of the reproductive disadvantage of afflicted males. Fig. 9-7 depicts a teenage boy with *Duchenne's muscular dystrophy*, the most common and severest type. It is known as a disease of young boys; the symptoms begin between the first and sixth years and result in confinement to wheelchairs by the ninth to twelfth years and death to over three fourths by the age of 20. It is not hard to "see" the lowered reproductive potential of such individuals to account for its extreme rarity among young girls. Evidence indicates that there are four different types of X-linked muscular dystrophy that form a multiple allelic series at a single locus.

A small number of abnormalities are inherited as X-linked dominant traits. Just as with autosomal dominants, one dose of such alleles is enough to produce the trait in either sex. However, a dominant allele in the hemizygous male ensures that he will pass the condition to each of his daughters but to none of his sons. Affected females who are heterozygous transmit the condition to half of their children of either sex, whereas females homozygous for the dominant allele will produce *all* affected children.

All evidence to date indicates that a form of vitamin D–resistant rickets, known technically as *hypophosphatemia*, is inherited as an X-linked dominant trait, since male-to-male transmission (father to son) has never been observed although all daughters of afflicted males appear to develop the condition. Also, a defect of the dentine of the teeth, which results in their rapid wear down to small stubs protruding from the gums, is caused by a dominant X-linked gene.

Perhaps the most clearcut example of an X-linked dominant trait is that of the antigen of the X_g blood group system. The genes for all other blood group systems are located on autosomes, but the particular location (exact chromosome number) of none is known. In the X_g blood group system the X stands for X-linked and the g for Grand Rapids, Michigan, the home of the person in whom the system was discovered. As in other blood groups, certain individuals possess a detectable antigen and others do not. The gene that produces the

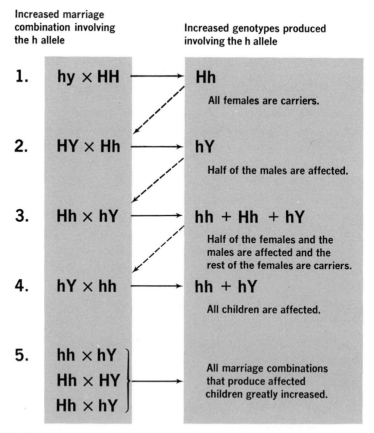

Fig. 9-6. The general steps in the progressive increase of treatment-dependent hemophiliacs (the *h* allele) in the population when the reproductive disadvantage of the victims is eliminated.

X_g-antigen is symbolized as X_g^a (a for antigen), and it dominates its inactive allele symbolized as X_g. Thus two phenotypes are possible, and these are designated as $X_{g(a+)}$ and $X_{g(a-)}$. Table 9-2 lists the genotypes and phenotypes possible for this sex-linked blood group system, and Fig. 9-8 demonstrates the transmission of the X_g groups to produce a typical X-linked dominant pattern of inheritance.

All X-linked genes also form a single linkage group with respect to crossing over, as discussed in the previous chapter. As a matter of fact, since crossing over that occurs in the female will be expressed directly in the phenotypes of

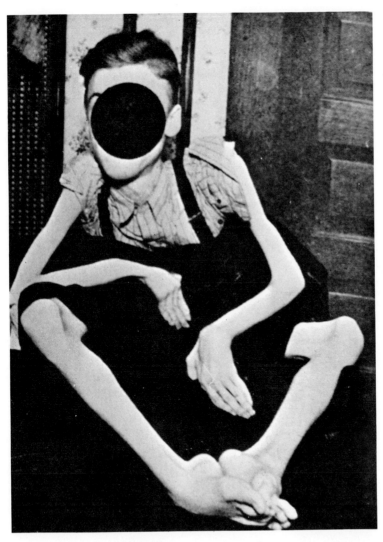

Fig. 9-7. A teenage boy with Duchenne's muscular dystrophy, an X-linked recessive trait rarely found in females. (From Winchester, A. M.: Human genetics, Columbus, Ohio, 1971, Charles E. Merrill Publishing Company.)

her hemizygous sons, the X chromosome is the only member of the human complement for which enough data have been collected to construct a partial chromosome map (refer to Chapter 8). The X chromosome seems to be about 175 map units long and is divided by a submetacentric centromere into a short arm of about 60 units and a long arm of more than 100 units. A tentative map of the X chromosome is presented in Fig. 9-9. It should be noted, however, that this map is probably incorrect in that, although the map distances given conform to the present data, the ordering to the right or left may be reversed in some

Table 9-2. The possible genotypes and phenotypes in the two sexes for the X-linked dominant X_g blood antigen

Males		Females	
Genotypes	Phenotypes	Genotypes	Phenotypes
X_g^a Y	$X_{g(a+)}$	X_g^a X_g^a	$X_{g(a+)}$
		X_g^a X_g	
X_g Y	$X_{g(a-)}$	X_g X_g	$X_{g(a-)}$

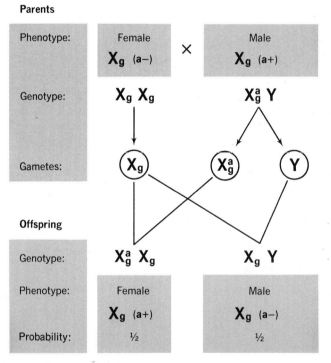

Fig. 9-8. A cross demonstrating the transmission of a dominant X-linked trait (the X_g antigen). Note that when a male has the trait and a female does not (as indicated here), *all* daughters inherit the condition but male-to-male transmission does not occur.

cases. Some researchers believe that the X_g locus is near the extreme end of the short arm and also a great distance from the G6PD-hemophilia complex. Therefore, X_g and the loci known to be close to it are placed in the short arm, and the other known linkage complex is placed in the long arm.

When you consider that this sketchy knowledge of gene loci on the X chromosome and the few autosomal linkages mentioned in the previous chapter that have no specific pair assignment represent our entire knowledge of gene locations in man, it is not difficult to realize the magnitude of the problem of mapping the human chromosomes. Such mapping would be a prerequisite to any so-called genetic surgery to correct a mutant gene (defective code) and thereby actually *cure* a genetic disease.

If a gene were located on the Y chromosome with no comparable locus on the X chromosome, the trait controlled by such a gene would appear in men only and would be transmitted consistently from male to male, that is, all the sons of an affected male would inherit their father's trait. This is known as a Y-linked, or *holandric*, pattern of inheritance. Since females never possess a Y chromosome, it is apparent that no locus controlling a critical developmental process or bodily function common to both sexes could possibly be Y-linked. Undoubtedly, the Y chromosome contains genes involved in male determination. The only distinguishable phenotype in man which consistently conforms to a Y-linked pattern of transmission is a trait known as *hypertrichosis pinnae auris*, or "hairy ears," which seems to be quite common in some parts of India.

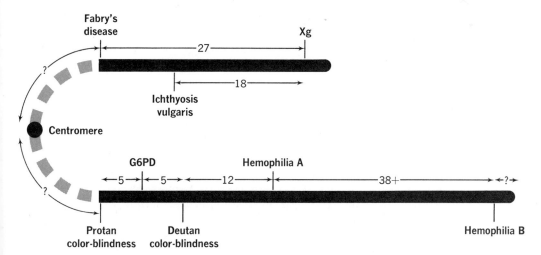

Fig. 9-9. Tentative map of the X chromosome. Loci included on the map but not discussed in the text are: *Fabry's disease,* a metabolic disorder causing serious kidney malfunction (recessive); *ichthyosis vulgaris,* a mild form of scaling of the skin (recessive); and G6PD, which stands for the enzyme glucose-6-phosphate-dehydrogenase, a lack of which produces an unusual form of anemia known as Favism which occurs when susceptible individuals eat the fava bean (recessive). Linkage data are not yet available for other X-linked loci.

The female as a mosaic

As the names X and Y imply, the sex chromosomes presented an enigma right from their discovery. The X chromosome was first described in 1891 from insect material as a peculiar chromatin body that lagged behind all others during anaphase I of spermatogenesis and then passed undivided to one pole during anaphase II. The uncertainty of its function resulted in it being named X, and the later discovery of its small "mate" in males lead to the naming of the Y chromosome.

After the role of the X and Y chromosomes in sex determination was established, the function of the X chromosome seemed more confusing than ever. Although the female had two doses of all genes on the X chromosome and the male only one of each, there seemed to be no significant qualitative difference in the products of X-linked genes in males and females. For many years geneticists searched for a mechanism to explain what had been termed "dosage compensation" to account for the equalization of X-linked gene products in the two sexes. Finally, in 1961 a theory hypothesized by several researchers throughout the world, but lead by the work of Dr. Mary Lyon in England and Dr. L. B. Russell in the United States, was developed that explained dosage compensation through X chromosome inactivation in the female. This theory, known generally as the *Lyon hypothesis*, has been well substantiated in all mammals and might properly be called the Lyon-Russell principle. Simply stated, it proposes that in each *somatic* cell of a normal female one of the two X chromosomes becomes functionally inactive early in the development of the embryo (by the sixteenth day). It is a matter of chance whether the X chromosome of maternal or paternal origin is inactivated. But once either one is inactivated in an embryonic cell, all progeny of that cell maintain the *same* inactive X chromosome. As a result, one group, or "patch," of cells may have one X inactivated, while an adjacent patch of tissue may have the other X inactivated. If the female is heterozygous at some particular X-linked locus, one patch of cells may exhibit the expression of one allele, while another patch may express the alternate allele's product. Thus, the female is a mosaic with respect to the expression of X-linked genes for which she is heterozygous. No X chromosome inactivation occurs in the female's germinal tissue, and no inactivation of the single X chromosome of the male occurs. However, all X chromosomes in excess of one become inactivated even when three or more are present (see Chapter 12). The inactivated X chromosome becomes a tightly coiled, deeply staining mass usually located next to the inner surface of the nuclear membrane and called a *Barr body* (after Dr. Murray Barr who first described them in cats). Fig. 9-10 schematically illustrates the dosage compensation effect and the production of mosaicism in females according to the Lyon-Russell principle.

The Lyon-Russell principle produces no contradictions of the facts of X-linked inheritance because heterozygosity at any X locus would mean that about half of the woman's cells would express the dominant allele and half the recessive allele. For example, in a female heterozygous for hemophilia, about half of the cells involved in producing AHF would express the dominant active allele to produce enough of the plasma component for normal or near-normal blood clotting to occur. However, only a small number of cells destined for any particular function is present at the early stage of development when inactivation takes place. Thus, it is possible that in rare instances, by chance alone, most or all of the active X chromosomes in the progenitors of the AHF-producing cells could contain the mutant allele. Such individuals would have extremely slow clotting time that approaches the symptoms of full hemophilia; these individuals are called "manifesting heterozygotes."

The application of the Lyon-Russell principle to methods of genetic counseling is exemplified by the progress made in the identification of carriers and preventative control by prenatal diagnosis of a severe neurological disease known as the *Lesch-Nyhan syndrome*. This X-linked recessive disease produces mental retardation associated with bizarre behavior characterized by spasticity and self-mutilation such as biting of the lips or fingers and banging of the head despite the pain produced. Biochemically an enzyme (gene product) known as

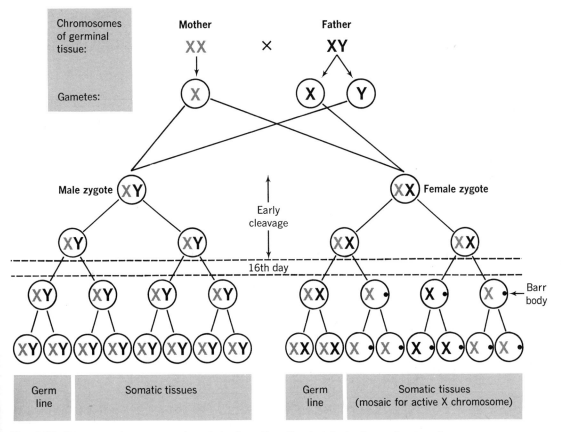

Fig. 9-10. Diagrammatic representation of the Lyon-Russell principle in the production of female mosaicism for X chromosome activity. Gray represents chromosomes of maternal origin, and black represents chromosomes of paternal origin.

HGPRT (which stands for hypoxanthine guanine phosphoribosyltransferase) is inactive, causing an upset in purine metabolism. This imbalance results in the mental impairment and behavioral disorder. Since the description of the disease in 1964, over 150 cases have been reported; no treatment of the neurological symptoms is yet known.

The Lesch-Nyhan syndrome is found almost entirely among boys who inherit it from their mother, who is a carrier. Using the technique of amniocentesis, it has been possible to obtain and culture fetal cells to test for HGPRT activity. In the first reported application of this procedure (in 1970) a female who had previously given birth to one normal son and one affected son was once again pregnant and concerned about producing a second affected child. Cells of this fetus were extracted and cultured. Karyotypes of the chromosome complement of the cells identified the fetus as male, and the assay for HGPRT activity was negative. The same enzyme assay performed on cultured somatic cells of the mother indicated two populations of cells, one with HGPRT activity and one without, as would be expected from the random inactivation of X chromosomes in a carrier female according to the Lyon-Russell principle. The pregnancy was terminated, and the prenatal detection of Lesch-Nyhan syndrome was confirmed in the aborted fetus.

Preventive control of this or any other recessive genetic disease cannot be very effective if carriers must first produce one affected child to be identified. Neither is it practical in terms of time, labor, or materials to perform complex, sophisticated cell culture and enzyme assay techniques on the whole population to identify carriers before they reproduce. Mass screening must be accomplished by much simpler methods, especially if we ever hope to assay for a large number of different defects.

A further step in this direction has been achieved through the development of a simple method for direct assay of HGPRT activity in single hair follicles obtained from the scalp. Hair follicles are especially useful in this case because each follicle develops from a very small number of cells. This gives a high probability of obtaining a "nonmosaic follicle," that is, one in which all cells have the same X-linked allele active. Thus, a female heterozygous for an X-linked gene may exhibit *three* classes of hair follicles—those consisting entirely of cells expressing only the normal allele, those of cells expressing only the mutant allele, or mosaic follicles in which some cells express one allele and some the other allele. In a sampling of single hair follicles plucked from different areas of the scalp, the enzyme activity in phenotypically normal women would either be constant and average, indicating homozygosity, or it would range from average to nil, indicating that the female was heterozygous. Such screening would make it possible to diagnose Lesch-Nyhan heterozygosity during early pregnancy to determine whether or not to offer a patient amniocentesis for fetal detection. The hair follicle, with its special attributes for determining X-linked gene activity, may very likely become an important diagnostic tool in screening for carriers of other defects such as hemophilia and progressive muscular dystrophy.

SEX-LIMITED AND SEX-INFLUENCED TRAITS

Sex-linked traits (from genes on the X or the Y chromosome) are characterized by the fact that the sex ratio of affected individuals is greatly skewed toward the male. However, there are traits in which the sex ratio among the phenotypes is abnormally unbalanced in one direction or the other and the governing genes are not sex-linked.

An inherited trait that appears in only one sex is said to be *sex-limited*. These traits are usually controlled by *autosomal* genes transmitted in a normal fashion by both sexes but phenotypically expressed in only one sex, often because the structure or function affected only develops in one sex or the other. The secondary sex characteristics are examples of sex-limited traits. In women the shape and size of breasts are inherited, although environmental factors such as physical activity, nursing, and diet also exert an influence. It is not uncommon to note that a woman may have breasts unlike her mother's but similar to those of her father's mother or one of his sisters. Beard pattern and texture are sex-limited traits expressed only in the male under normal conditions. Thus, a man with a light, soft beard may have a son who develops a heavy, wiry beard because of genes inherited through the mother. It is usually the hormonal differences between the sexes that influence the expression of sex-limited genes. With respect to the beard, for example, it has been found that women have the same number of hairs on the face as men, but they do not grow as in men. The stimulation of male hormones, however, can produce beard development in females. Of course the basic structure influenced by the genes must also be present in order to observe their effect. Testicular feminization, discussed earlier in the chapter, is sex-limited to males because females do not develop testes. Fig. 9-11 presents a generalized model of the sex-limited phenomenon.

Finally, there are autosomally inherited genes that produce phenotypic expressions in *both* sexes but which are much more prevalent in one sex than the other. Traits produced in this manner are said to be *sex-influenced* because sex

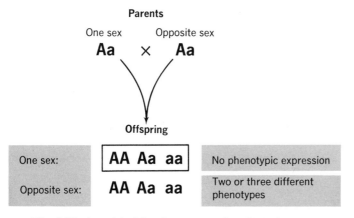

Fig. 9-11. A model of the phenomenon of sex-limited traits.

Nomenclature

| **A** | = | Normal condition |
| **A'** | = | Mutant condition |

Parents

AA' × AA'

Possible progeny genotypes

A' dominant

| One sex: | $^1/_4$ AA | $^2/_4$ AA' | $^1/_4$ A'A' | = | 3 mutant phenotypes to 1 normal |
| Opposite sex: | $^1/_4$ AA | $^2/_4$ AA' | $^1/_4$ A'A' | = | 1 mutant phenotype to 3 normals |

A' recessive

Fig. 9-12. A model of the phenomenon of sex-influenced inheritance.

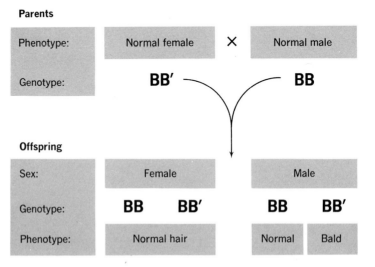

Parents

| Phenotype: | Normal female | × | Normal male |
| Genotype: | **BB'** | | **BB** |

Offspring

Sex:	Female		Male	
Genotype:	**BB**	**BB'**	**BB**	**BB'**
Phenotype:	Normal hair		Normal	Bald

Fig. 9-13. The transmission of baldness from a normal female to half her sons but to none of her daughters, according to a sex-influenced pattern of inheritance.

controls the dominant-recessive relationship of the alleles. The *same* allele that is dominant in one sex acts recessive in the other sex, and vice versa. When both parents are heterozygous for a sex-influenced trait, it will be expected in a 3:1 ratio in the sex in which it acts dominant and in a 1:3 ratio in the sex in which it acts recessive.

A model of this phenomenon is presented in Fig. 9-12. To avoid confusion with respect to capital and lower case letters representing dominant and recessive alleles that would be reversed in one of the sexes, symbols with and without prime notation are used to distinguish sex-influenced alleles; the prime mark represents the mutant form. Common baldness is inherited in this manner with the baldness allele *(B')* acting dominant over the allele for normal hair pattern *(B)* in males, but recessive to the normal allele in females. Thus, males with a single dose of *B'* will become bald, while females must be homozygous *B'B'* to develop the phenotype. Baldness also displays variable expression, especially between the sexes, where females exhibit extreme thinning of the hair but rarely complete balding. This mode of inheritance explains why it is erroneous for a male to believe that if his father was not bald he cannot become bald either. As Fig. 9-13 illustrates, a son may inherit baldness from his normal mother. A sex-influenced trait that appears to act dominant in females and recessive in males is a nonpainful swelling of the terminal finger joints known as *Heberden's nodes*.

10

Pedigrees and family predictions

In the preceding three chapters the major modes of hereditary transmission were discussed, with the exception of continuous patterns produced by the action of many genes affecting complex traits, which will be discussed in the next chapter. In this chapter we shall consider the methods of detecting and tracing various hereditary patterns within families and extend our knowledge of genetic predictions at the family level.

It has already been mentioned (Chapter 7) that testcrosses or any prescribed matings are not only unacceptable but really impractical as a means of collecting genetic data in the human species. The common method of genetic analysis in man involves the construction of a *pedigree*, or family history, indicating all the information available about the trait under consideration back through as many generations as possible. From a pedigree it may be possible to determine the mode of heredity of a previously unknown trait or to make predictions concerning the appearance of a known trait in future progeny. However, since many family pedigrees are small, the data are often inconclusive. Also, because many human traits are genetically complex and involve various kinds of interactions (genetic and environmental), it is not always possible to associate a phenotype with the action of a single gene conforming to a particular hereditary pattern. Nevertheless, for many traits pedigrees are a valuable tool for the human geneticist and an important source of information for genetic counseling.

PEDIGREE CONSTRUCTION

A pedigree is constructed by diagramming all known relationships, contemporary and ancestral, of a particular individual with respect to some specific trait, or traits, to be studied. The individual with whom the construction of a pedigree is initiated is known as the *proband* or *propositus*. Fig. 10-1 presents a set of symbols commonly used in pedigree construction that will be applied in this text. Fig. 10-2 illustrates how these symbols are combined to produce a pedigree. The pedigree depicted here is hypothetical; it attempts only to utilize most symbols at once for illustrative purposes. Note also that each generation is assigned a roman numeral, and each individual within a generation is assigned an arabic numeral. This makes the identification of any particular individual easily accomplished by his or her roman and arabic designation, for example, I-2 or III-7, and so forth. Additional information, such as age of death or date of birth, is sometimes given to the right of a symbol when it might contribute to the analysis being conducted. Various modifications and

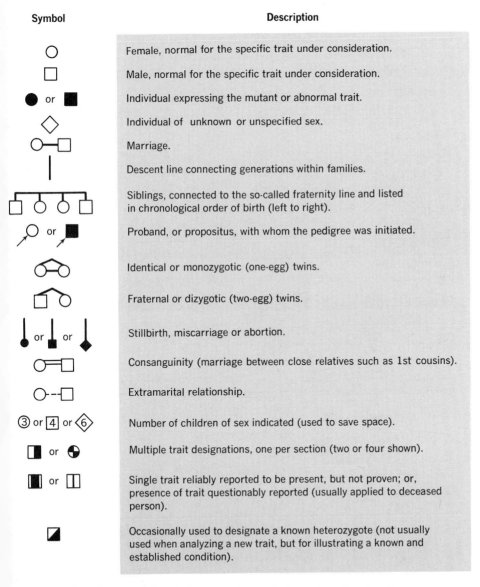

Symbol	Description
	Female, normal for the specific trait under consideration.
	Male, normal for the specific trait under consideration.
or	Individual expressing the mutant or abnormal trait.
	Individual of unknown or unspecified sex.
	Marriage.
	Descent line connecting generations within families.
	Siblings, connected to the so-called fraternity line and listed in chronological order of birth (left to right).
or	Proband, or propositus, with whom the pedigree was initiated.
	Identical or monozygotic (one-egg) twins.
	Fraternal or dizygotic (two-egg) twins.
or or	Stillbirth, miscarriage or abortion.
	Consanguinity (marriage between close relatives such as 1st cousins).
	Extramarital relationship.
③ or ④ or ⑥	Number of children of sex indicated (used to save space).
or	Multiple trait designations, one per section (two or four shown).
or	Single trait reliably reported to be present, but not proven; or, presence of trait questionably reported (usually applied to deceased person).
	Occasionally used to designate a known heterozygote (not usually used when analyzing a new trait, but for illustrating a known and established condition).

Fig. 10-1. A set of symbols commonly used in the construction of pedigrees.

additions to the symbols and patterns indicated here also exist, but they do not substantially alter the basic analysis of a pedigree.

PEDIGREE ANALYSIS

In analyzing a pedigree for the purpose of determining the possible mode of inheritance of a specific trait (if it is indeed inherited in a simple manner), the general procedure is to progress through each generation, beginning with the first, "testing" whether or not a particular mode of hereditary transmission of the trait would be possible. The objective is to make an assumption such as, "the trait is inherited as an autosomal recessive." Then one must determine from the pattern of phenotypes in the pedigree if the hypothesis under consideration could possibly "fit" the pedigree or if it definitely could not fit. The modes of transmission that can be analyzed with relative ease and accuracy in this manner are:
1. Autosomal dominance
2. Autosomal recessiveness
3. X-linked dominance
4. X-linked recessiveness
5. Y-linkage
6. Sex-limitation

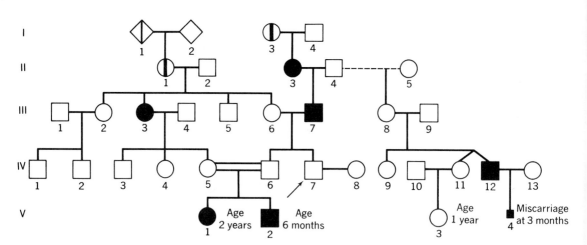

Fig. 10-2. Hypothetical pedigree illustrating the use of the common symbols and the patterns produced. Note that individual IV-7 is the proband. Also note that individuals III-4 and IV-10 are not brothers of their wives for they are not connected to the fraternity line. An interpretation of generation I would be that all were deceased for some time before the pedigree was constructed and that one great-grandparent of the proband on each side of his family is thought to have possessed the trait in question. In one case a paternal great-grandmother is reliably thought to have been affected, while on the maternal side the presence of the trait is questionable, and it is not certain which great-grandparent might have had it.

7. Sex-influence
   A. Acting dominant in males and recessive in females
   B. Acting dominant in females and recessive in males

Testing each of the eight possibilities (considering 7A and 7B separately) as an independent hypothesis produces a series of yes and no answers. A no answer definitely excludes the simple form of the hypothesis (ignoring penetrance, variable expression, or other modifying effects), whereas a yes answer means that that particular simple mode fits the pedigree pattern, but does not confirm it as the correct mode. A yes answer to a hypothesis fitting a particular pedigree pattern really means "yes, it is possible" or "yes, because no definite *impossibility* is found in the pedigree." In many small pedigrees it is not unusual to come up with more than one yes answer. But obviously, if the trait is simply inherited, it is not being transmitted in several ways at the same time; such a pedigree is said to be *inconclusive*. In Fig. 10-3 two very similar pedigrees are presented. In testing the eight hypotheses for simple modes of inheritance, we find that pedigree *A* in Fig. 10-3 is inconclusive because the trait could be inherited as an autosomal recessive, an X-linked recessive, a sex-limited character, or it could be governed by a sex-influenced allele acting dominant in males and recessive in females. (It should be noted that when testing the various hypotheses, it must be assumed that normal phenotypes may be carriers of recessive alleles when their genotype is not known for certain.) On the other hand the same analyses applied to pedigree *B* reject all but autosomal recessive inheritance as the mode of transmission; this is a conclusive pedigree. If the same trait is under consideration in pedigrees *A* and *B* of Fig. 10-3, then the confirmation of its mode of inheritance in *B* must apply to *A* also.

Accurate phenotypic identification and collective analyses of several inconclusive pedigrees can often establish the mode of transmission of an inherited trait. This makes it possible to apply the collective genetic knowledge on the mode of inheritance to each specific pedigree and achieve a predictive power

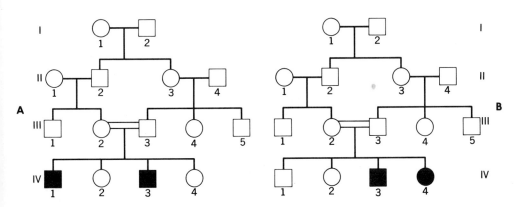

Fig. 10-3. Two very similar pedigrees in which **A** is inconclusive and **B** is conclusive regarding the mode of inheritance.

that would not be possible from the information contained within any single pedigree.

Fig. 10-4 depicts three pedigrees involving three unrelated families that have the *same* trait under investigation. These pedigrees are accompanied by a tabularized analysis of each pedigree with respect to the eight hypotheses for simple genetic transmission. It can be seen in the analyses that, although each pedigree is inconclusive in itself, collectively only one mode of transmission fits all three consistently. Thus, the trait seems to be inherited as an autosomal recessive condition, and the other "yes" fits are simply artifacts produced within small pedigrees. Further investigations and the eventual discovery of a single conclusive pedigree could confirm the hereditary nature of the condition, which, in this case, could have ranged from blue eyes to albinism or forms of mental retardation.

Once the mode of inheritance is established, it is possible to gain additional knowledge about the genotypes of normal individuals and to make predictions regarding future occurrences of the trait through genetic counseling. For example, if individuals III-2 and III-4 in Fig. 10-4 were to marry, the probability that they would produce an affected child would be ¼ because both of these persons must be heterozygous for the recessive allele since each had a homozy-

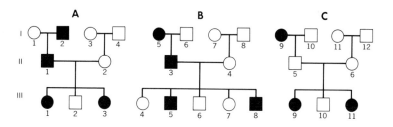

Analysis

Possible mode of inheritance	Pedigree (A)	Pedigree (B)	Pedigree (C)
Autosomal dominance	Yes	Yes	No
Autosomal recessiveness	Yes	Yes	Yes
X-linked dominance	Yes	No	No
X-linked recessiveness	Yes	Yes	No
Y-linked	No	No	No
Sex-limited	No	No	Yes
Sex-influenced-male dominance	Yes	Yes	No
Sex-influenced-female dominance	No	No	Yes

Fig. 10-4. Three pedigrees involving the same trait in unrelated families and the analysis showing that, although each pedigree is itself inconclusive, collectively autosomal recessiveness is the only mode of inheritance that consistently fits all three.

gous parent (individuals II-1 and II-3). On the other hand, if individuals III-7 and III-10 were to marry, the best prediction possible regarding their production of an affected child is $\frac{1}{6}$. This is a lower probability than the first case but is also less accurate because it is an overall prediction that will be *wrong* for every specific case. The reason that the prediction in this situation is less accurate than the first situation stems from the fact that less is known about the genotype of III-10 than is known about III-7 (or III-2 and III-4), who must be heterozygous because of her homozygous father (II-3). The parents of III-10 (II-5 and II-6) must both be heterozygotes. This is known because although they are both normal, they produced two affected children (III-9 and III-11). Therefore, we know that III-10 is $A\_\_$ (using A and a to represent the alleles in question), but we do not know whether his genotype is AA or Aa; it could be either. It is certain that his genotype is not aa, however, because his phenotype is normal. Since III-7 is definitely heterozygous, only if III-10 is heterozygous also will there be any probability of producing an affected child *(aa)*. But III-10 may or may not be heterozygous. If he is, the probability of producing an affected child is $\frac{1}{4}$; if he is not, the probability is zero. To proceed further, we must determine the probability of III-10 being heterozygous. In a simple Mendelian cross between two carriers of an autosomal recessive allele, the classical genotypic ratio of $1AA : 2Aa : 1aa$ would be expected and the probability of the Aa genotype would be $\frac{2}{4}$ or $\frac{1}{2}$. However, our pedigree tells us that phenotypically III-10 is normal, which means that the mutant phenotype (aa genotype) can be excluded from our consideration. This leaves $1AA : 2Aa$ (the phenotypically normal portion of the genotypic ratio) as the total set of genotypic possibilities for III-10. Of the three possibilities, two are heterozygous and one is homozygous. Thus, $\frac{2}{3}$ is the probability that III-10 is heterozygous and $\frac{1}{3}$ that he is homozygous. This is a higher probability than the $\frac{1}{2}$ derived from the entire genotypic ratio because the phenotypic input provided additional information that increased the predictive power. Since we cannot be sure of the genotype of III-10, the overall prediction must account for the fact that the probability that III-10 is Aa is $\frac{2}{3}$. The complete probability statement is 1 (the probability that III-7 is Aa) times $\frac{2}{3}$ (the probability that III-10 is Aa) times $\frac{1}{4}$ (the probability that aa will be produced from $Aa \times Aa$ parents). Thus, $1 \times \frac{2}{3} \times \frac{1}{4} = \frac{2}{12} = \frac{1}{6}$. Note that $1 \times \frac{2}{3} = \frac{2}{3}$ is simply the probability that the genotypic combination is $Aa \times Aa$, which is the only way of producing an affected child in this situation. The other possibility is $Aa \times AA$ with a probability of $1 \times \frac{1}{3} = \frac{1}{3}$, but it will produce no affected children. Hence, the overall probability of $\frac{1}{6}$ is an average that would materialize from a great number of situations such as this. However, in any particular case the true probability will either be $\frac{1}{4}$ or zero. Only if a test for heterozygosity, such as an enzyme assay, were available could the $\frac{1}{4}$ or zero predictions be distinguished.

The following pedigrees, taken from actual case studies, constitute a series that illustrates most of the modes of inheritance considered in a basic approach to pedigree analysis. In many cases examples not previously considered in this text are used to increase our repertoire of inherited traits.

Pedigrees of autosomal dominant traits can sometimes trace the transmission of the dominant allele back to the parental combination that gave rise to the new mutant form. This is possible for fully penetrant conditions because the dominant allele is never masked and never skips generations. Fig. 10-5 illustrates one such situation concerning a rare abnormality in man known as *piebaldness,* a spotting of the body caused by a mixture of pigmented and unpigmented skin patches. The mutation to the piebald condition that produced II-4 apparently arose in the egg or the sperm of I-1 or I-2. They had 8 other normal children. II-4 had 15 children, of whom 8 were piebald and 7 normal. Five of the normal children married and produced 15 normal children and 4 normal grandchildren. All 5 of the piebald children of generation III who reproduced transmitted the trait to their children who, in turn, were transmitting it into generation V at the time the pedigree was constructed. Note that once established, dominant traits do not skip generations and nearly a 1:1 ratio of affected to normal is observed. In generation III, 8 piebald to 7 normal children is as close to 1:1 as is possible for 15 children. It should also be noted, however, that among the offspring of affected persons there are 24 piebald to only 11 normal in the entire pedigree. We might suspect that this is a quite large deviation from random segregation and zygote production. However, in this particular case chance alone was probably not operating because many of the people with this trait joined circuses and freak shows, appearing under such billings as "The Tiger Lilies" and "The Striped Graces." As these people became scattered over the United States and Europe, they lost contact with their family and, although piebald family members were traced accurately, the data on normal individuals are less reliable. It is very likely therefore that, in generation IV especially, some undiscovered normal siblings are missing from the pedigree, causing the ratio to be biased toward piebaldness.

Many rare autosomal recessive abnormalities appear in families where intermarriages (consanguinity) produce homozygosity and the appearance of the abnormal phenotype because both carriers possess the *same* recessive allele as

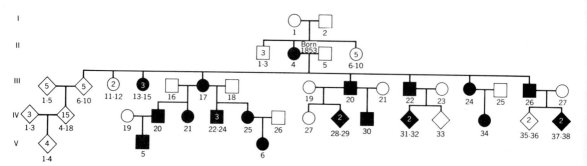

Fig. 10-5. Pedigree illustrating autosomal dominant transmission of piebalding. (Based on Keeler, C. E.: Heredity of congenital white spotting in Negroes, J. A. M. A. **103**:179, 1934.)

a result of its common descent to them through their common ancestry. In other words, the probability that two parents will be carriers for the recessive allele of a rare trait is greatly enhanced when both could have received it from the same source of origin. This will be elaborated upon further in Chapter 13. A pedigree illustrating the appearance of an autosomal recessive condition known as *microcephaly* ("small head") is presented in Fig. 10-6. This inherited form of microcephaly is one of the most common genetic forms, but there are other genetic causes and also environmental causes of this condition. We see in this pedigree that in both families that produced affected children in generation VI, the parents represent a consanguineous marriage (each related as third cousins). We can also observe that each of the four individuals in generation V can be traced back to a common ancestral couple in the 1700s (generation I).

The two types of X-linked recessive colorblindness, deutan and protan, are indicated simultaneously in the three independent and unrelated pedigrees in Fig. 10-7. Considering either trait independently provides a typical example of X-linked recessive transmission. For instance, considering only the deutan phenotype in pedigree A, the affected mother (I-1) must be homozygous and therefore transmits her trait to all sons, but no daughters. When only the protan phenotype is considered in pedigree A, the affected father (hemizygous)

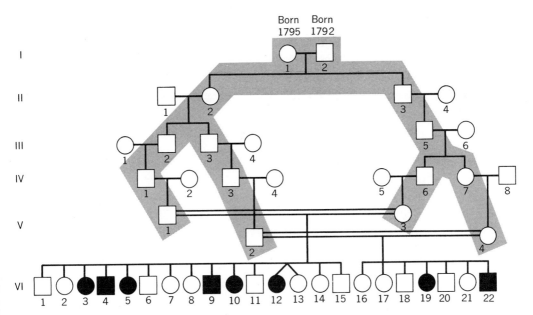

Fig. 10-6. Pedigree illustrating autosomal recessive transmission of microcephaly expressed in the children of two families representing the intermarriage of close relatives. The shaded pathways trace the probable transmission from a common ancestral pair in generation I to the parents in generation V. (Based on McKusick, Victor A.: Some principles of medical genetics. In Bartalos, Mihaly, editor: Genetics in medical practice, Philadelphia, 1968, J. B. Lippincott Company.)

transmits his trait to none of his sons or daughters, although the daughter (II-2) would be a carrier. Finally, when *both* types of colorblindness are considered simultaneously, we find that II-2, II-5, and II-6 are carriers for *both* types of colorblindness and that I-5 must also be a double hetereozygote because she produced both deutan and protan sons. This latter case and the fact that II-2, II-5, and II-6 are not affected, even though both parents are colorblind, provide evidence that the the deutan and protan genes occupy two different loci on the X chromosome (Fig. 9-9).

The inheritance of a sex-influence trait, such as baldness, is difficult to prove conclusively within a single pedigree. In most pedigrees of sex-influenced transmission it is common to find that autosomal dominant or autosomal recessive or X-linked recessive modes of transmission also fit. The two pedigrees of "skull-cap pattern" baldness presented in Fig. 10-8 taken together are conclusive for transmission as a sex-influenced trait inherited as a dominant condition in males and recessive in females (see Fig. 9-12 for model). In pedigree A the marriage between III-8 and III-9, both of whom are affected, and their production of a normal daughter (IV-8) contradicts both autosomal and X-linked recessiveness as possible modes of transmission. And, in pedigree B, the production of affected children by normal parents in two instances, II-10 married to II-11, and III-12 married to III-13, eliminates autosomal dominance (and X-linked dominance) as the mode of transmission. The fact that all affected individuals in this pedigree are males also substantiates male-dominant sex-influenced transmission. It should also be noted that father to son transmission is common, but not absolute, whereas affected females produce *all* sons with the trait.

A vivid picture of genetic transmission is obtained in cases where codomi-

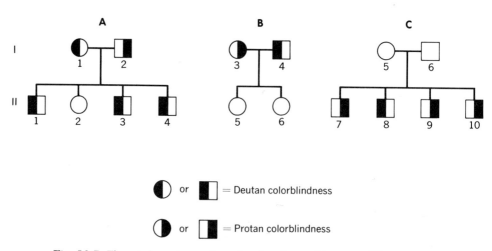

Fig. 10-7. Three independent and unrelated pedigrees illustrating X-linked recessive transmission of the deutan and protan types of colorblindness. (Based on Stern, C.: Principles of human genetics, San Francisco, 1960, W. H. Freeman and Company.)

nance exists between autosomal alleles. An example of this situation is provided by *Cooley's anemia* (Mediterranean anemia). This is a relatively common disease that occurs in a band around the Old World encompassing the Mediterranean, Middle East, India, and the Orient. The pattern of transmission is exactly the same as that of sickle cell anemia and sickle cell trait (Fig. 7-3). However, Cooley's anemia appears to be caused by changes in the amount of normal hemoglobin produced rather than structural change in a polypeptide. The anemia becomes apparent in early childhood and progressively worsens, causing death by the age of 10 to the majority of those afflicted. The spleen and liver enlarge, often to massive proportions, and the skin assumes a characteristic muddy tinge. Many of the red blood cells are destroyed in the marrow soon after their production, and those that enter the blood to circulate rapidly disintegrate. Survival to adulthood is almost unknown, and reproduction involving affected persons practically never occurs.

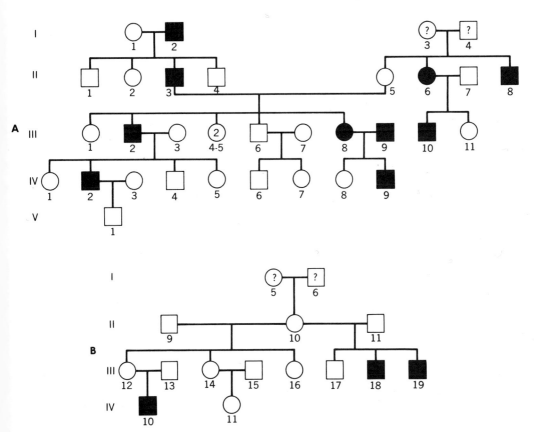

Fig. 10-8. Two pedigrees illustrating the transmission of "pattern" baldness as a sex-influenced trait inherited as a dominant condition in males and recessive in females. (Based on Gates, R.: Pedigrees of Negro families, New York, 1949, The Blakiston Company.)

Some 50 years ago, when Cooley's anemia was first recognized in children of Mediterranean ancestry, Italian physicians, studying a far less severe form of anemia that was widespread in their country, initiated extensive research into the transmission of these anemias. It was ultimately revealed that Cooley's anemia is the result of a double dose of a gene (homozygote) that in single dose (heterozygote) seemed to produce the milder form of anemia. The disease was named *thalassemia,* derived from the Greek word for sea, *thalassa,* signifying its discovery in persons of Mediterranean stock. The homozygous condition is now called *thalassemia major* (Cooley's anemia), and the heterozygous condition is known as *thalassemia minor.* Tracing thalassemia major and minor through pedigrees has confirmed this genetic interpretation, as illustrated in Fig. 10-9. With the advent of biochemical analyses to detect carriers of autosomal recessive disorders and the possibility of public mass screening for diseases such as sickle cell anemia, pedigrees, such as Fig. 10-9, that include the designation of carriers will become much more common and will provide for greater accuracy in determining probabilities to aid in genetic counseling.

Counseling, even with the aid of detailed pedigree data, is not always a simple matter, however. An extraordinarily perplexing situation arises in the case of an autosomal dominant disease known as *Huntington's chorea.* The disease was named after Dr. George Huntington who, in 1872, described cases of chorea (nervous disorder characterized by spasmodic twitching) among inhabitants of Long Island. Although the principles of Mendelian genetics were not yet acknowledged in the scientific community, Dr. Huntington wrote:

> And now I wish to draw your attention more particularly to a form of the disease which exists, so far as I know almost exclusively on the east end of Long Island. It is peculiar in itself and seems to obey certain laws. The hereditary character as I shall call it is confined to certain and particularly a few families and has been transmitted to them, an heirloom from generations away back in the dim past. It is spoken of by those in

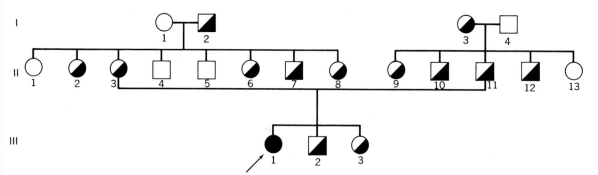

Fig. 10-9. Pedigree illustrating the transmission of autosomal codominant alleles responsible for thalassemia major and minor. Note that the propositus with thalassemia major (III-1) is traced to parents who were both heterozygous (thalassemia minor) and themselves received an abnormal allele from one of their parents, I-2 and I-3. (Based on Valentine, W. N., and Neel, J. V.: Hematologic and genetic study of transmission of thalassemia [Cooley's anemia; Mediterranean anemia] Arch. Intern. Med. **74**:185, 1944.)

whose veins the seeds of the disease are known to exist with a kind of horror, and not at all alluded to except through dire necessity, when it is mentioned as *that disease* . . . . There are three marked peculiarities of this disease: (1) Its hereditary nature. (2) A tendency to insanity and suicide. (3) Its manifesting itself as the grave disease only in adult life. (From Porter, I. H.: Heredity and disease, New York, 1968, McGraw-Hill Book Company, pp. 89-90.)

Even today this disease, characterized by spasmodic movements and progressive mental deterioration, presents an enigma in most counseling situations. Although an autosomal dominant gene causes the disease, the symptoms do not appear until well after reproductive age has been reached. The average age of onset is 35, but the disease may appear as early as age 20 or as late as age 50. Death usually occurs 12 to 15 years after onset of the symptoms. In addition to the long period of progressive physical and mental deterioration and the agony this produces for the afflicted person and his or her family, the genetic situation is frustrating because there is no way of determining who in a family at risk does or does not possess the abnormal allele until the symptoms appear. By that time the afflicted person may have produced several children, each with a $\frac{1}{2}$ chance of receiving the abnormal allele. The anxiety and frustration of having a parent suffer and die of Huntington's chorea and knowing that there is a 50% chance of having it places a person under extreme psychological stress when considering marriage and the rearing of a family, both of which usually occur before the average age of onset. If the individual develops Huntington's chorea, then his or her children have a 50% chance of receiving the abnormal allele and the cycle starts all over again. There is yet no sure way of identifying afflicted individuals before the onset of symptoms, and there is no sufficient treatment available for the disease.

However, recent attempts have been made to detect Huntington's chorea in the presymptomatic state, that is, prior to any overt signs of the disease and preferably before the child-bearing years. It appears that injections of L-dopa or levodopa, a modified amino acid found as a intermediary in melanin production, can evoke manifestations of the disease in young people who possess the Huntington's gene. The symptoms rapidly disappear when injections are discontinued. But for those who test positively, it is a grim preview of their future.

If this test is confirmed as a diagnostic measure for Huntington's chorea, a whole new set of problems arise such as the ethical question of using a drug to provoke symptoms to diagnose an incurable disease, the moral question of what to do once "carriers" are detected (should they be sterilized to prevent transmission of the disease?), and the serious problems associated with counseling young people diagnosed to have a fate of dementia and death from this disease.

Without an established clinical presymptomatic diagnosis available as yet, the variability in age of onset complicates genetic counseling for Huntington's chorea because in many cases accurate probabilities cannot be determined, and an overall prediction is the best that can be given. Before any symptoms develop in a person who had a parent with the disease, the probability is ½ that he or she received it from that parent. (All afflicted individuals are assumed to be heterozygous for the dominant allele, and the homozygous dominant genotype is assumed to be lethal at an early stage in embryonic development—a situation quite common for severe dominant defects.) Therefore, before diagnosis is possible, the probability that the person in question will have an affected child is ½ (the probability that the parent in question has the abnormal allele) × ½ (the probability that the child will receive the allele if the parent had it) = ¼. For such rare diseases, a spouse chosen at random is assumed to be homozygous normal unless his or her pedigree indicates the presence of the disease. The probability of ¼ in this case is only an overall chance because once diagnosis of the parent is certain (by age 50), the probability for the child becomes either ½ (if the parent had the disease) or zero (if the parent did not develop the disease).

Fig. 10-10 presents a pedigree of Huntington's chorea in which four sets of children in generation IV (families A, B, C, and D) have different probabilities of being affected, depending on the amount of information available to make the predictions. In family A the situation is clearcut; the mother (III-2) has developed the symptoms of Huntington's chorea, thereby making the probability of being affected ½ for each of her children. In family B the parent at risk is the father (III-6) who does not have the symptoms but at age 33 may still develop them. What is the father's chance of having the abnormal allele? Tracing back to generation II we find that II-3 was at risk (member of the family lineage in which the abnormal allele is being transmitted), but he was killed at age 30 not having shown the symptoms but too young at death for an accurate diagnosis to be made. We can be quite sure that II-3 did possess the mutant allele, however, because his daughter III-4 has the symptoms and must have inherited the disease from her father. Thus, assuming that II-3 had the

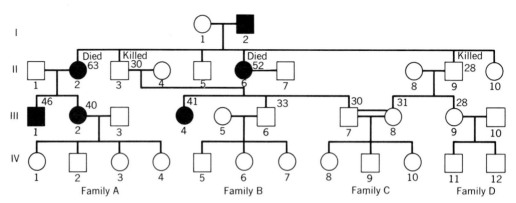

Fig. 10-10. Pedigree illustrating the transmission of an autosomal dominant allele with variable age of onset responsible for Huntington's chorea. Numbers to the upper right of symbols indicate age. (Based on Roderick, G. W.: Man and heredity, New York, 1968, Macmillan and Company.)

abnormal allele, the probability is $\frac{1}{2}$ that III-6 received it and $\frac{1}{2}$ that his children would receive it if he has it, giving $\frac{1}{2} \times \frac{1}{2} = \frac{1}{4}$ as the best overall probability of the children in family B being affected.

Skipping family C for the moment and considering family D, the situation is different again. Here parent III-9 is at risk, but her father, II-9, at risk in generation II was killed at age 28, not having developed any symptoms at the time of death. Contrary to the situation in family B, II-9 produced no children yet diagnosed with the disease; the best we can predict in this case is that the probability was $\frac{1}{2}$ that II-9 received the mutant allele from his father, I-2. There would also be a $\frac{1}{2}$ probability of III-9 receiving the allele if II-9 had it and another $\frac{1}{2}$ chance that the children IV-11 and IV-12 would inherit the allele if III-9 had it. Thus, $\frac{1}{2}$ (probability that II-9 had it) $\times$ $\frac{1}{2}$ (probability that III-9 has it) $\times$ $\frac{1}{2}$ (probability that IV-11 or IV-12 have it) $= \frac{1}{8}$ as the best overall estimate for family D.

Getting back to family C, the situation is complicated by the fact that the parents in generation III represent a consanguineous marriage in which either one or both could have received the abnormal allele through descent in their common lineage from I-2. The children in family C have both parents at risk and could receive the abnormal allele from either parent. Thus, three possibilities exist concerning the genotypes of III-7 and III-8 that could produce affected children in generation IV. Using the symbol *Ht* for the dominant mutant allele and *ht* for the recessive normal allele, we must first determine the probabilities of III-7 and III-8 being *Htht* or *htht*. The probability that III-7 is *Htht* is $\frac{1}{2}$ for the same reason as his brother, III-6, previously discussed. The probability that III-8 is *Htht* is $\frac{1}{4}$ (and $\frac{3}{4}$ for *htht*) for the same reason as her sister, III-9, previously discussed. Table 10-1 lists the three genotypic combinations between III-7 and III-8 that could lead to affected children, the probabilities of these combinations, the probabilities of affected children from these combinations, and the total probability for each case. Finally, since each combination represents a mutually exclusive event and each event may produce affected children, the overall probability is the sum

Table 10-1. Genotypic combinations and probabilities for the production of affected children in family C of Fig. 10-9

Possible parental genotypes III-7 III-8	Probability of parental genotype III-7 III-8	Probability of an affected child (*Ht ht*)	Total probability of an affected child
Ht ht $\times$ *Ht ht*	$\frac{1}{2} \times \frac{1}{4}$	$\times$ $\frac{2}{3}$ $=$	$\frac{2}{24} = \frac{4}{48}$
Ht ht $\times$ *ht ht*	$\frac{1}{2} \times \frac{3}{4}$	$\times$ $\frac{1}{2}$ $=$	$\frac{3}{16} = \frac{9}{48}$
ht ht $\times$ *Ht ht*	$\frac{1}{2} \times \frac{1}{4}$	$\times$ $\frac{1}{2}$ $=$	$\frac{1}{16} = \frac{3}{48}$
	Overall probability (sum of mutually exclusive events)		$\frac{16}{48} = \frac{1}{3}$

of the three possible ways an affected child can be obtained. Note in the first combination in the table that the probability of an *Htht* child is given as $\frac{2}{3}$; this assumes that *HtHt* is lethal in embryonic development and that of the children that may be born two of three would be *Htht* and one of three *htht*. A comparison of families B and D to family C illustrates the effect of consanguinity in increasing the chances of defects in the offspring.

FAMILY PREDICTIONS

After the student has become acquainted with 3:1 ratios for recessive traits or 50:50 chances for sex determination and other genetic probabilities, it often seems confusing that within some families these ratios are not observed or may even be turned around. For example, if two parents are heterozygous for blue eyes and we expect a $\frac{1}{4}$ probability of blue-eyed children and a $\frac{3}{4}$ probability of dark-eyed children, how do we explain the occurrence in some families of three children with blue eyes and only one child with dark eyes or four in a row with blue eyes? Do such observations contradict the principles of genetic transmission? Do families of five children consisting of all boys or all girls threaten the meiotic basis of X and Y chromosome segregation? Actually, the situations described here substantiate rather than contradict the basic principles of heredity because they represent predictable chance combinations of independent events. Whenever there is an either-or situation genetically, for example, blue or dark eyes, normal or albino pigmentation, male or female sex, all combinations of such events within a limited sample size (family) are likely to occur. However, the probability of some combinations will be greater than others; the most likely combination is always that closest to the genetic prediction. The probabilities of the various combinations of simple genetic events are easily calculated algebraically through the binomial expansion. This simple procedure explains many apparent aberrant family combinations.

The binomial expansion is written as $(a + b)^n$ where a and b represent the either-or events and n equals the power of the expansion. In our case the power of the expansion is the number of individuals in the sample, or simply the family size. The family size, n, determines the number of combinations of the two events, a and b, that are possible. These combinations are obtained in an orderly progressive manner by beginning with all a events and no b events and then decreasing a by one and increasing b by one until all b and no a is obtained. For example in a family of four children, if a = boys and b = girls, the expansion of $(a + b)^4$ would describe all possible combinations of boys and girls as $a^4 + a^3b + a^2b^2 + ab^3 + b^4$ (b^0 and a^0 are not written). The middle term, a^2b^2, represents two boys and two girls, the theoretical expectation, and all other combinations are deviations from the expected combination with a^4 (all boys) and b^4 (all girls) representing the greatest deviations. (Note also that the sum of the exponents always equals the family size, n.) It was already mentioned that the genetically predicted combination is the most likely to occur, and logically, the combination representing the greatest deviation would be expected to have a lower probability of occurrence. But just what proportions of each combination should be expected? These proportions are obtained by calculating coefficients for each term in the expansion. These coefficients represent the proportion of each combination expected relative to the most extreme combination (in our example a^4 and b^4). When plotted, the coefficients describe a normal, or bell-shaped, distribution curve. The coefficients can be easily obtained by either of two methods. The coefficient of the first term in any expansion (a^n) is always one. The coefficient of the next term can be calculated by multiplying the coefficient of the preceding term by the exponent of a in that term and dividing by the number of the first term in the serial order of terms (the ordinal number of the first term). Thus in our example of $(a + b)^4$, the first term would have a coefficient of 1. The coefficient of the second term (a^3b) would equal:

$$\frac{\text{Coefficient of preceding term} \times \text{Exponent of a in preceding term}}{\text{Ordinal number of preceding term}} = \frac{4 \times 1}{1} = 4$$

The coefficient of the third term (a^2b^2) would be $\dfrac{4 \times 3}{2} = 6$, that of the fourth term (ab^3) would be $\dfrac{6 \times 2}{3} = 4$, and that of the final term (b^4) would be $\dfrac{4 \times 1}{4} = 1$.

Note that the coefficients form a symmetrical series, 1-4-6-4-1, with the proportion of a^2b^2 combinations expected to be 6 times greater than a^4 or b^4.

Expansions of binomials for n = 1 to 6 are given in Table 10-2; the coefficients can be checked by the above procedure. The coefficients are graphically illustrated by Pascal's triangle (named for the seventeenth century French mathematician and philosopher who was one of the earliest men to deal with the concept of probability) shown in Fig. 10-11, where each coefficient is simply calculated as the sum of the two coefficients immediately above and to the left beginning with 1 as the coefficient for a and b when n = 1. Note that the coefficient of the second and next to last terms in each expansion is always equal to the power of the expansion (n).

Table 10-2. Expansion of the binomials $(a + b)^n$ for n equals 1 to 6

Power of the expansion (n)	Ordinal number of expansion terms						
	1	2	3	4	5	6	7
1	1 a	+ 1 b					
2	1 a^2	+ 2 ab	+ 1 b^2				
3	1 a^3	+ 3 a^2b	+ 3 ab^2	+ 1 b^3			
4	1 a^4	+ 4 a^3b	+ 6 a^2b^2	+ 4 ab^3	+ 1 b^4		
5	1 a^5	+ 5 a^4b	+ 10 a^3b^2	+ 10 a^2b^3	+ 5 ab^4	+ 1 b^5	
6	1 a^6	+ 6 a^5b	+ 15 a^4b^2	+ 20 a^3b^3	+ 15 a^2b^4	+ 6 ab^5	+ 1 b^6

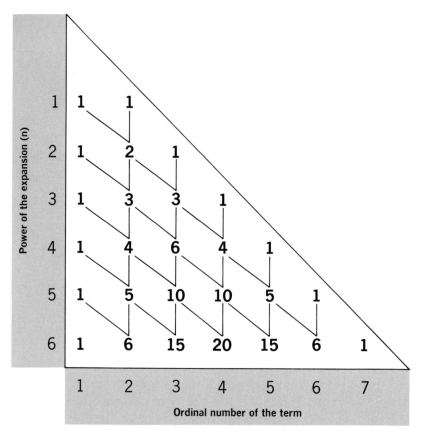

Fig. 10-11. Pascal's triangle illustrating a method of addition for obtaining the coefficients of a binomial expansion.

Once the terms of the expansion and the coefficients are obtained for a certain family size, the probability of any particular combination occurring can be obtained by substituting the proper genetic probability for events a and b. For example, assuming a 50:50 sex ratio, the probability of a boy (a) is $\frac{1}{2}$ and the probability of a girl (b) is $\frac{1}{2}$. Substituting these probabilities for a and b in the proper expansion term and performing the arithmetic produces the probability for the particular term. If we want to know the probability of obtaining the expected ratio of two boys and two girls in a family of four, we find in Table 10-2 under n = 4 that the complete expansion term that describes this combination is $6\,a^2b^2$ (remember, the coefficient is important because it gives the relative proportion for the term). Substituting the genetic probabilities, the term becomes $6(\frac{1}{2})^2(\frac{1}{2})^2 = 6(\frac{1}{4})(\frac{1}{4}) = \frac{6}{16}$. A probability of $\frac{6}{16}$ for a particular combination in a binomial expansion means that for every 16 families of 4 children, 6 families are expected to consist of 2 boys and 2 girls; the rest will be of different combinations. Thus, not even half of all 4-children families are expected to meet the 50:50 sex ratio.

If we asked how likely it is to have 5 girls and 1 boy in a family of 6, the proper expansion term, under n = 6, is $6ab^5$ which, by substituting the genetic probabilities of the events, gives $6(\frac{1}{2})(\frac{1}{2})^5 = \frac{6}{64}$ as the probability for that combination. This means that for every 64 families of 6 children, only 6 are expected to have 5 girls and 1 boy. On the other hand, $20a^3b^3 = 20(\frac{1}{2})^3(\frac{1}{2})^3 = \frac{20}{64}$ are expected to meet the 50:50 ratio.

In many cases the probabilities of events a and b do not equal $\frac{1}{2}$. Different genetic probabilities for the events obviously produce different probability statements for the various combinations of those events. However the procedure in the use of the binomial expansion remains the same. For example, if two parents are carriers for albinism (autosomal recessive trait), the probability of a normally pigmented child is $\frac{3}{4}$ and that of an albino child is $\frac{1}{4}$. Calling normal pigmentation event a and albinism event b, we might ask what is the probability of producing the expected 3 normal : 1 albino ratio in a family of 4 children? The proper binomial expansion term in this case is $4\,a^3b$ (under n = 4 in Table 10-2). When we substitute the proper genetic probabilities of $\frac{3}{4}$ for event a and $\frac{1}{4}$ for event b, this expression becomes $4(\frac{3}{4})^3(\frac{1}{4}) = 4\,(\frac{27}{64})(\frac{1}{4}) = \frac{108}{256}$. This means that 108 of every 256 families of 4 children produced by parents who are both heterozygous for albinism would be expected to consist of 3 normally pigmented children and 1 albino child. It should be apparent that in this case it would not be a very great deviation from the expected 3 normal : 1 albino if all 4 were normal. But it would be a very great deviation if all 4 children were to be albino. This is verified by the predictions from the binomial as follows. In the first case the probability of 4 normal children is derived from the term $1a^4$ by substituting $\frac{3}{4}$, the genetic probability for normal pigmentation in this cross, giving $1(\frac{3}{4})^4 = \frac{81}{256}$ as the probability for this combination, while the probability of 4 albino children is derived from the term $1b^4$ by substituting $\frac{1}{4}$, the genetic probability for an albino child in this cross, giving $1(\frac{1}{4})^4 = \frac{1}{256}$ as the probability for this combination. One of 256 is a quite low probability, whereas 81 of 256 is much more likely and not greatly lower than the probability of 108 of 256, which is the expectation for the "perfect" 3:1 ratio.

The expression $4ab^3 = 4(\frac{3}{4})(\frac{1}{4})^3 = \frac{12}{256}$ tells us that in 12 of every 256 families of the type in question the 3:1 ratio can actually be expected to be turned around to 1 normal child and 3 albinos.

An alternative to expanding the entire binomial when only the probability of one particular combination is desired is provided by the formula

$$\frac{n!}{s!\ t!}\ (p)^s\ (q)^t$$

where

    n = the power of the expansion (family size)
    p = the genetic probability of event a
    s = the number of times event a occurs
    q = the genetic probability of event b
    t = the number of times event b occurs.

Note that the symbol n! (n factorial) means that all whole numbers from n down to 1 are serially multiplied together to give n!. This holds true for whatever number precedes the ! symbol. Also, O! $= 1$ and p or q to the zero power $= 1$. Actually $\frac{n!}{s!t!}$ is simply the generalized formula for obtaining any particular coefficient, and $(p)^s (q)^t$ is the generalized formula for any specific combination of events a and b, for example, a^3b^2. To illustrate the function of the formula, let us apply it directly to the last example of predicting the probability of 1 normally pigmented child : 3 albino children, when both parents are carriers for albinism. The pertinent data are:

Event a $=$ normal pigmentation
p $=$ genetic probability of a $= \frac{3}{4}$
s $=$ number of times event a occurs $= 1$
Event b $=$ albinism
q $=$ genetic probability of b $= \frac{1}{4}$
t $=$ number of times event b occurs $= 3$
n $=$ family size $= 4$

Thus,

$$\frac{n!}{s!t!} (p)^s (q)^t = \frac{4\cdot3\cdot2\cdot1}{1\cdot3\cdot2\cdot1} (\tfrac{3}{4})^1 (\tfrac{1}{4})^3 = \tfrac{4}{1} \cdot \tfrac{3}{4} \cdot \tfrac{1}{64} = \tfrac{12}{256}$$

The formula can also be expanded to cover 2 or more genetic factors at one time. For example, if we consider albinism and sex together, we might ask the probability of a family of 8 consisting of 3 boys with normal pigmentation : 1 albino boy : 3 girls with normal pigmentation : 1 albino girl, from parents who are both heterozygous for albinism. This would be the "perfect" ratio for these two traits together. To extend the formula, we simply add the proper number of events, their frequencies of occurrence, and their genetic probabilities. In this case the formula will be:

$$\frac{n!}{s!\, t!\, u!\, v!} (p)^s (q)^t (x)^u (y)^v$$

Where:

Event a $=$ a boy with normal pigmentation
p $=$ genetic probability of a $= \frac{1}{2}$ (sex probability) $\times \frac{3}{4}$ (trait probability) $= \frac{3}{8}$
s $=$ number of times event a occurs $= 3$
Event b $=$ an albino boy
q $=$ genetic probability of b $= \frac{1}{2} \times \frac{1}{4} = \frac{1}{8}$
t $=$ number of times event b occurs $= 1$
Event c $=$ a girl with normal pigmentation
x $=$ genetic probability of c $= \frac{1}{2} \times \frac{3}{4} = \frac{3}{8}$
u $=$ number of times event a occurs $= 3$
Event d $=$ an albino girl
y $=$ genetic probability of d $= \frac{1}{2} \times \frac{1}{4} = \frac{1}{8}$
v $=$ number of times event d occurs $= 1$
n $=$ family size $= 8$

Thus,

$$\frac{n!}{s!t!u!v!} (p)^s (q)^t (x)^u (y)^v = \frac{8\cdot7\cdot6\cdot5\cdot4\cdot3\cdot2\cdot1}{3\cdot2\cdot1\cdot1\cdot3\cdot2\cdot1\cdot1} (\tfrac{3}{8})^3 (\tfrac{1}{8})^1 (\tfrac{3}{8})^3 (\tfrac{1}{8})^1 =$$

$$\frac{1120}{1} \cdot \frac{27}{512} \cdot \frac{1}{8} \cdot \frac{27}{512} \cdot \frac{1}{8} = \frac{816,480}{16,777,216} \approx \frac{1}{20}$$

This tells us that about 1 out of every 20 families that fit these criteria would consist of the theoretical genetic ratio. The vast majority would deviate to a greater or lesser extent, resulting in literally every combination possible.

Understanding the implications of chance combinations within family groups, as described by the binomial expansion, explains why most families do not display the theoretical genetic ratios predicted by the various modes of hereditary transmission. One should recognize that genetic ratios are averages for the whole population of families conforming to a certain set of criteria (large sample size) and that chance combinations within very small samples of that population (individual families) occur at predictable frequencies (according to the binomial expansion). Thus, the existence of seemingly aberrant and contradictory family combinations can be appreciated as normal components of a larger genetic system.

In Chapter 13 we shall move beyond specifically defined family groups, such as all families of 5 children with heterozygous parents, and so forth, to a consideration of gene action within an entire population. Before we reach the populational level, however, we must consider one more very important mode of hereditary transmission, polygenic traits (Chapter 11). We shall also study a significant set of abnormalities at the familial level caused by mutational events concerning chromosome number and structure (Chapter 12).

11

Complex characteristics and complex problems

Many important behavioral characteristics and physical attributes have a genetic basis that is more complex than the modes of inheritance previously discussed and which is not conducive to simple pedigree analysis. These complex characteristics show continuous variation within the population rather than sharply contrasting qualitative alternatives in phenotype. This continuous variation results from the action of several loci influencing the expression of a particular trait, often coupled with a significant environmental interaction. Characteristics such as stature, skin color, and intelligence are controlled by this mode of inheritance known as *quantitative inheritance* (as opposed to qualitative inheritance) or *polygenic inheritance* (many genes). Polygenic characteristics are much more difficult to study and analyze than the qualitative traits we have previously discussed. And since sophisticated statistical techniques beyond the scope of this text are often required, the objective of this chapter will be to simply introduce this important mode of inheritance and to discuss a few specific examples in man. It is hoped that an appreciation of this genetic mechanism and the so-called nature-nurture problem related to such important characteristics as human intelligence can be achieved.

THE MECHANISM OF POLYGENIC INHERITANCE

When a number of loci control the expression of a particular characteristic, the Mendelian principles of segregation and independent assortment of alleles predict a continuous distribution of phenotypes, especially if the following conditions (assumptions) prevail:

1. Each locus consists of an active and inactive allele with respect to contributing to the expression of the phenotype.
2. The alleles at each locus are codominant.
3. The effect of the active allele at each locus is small and equal to that of

232

every other active allele in the system, and the effect of all alleles on the phenotype is cumulative in a linear manner.

4. The loci assort independently, that is, they are not linked.

These assumptions allow the simplest interpretation of polygenic inheritance. If dominance and recessiveness exist among alleles, or if the effects are not equal among alleles, or if some alleles are linked, analysis and interpretation becomes more complex and difficult. We will limit our discussion to a context in which the above conditions are fulfilled.

Let us oversimplify and assume that stature (height) is controlled by three loci and that each locus has an active allele that adds 3 inches to the potential for height over and above a basic or fundamental height of 4'9" achieved by normal development without the action of any "stature" alleles. If we symbolize the stature gene as S, with S representing the active form and s the inactive form, the possible genotypic combinations of the alleles at three loci in the diploid state would range from *ss ss ss* (no active alleles) giving the shortest normal stature possible (the basic 4'9") to *SS SS SS* (all active alleles) giving the tallest normal stature possible (6'3" in our model; $6 \times 3"$ per active allele beyond the basic 4'9"). In reality there may be dozens of loci with active alleles contributing only a fraction of an inch to stature potential and with the range of height extended downward and upward. It should also be noted that poor nutrition (suboptimal environment) can prevent the genetic potential from being reached, whereas proper nutrition (optimal environment) can not extend stature beyond the maximum genetic potential. This environmental interaction obviously blurs the full phenotypic expression of the genotype so that phenotypic observations may not be a true reflection of the genotype. Our model will assume optimal and equal environmental factors and equal frequencies of active and inactive alleles at each locus within the population as a whole; thus, the probability of S or s in the gene pool is $\frac{1}{2}$ in each case. The binomial expansion discussed in the previous chapter illustrates the concept of continuous variation for polygenic traits.

In our three-locus model each diploid genotype would consist of six alleles. If event a of the binomial is considered to represent the occurrence of S alleles, and event b the occurrence of s alleles, then the expansion of the binomial $(a+b)^6$ describes the possible genotypic combinations of S and s, and the coefficient of each term indicates its relative frequency in the distribution of genotypes (see Table 10-2). Fig. 11-1 represents the plot of these relative genotype frequencies, depicting a normal, bell-shaped distribution. Because of the effect of environmental interactions, each phenotype in this distribution gradually merges into the next, producing continuous variation for stature, rather than just tall, medium, and short heights. The most frequent phenotypes would be those nearest the average for the system (5'6" in our model); the frequencies decline as the extremes at each end of the range are approached. If the frequencies of the alternate alleles in the system were not equal, the phenotypic distribution would become skewed in the direction of the more frequent allelic forms. Note also in Fig. 11-1 that the genotypes are only described according to the number of S and s alleles present and not their precise arrangement by locus and chromosome pair. (The alleles of three independently assorting loci would occupy three pairs of chromosomes.)

Although it is the cumulative *number* of active and inactive alleles that determines phenotypic potential for the individual, it is the *arrangement* of the alleles on specific chromosome pairs that influences the genetic transmission of the trait. If the identification of "stature" genes in our model is refined so that S_1 and s_1 represent alleles at one locus, S_2 and s_2 represent alleles at another locus, and S_3 and s_3 represent alleles at a third locus, a normal diploid genotype would specifically consist of a double dose of each locus *(11 22 33)* and never some other mixture. This is equivalent to considering three loci located on any three autosomal chromosome pairs undergoing typical meiotic segregation and independent assortment. For example, the phenotypic potential (height) controlled by four active and two inactive alleles (5′9″) could be produced by any one of six genotypes, listed in Table 11-1.

These would form different kinds of gametes (haploid genotypes) during meiosis and they would, as a result, have the possibility of producing different genotypic (and phenotypic) ratios in their offspring. Fig. 11-2 illustrates this point by depicting three crosses (*A, B,* and *C*) between parents who both have a 5′9″ phenotypic potential (genotype with four active and two inactive alleles), but whose specific genotypes, that is, actual allelic arrangements, differ.

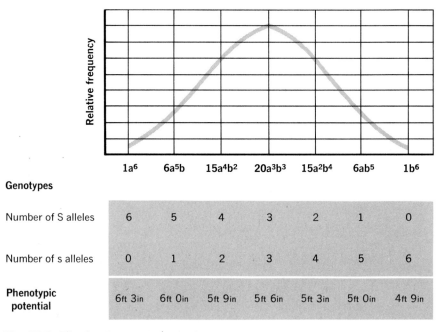

Genotypes	$1a^6$	$6a^5b$	$15a^4b^2$	$20a^3b^3$	$15a^2b^4$	$6ab^5$	$1b^6$
Number of S alleles	6	5	4	3	2	1	0
Number of s alleles	0	1	2	3	4	5	6
Phenotypic potential	6ft 3in	6ft 0in	5ft 9in	5ft 6in	5ft 3in	5ft 0in	4ft 9in

Fig. 11-1. The distribution of "stature" alleles in a three-locus (six allele) model with the potential phenotype effect of each active allele *(S)* an increase of 3 inches in height.

Note that the types and probabilities of offspring are different in each case and that the range of tallest to shortest also varies, depending on the particular genotypic combination of the parents. Since hormonal influence is known to be an important environmental factor that affects stature differentially between the sexes, phenotypic potentials in females are consistently expressed at lower levels (shorter actual height) then comparable potentials in males. However, this developmental difference in no way affects genetic transmission of the "stature" alleles and can be ignored in our simplified model.

The three different sets of progeny predictions in Fig. 11-2 resulting from phenotypically similar parental combinations (all 5′9″) point up one of the major problems in making genetic predictions involving polygenic traits. Even if the phenotypes of the parents provided a perfect reflection of their genotypes (without the blurring of environmental influences), accurate progeny predictions could not be made because there is no way to determine the *specific allelic arrangement* in the genotype. Therefore there is no way of predicting the types of gametes and zygotic combinations possible. In other words, there is no way of distinguishing situations such as *A*, *B*, and *C* in Fig. 11-2 from one another, and this is the critical information necessary to calculate accurate probabilities. This problem can be reiterated by considering the model for another polygenic characteristic—*skin color*.

The amount of melanin pigment distributed in the skin determines its color. Just as there is a basic level of stature achieved through the normal development process which then may be augmented by active "stature" alleles to increase height, there is a basic level of melanin production. This process produces the minimal level of pigmentation (distinct from albinism that is lack of pigment due to a block of any pigment production) which then may be augmented by active "pigment" alleles that increase melanin deposition to produce progressively darker skin colors according to the number of active alleles present. Current estimates range from four to six as the number of loci involved in the determination of skin color.

We shall use a four-locus model in our consideration of the characteristic, with *P* symbolizing active pigment-increasing alleles and *p* representing the inactive forms. Thus, the lightest (nonalbino) skin color genotype would be

Table 11-1. The 6 different genotypes having 4 active and 2 inactive "stature" alleles that will produce different kinds of gametes (haploid genotypes) when meiosis takes place

1. S_1S_1	S_2S_2	s_3s_3
2. S_1S_1	s_2s_2	S_3S_3
3. s_1s_1	S_2S_2	S_3S_3
4. S_1s_1	S_2s_2	S_3S_3
5. S_1s_1	S_2S_2	S_3s_3
6. S_1S_1	S_2s_2	S_3s_3

A $S_1S_1 \ S_2S_2 \ s_3s_3 \times s_1s_1 \ S_2S_2 \ S_3S_3$

Gametes

all $S_1S_2s_3$

Gametes

all $s_1S_2S_3$

Offspring

Genotypes

all $S_1s_1 \ S_2S_2 \ S_3s_3$

Phenotypic range

5ft 9in

B $S_1S_1 \ S_2S_2 \ s_3s_3 \times S_1S_1 \ S_2s_2 \ S_3s_3$

Gametes

all $S_1S_2s_3$

Gametes

¼ $S_1S_2S_3$

¼ $S_1S_2s_3$

¼ $S_1s_2S_3$

¼ $S_1s_2s_3$

Offspring

Genotypes

¼ $S_1S_1 \ S_2S_2 \ S_3s_3$

¼ $S_1S_1 \ S_2S_2 \ s_3s_3$

¼ $S_1S_1 \ S_2s_2 \ S_3s_3$

¼ $S_1S_1 \ S_2s_2 \ s_3s_3$

Phenotypic range

6ft 0in

5ft 9in

5ft 9in

5ft 6in

Fig. 11-2. Three crosses between parents with potential phenotypes of 5'9" (four active and two inactive "stature" alleles), illustrating the different outcomes possible among their offspring depending on the particular genotype combination involved.

C $S_1S_1\ S_2s_2\ S_3s_3 \times S_1s_1\ S_2s_2\ S_3S_3$

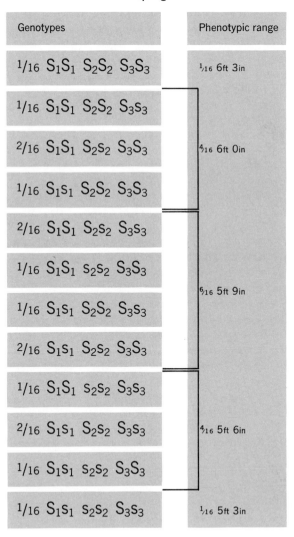

Gametes

1/4 $S_1S_2S_3$
1/4 $S_1S_2s_3$
1/4 $S_1s_2S_3$
1/4 $S_1s_2s_3$

Gametes

1/4 $S_1S_2S_3$
1/4 $S_1s_2S_3$
1/4 $s_1S_2S_3$
1/4 $s_1s_2S_3$

Offspring

Genotypes	Phenotypic range
1/16 $S_1S_1\ S_2S_2\ S_3S_3$	1/16 6ft 3in
1/16 $S_1S_1\ S_2S_2\ S_3s_3$	4/16 6ft 0in
2/16 $S_1S_1\ S_2s_2\ S_3S_3$	
1/16 $S_1s_1\ S_2S_2\ S_3S_3$	
2/16 $S_1S_1\ S_2s_2\ S_3s_3$	6/16 5ft 9in
1/16 $S_1S_1\ s_2s_2\ S_3S_3$	
1/16 $S_1s_1\ S_2S_2\ S_3s_3$	
2/16 $S_1s_1\ S_2s_2\ S_3S_3$	
1/16 $S_1S_1\ s_2s_2\ S_3s_3$	4/16 5ft 6in
2/16 $S_1s_1\ S_2s_2\ S_3s_3$	
1/16 $S_1s_1\ s_2s_2\ S_3S_3$	
1/16 $S_1s_1\ s_2s_2\ S_3s_3$	1/16 5ft 3in

Fig. 11-2, cont'd. For legend see opposite page.

$p_1p_1\ p_2p_2\ p_3p_3\ p_4p_4$ and the darkest skin color genotype would be $P_1P_1\ P_2P_2$ $P_3P_3\ P_4P_4$. In a cross between individuals of the lightest and darkest skin colors all offspring would be of intermediate color consisting of four active and four inactive alleles arranged as a quadruple heterozygote, $P_1p_1\ P_2p_2\ P_3p_3\ P_4p_4$.

This specific genotype formed by the gametes of fully homozygous parents of the lightest and darkest genotypes is technically known as a *mulatto*. Intermarriage of mulattos is the *only* parental combination involving skin colors other than the lightest or darkest for which genetic predictions can be accurately made. In this case the entire phenotypic range of skin colors is possible among the progeny in expected frequencies that form a normal distribution as shown in Fig. 11-3 (derived from expansion of the binomial $[a+b]^8$). Note that most children would be expected to be similar to or only slightly lighter or darker than the intermediate mulatto parents. The children of mulattos more generally represent all the genotypes and skin colors possible in a randomly breeding population with equal frequencies of each allele in the population as a whole.

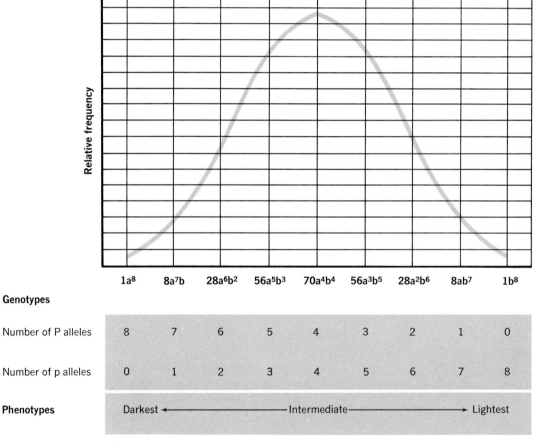

Genotypes	$1a^8$	$8a^7b$	$28a^6b^2$	$56a^5b^3$	$70a^4b^4$	$56a^3b^5$	$28a^2b^6$	$8ab^7$	$1b^8$
Number of P alleles	8	7	6	5	4	3	2	1	0
Number of p alleles	0	1	2	3	4	5	6	7	8
Phenotypes	Darkest ←				Intermediate			→	Lightest

Fig. 11-3. The distribution of skin color genotypes expected from a cross involving two mulattos.

Any crosses involving these children present the same problem regarding genetic predictions as that illustrated in Fig. 11-2 for stature. In other words without knowledge of the specific allelic arrangement within any particular genotypic category (4 active, 4 inactive, and so forth), no progeny predictions or even phenotypic ranges among progeny can be ascertained for any particular parental combination. At best only average expectations for all similar combinations within an entire, large population (particular race, for example) might be made. This possibility will be pursued later. But first let us consider one more illustration of the complex problem of progeny predictions associated with the polygenic mode of inheritance.

If a "moderately dark-skinned" person (5 active and 3 inactive alleles) married a "moderately light-skinned" person (3 active and 5 inactive alleles), each might have a genotype consisting of any one of the 16 different allelic arrangements listed in Table 11-2. These combinations would produce different arrays of gametes during meiosis. Fig. 11-4 illustrates the outcome of 2 of the possible 256 (16×16) specific genotypic combinations that could constitute the mating between these two individuals. Cross *A* represents a genotypic combination that produces the smallest phenotypic range of offspring possible between parents of the phenotypes described. Note that half of the progeny are expected

Table 11-2. The 16 different specific allelic arrangements possible in the genotypes of *A* individuals with 5 active and 3 inactive alleles and *B* individuals with 3 active and 5 inactive alleles

A Possible allelic arrangements in the genotype with 5 active and 3 inactive alleles				B Possible allelic arrangements in the genotype with 3 active and 5 inactive alleles			
*P_1P_1	P_2p_2	P_3p_3	P_4p_4	P_1p_1	P_2p_2	P_3p_3	p_4p_4
P_1p_1	P_2P_2	P_3p_3	P_4p_4	P_1p_1	P_2p_2	p_3p_3	P_4p_4
P_1p_1	P_2p_2	P_3P_3	P_4p_4	P_1p_1	p_2p_2	P_3p_3	P_4p_4
P_1p_1	P_2p_2	P_3p_3	P_4P_4	*p_1p_1	P_2p_2	P_3p_3	P_4p_4
P_1P_1	P_2P_2	P_3p_3	p_4p_4	P_1P_1	P_2p_2	p_3p_3	p_4p_4
†P_1P_1	P_2P_2	p_3p_3	P_4p_4	P_1P_1	p_2p_2	P_3p_3	p_4p_4
P_1P_1	P_2p_2	P_3P_3	p_4p_4	†P_1P_1	p_2p_2	p_3p_3	P_4p_4
P_1P_1	p_2p_2	P_3P_3	P_4p_4				
P_1P_1	P_2p_2	p_3p_3	P_4P_4	P_1p_1	P_2P_2	p_3p_3	p_4p_4
P_1P_1	p_2p_2	P_3p_3	P_4P_4	p_1p_1	P_2P_2	P_3p_3	p_4p_4
				p_1p_1	P_2P_2	p_3p_3	P_4p_4
P_1p_1	P_2P_2	P_3P_3	p_4p_4				
p_1p_1	P_2P_2	P_3P_3	P_4p_4	P_1p_1	p_2p_2	P_3P_3	p_4p_4
P_1p_1	P_2P_2	p_3p_3	P_4P_4	p_1p_1	P_2p_2	P_3P_3	p_4p_4
p_1p_1	P_2P_2	P_3p_3	P_4P_4	p_1p_1	p_2p_2	P_3P_3	P_4p_4
P_1p_1	p_2p_2	P_3P_3	P_4P_4	P_1p_1	p_2p_2	p_3p_3	P_4P_4
p_1p_1	P_2p_2	P_3P_3	P_4P_4	p_1p_1	P_2p_2	p_3p_3	P_4P_4
				p_1p_1	p_2p_2	P_3p_3	P_4P_4

*Indicates genotypes used in Fig. 11-4 cross B.
†Indicates genotypes used in Fig. 11-4 cross A.

to be intermediate and that the range of skin colors among the progeny would not exceed that of the parents in either direction. On the other hand, cross *B* represents a genotypic combination that produces the widest phenotypic range possible between parents of the phenotypes under consideration. Note especially that in this case only $^2\!\!/_{64}$, or less than a third, of the progeny are expected to be intermediate (compared to $\frac{1}{2}$ in cross *A*). Notice also that the range of skin colors among the progeny transcends significantly that of the parents in both directions with a total of $^7\!\!/_{64}$ (about $\frac{1}{8}$) expected to be darker than the darkest-skinned parent and a like number expected to be lighter than the lightest-skinned parent. These two examples should make it apparent that no generalizations about the skin color of children can accurately be made when the parental combination involves genotypes other than the lightest, the darkest, or true mulattos.

The variation in skin color within the human species encompasses the entire phenotypic range; within several races, including the Negro and Caucasian races, the range of variation is very broad. But we might ask how differences in skin color characterizing some races as "dark" and others as "light" can be explained within the context of a polygenic system. The concept of race will be discussed in more detail in Chapter 13. For the present discussion suffice it to note that the major distinction between races concerns differences

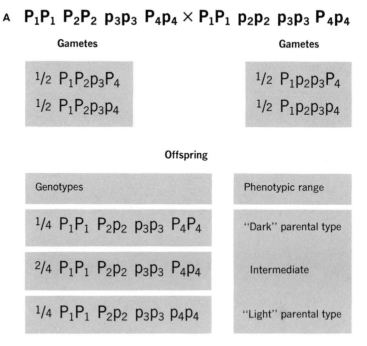

Fig. 11-4. The genetic predictions from two of the possible 256 genotypic combinations in the marriage of a "moderately dark-skinned" person to a "moderately light-skinned" person, illustrating **A,** the smallest and **B,** the widest phenotypic ranges among the expected offspring. (See Table 12-2 and text for further information.)

in the *frequency* of various alleles. Recall that our idealized examples have considered allelic frequencies to be equal within the population as a whole (50:50 at each locus), thereby generating a bell-shaped, symmetrical distribution of phenotypes. However, it was also mentioned that if the allele frequencies are not equal, the distribution will be skewed in the direction of the phenotypes produced by the more frequent allelic forms. The most important factor affect-

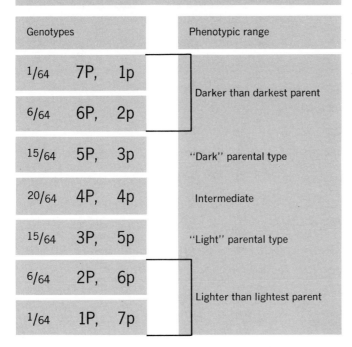

B P_1P_1 P_2p_2 P_3p_3 P_4p_4 $\times$ p_1p_1 P_2p_2 P_3p_3 P_4p_4

Gametes

Eight different kinds with ⅛ probability each ranging from

$P_1P_2P_3P_4$ to $P_1p_2p_3p_4$

Gametes

Eight different kinds with ⅛ probability each ranging from

$p_1P_2P_3P_4$ to $p_1p_2p_3p_4$

Offspring

64 combinations of gametes (8 X 8) producing 27 different allelic arrangements, which can be generated by a genotypic tree diagram. Listed below are the "general" genotypes according to the number of active and inactive alleles present without regard to their specific arrangement—this provides phenotypic range.

Genotypes		Phenotypic range
1/64	7P, 1p	Darker than darkest parent
6/64	6P, 2p	
15/64	5P, 3p	"Dark" parental type
20/64	4P, 4p	Intermediate
15/64	3P, 5p	"Light" parental type
6/64	2P, 6p	Lighter than lightest parent
1/64	1P, 7p	

Fig. 11-4, cont'd. For legend see opposite page.

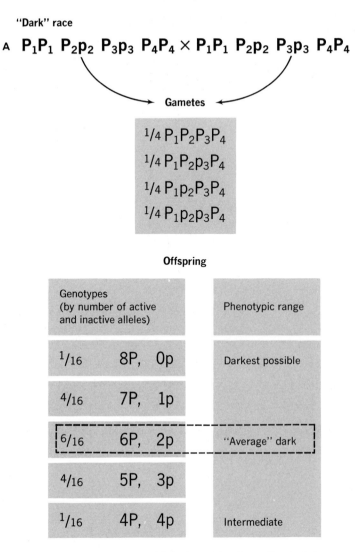

Fig. 11-5. Representative crosses of established **A,** "dark" and **B,** "light" races resulting from the fixation of allelic frequencies and some loci (Loci 1 and 4), presumably through natural selection. (The allelic arrangement for each "general" genotype listed can be generated by a genotypic tree diagram.) The most commonly expected genotype and phenotype are outlined; the other types form a normal distribution in either direction.

ing allelic frequencies within any particular population (race) is natural selection. Natural selection favored dark skin in tropical areas as protection against constant exposure to the sun. Light-skinned individuals become "tan" when exposed to the sun as a physiological adaptation of the body for protection; this supports the selective advantage of darker skin in sunny environments. Thus, as pigment-inducing alleles increase in frequency, some loci might consist of nearly all active alleles, or by chance some loci might become fixed and contain *all* active alleles with only occasional mutations producing the inactive form at that particular locus. (The reverse is true where light skins, inactive

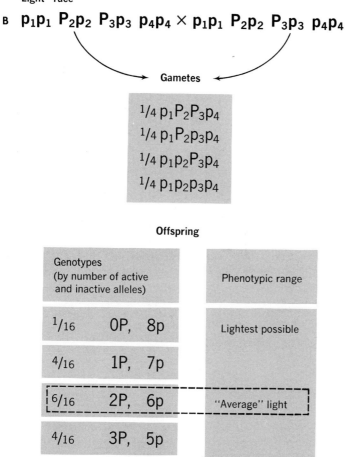

"Light" race

B p_1p_1 P_2p_2 P_3p_3 p_4p_4 × p_1p_1 P_2p_2 P_3p_3 p_4p_4

Gametes

1/4 $p_1P_2P_3p_4$
1/4 $p_1P_2p_3p_4$
1/4 $p_1p_2P_3p_4$
1/4 $p_1p_2p_3p_4$

Offspring

Genotypes (by number of active and inactive alleles)		Phenotypic range
1/16	0P, 8p	Lightest possible
4/16	1P, 7p	
6/16	2P, 6p	"Average" light
4/16	3P, 5p	
1/16	4P, 4p	Intermediate

Fig. 11-5, cont'd. For legend see opposite page.

alleles, are selected.) These frequency differences skew the phenotypic distributions within populations. When two populations have been geographically isolated for a long period of time with different selection pressures operating on an obvious characteristic like skin color, the distinctions that evolve define the categories as races. Most racial differences are subtle, and all are a matter of frequency and distribution, not of entirely new or different genetic systems (see Chapter 13 for further discussion). In Fig. 11-5 two crosses are presented as representative of *A*, a "dark" race, and *B*, a "light" race established through the fixation of active or inactive pigment-increasing alleles and the subsequent restriction in phenotypic range produced by unequal allelic frequencies. Note that all marriage combinations between progeny within a "race" would still tend to produce offspring with a phenotypic range characteristic of the "race" because of the fixed alleles that ensure certain limits. However, any crossmating ("interracial" marriages) would "hybridize" the fixed limits and restore a broader phenotypic range in future generations. This latter situation is illustrated in Fig. 11-6. (Both Figs. 11-5 and 11-6 are selected examples that exaggerate the situation for emphasis.) If the entire human species were to become an essentially single random breeding population, the complete range of skin colors would undoubtedly be maintained indefinitely, except that it would no longer be a "racial" characteristic—it would simply be another highly variable polygenic trait in our species. Only if selection pressures were re-established in one direction or the other would this variability be lost in a single cosmopolitan human population.

Following is a list of some other human characteristics that are transmitted through the polygenic mode of inheritance.

1. Anencephaly (absence of brain)
2. Blood pressure
3. Cleft palate (talipes equinovarus)
4. Club foot
5. Congenital dislocated hip
6. Diabetes mellitus
7. Harelip
8. Fingerprints (ridge count)
9. Pyloric stenosis (constricted stomach valve)
10. Spina bifida (cleft spine)
11. Intelligence

The last of these, intelligence, will be discussed in some detail in the remainder of this chapter first, because it is a very important characteristic subject to all the problems previously discussed concerning genetic predictions involving polygenic traits and secondly, because the confounding effect of environmental factors, the "nature-nurture" problem, is exemplified in this characteristic.

INTELLIGENCE: THE NATURE-NURTURE PROBLEM

It is quite well established that there is an important hereditary component to intelligence potential which interacts significantly with an individual's entire environmental milieu to produce some degree of realization of the potential. This end product is what we label as an individual's intelligence and what we attempt to measure with various intelligence tests. The influence of environmental factors on the achievement of some particular intelligence level within a genetically predetermined range has produced a host of problems and controversies concerning the very characteristics responsible for the "success" of the human species, including the capability for us to discuss the subject in this text!

Intelligence is a composite characteristic produced from the combination and interaction of many traits and capacities such as abstract thinking, memory, mathematical ability, and language acuity, along with "intuition," "creativity," and "common sense." These various components of intelligence are developed

P_1P_1 P_2p_2 P_3p_3 P_4P_4 $\times$ p_1p_1 P_2p_2 P_3p_3 p_4p_4

Gametes: same as Figure 11-5 (A) and (B)

Offspring

Genotypes (by number of active and inactive alleles)	Phenotypic range
$1/16$ 6P, 2p	"Average" dark
$4/16$ 5P, 3p	
$6/16$ 4P, 4p	Intermediate
$4/16$ 3P, 5p	
$1/16$ 2P, 6p	"Average" light

Future generations (assuming random intermarriages of offspring above)

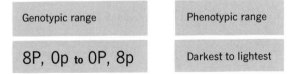

Genotypic range	Phenotypic range
8P, 0p to 0P, 8p	Darkest to lightest

Fig. 11-6. Example of the restoration of phenotypic variability following "interracial" marriages. The most commonly expected genotype and phenotype are outlined; compare this to Fig. 11-5.

to greater or lesser degrees somewhat independently so that certain individuals seem to be very "smart" in some ways but quite "dumb" in others, for example, the mathematical "wizard" who has no "common sense"! There is no doubt that a great many genes contribute to the human intelligence potential. In fact, each component of intelligence is probably controlled by a polygenic system in which the final product in the individual represents an hierarchical polygenic complex consisting of genes that govern brain structure and function, sensory organ functions, and a myriad of particular processes related to general and specific mental traits and abilities.

Most objective data on intelligence are available in terms of *intelligence quotient* (IQ) which is conventionally obtained by dividing mental age (as determined by a standardized test) by chronological age, and multiplying the quotient by 100, thus making a score of 100 normal by definition. Later in this chapter we will consider the problems related to the use of IQ scores for genetic interpretations about *specific* individuals, but first we shall consider several lines of evidence that support the genetic basis for intelligence potential.

Since pedigree analyses are not conducive to the study of complex polygenic traits that are subject to strong environmental interactions, other techniques must be used to gain genetic data. Some of the most fruitful data on intelligence have come from *twin studies*. Twins occur about once in every 100 births and are sometimes called experiments in nature because they produce a basis for comparison not available in most studies of human heredity. *Identical*, or *monozygotic* (MZ) twins are produced by the division and separation of a single fertilized egg during a very early embryonic stage. (Siamese twins result when separation is not complete.) Monozygotic twins are genotypically identical and phenotypically very similar; they are always the same sex. On the other hand, *fraternal*, or *dizygotic* (DZ) twins are produced when two separate eggs (multiple ovulation) are fertilized by two different sperms and each zygote successfully implants and completes development. Such twin pairs are genotypically and phenotypically no more similar than any two *sibs*, or *siblings* (offspring of the same parents); they may be of the same or opposite sexes. About two of every three twin births are dizygotic. Multiple births of three or more are quite infrequent and may result from a single fertilized egg (identicals), multiple ovulation (fraternals), or a combination of identical and fraternal siblings. There is some evidence that dizygotic twinning may have a genetic component in some way related to causing multiple ovulation, whereas monozygotic twinning appears to result from chance developmental accidents.

Twins provide controls over the two important variables in the nature-nurture problem—heredity and environment. In identical twins heredity is held constant (identical genotypes), and nearly all variation can be considered environmental, whereas fraternal twins represent hereditary variation with environment held relatively constant. Twins reared apart provide an additional degree of environmental variation to compare to those reared together. Twin groups can also be compared to normal sib pairs in which age differences are another environmental factor and to parent-child, parent–adoptive child, and unrelated pair groups, all of which represent different combinations of genetic and environmental relationships.

It was pointed out in the introduction to this chapter that rather sophisticated statistical techniques, beyond the scope of this text, are often needed to analyze polygenic inheritance. Some of the statistics that provide valuable basic information concerning the role of heredity in intelligence are not highly complex procedures and may be found in any introductory statistics textbook. We shall consider the analysis of IQ scores by a few of these procedures which will be defined for interpretation but not derived and formulated mathematically.

Table 11-3. Average pair differences for Binet IQ scores among 4 categories of pair groups

Identical twins reared together (50 pair)	Identical twins reared apart (19 pair)	Fraternal twins reared together (52 pair)	Paired sibs (like sexes) reared together (47 pair)
5.9	8.2	9.9	9.8

(Data from Newman, Freeman, and Holzinger: Twins: a Study of heredity and environment, Chicago, 1937, University of Chicago Press.)

The simplest statistic to shed light on the nature-nurture relationship to intelligence is the average difference in IQ scores for various categories of paired individuals. Table 11-3 presents average pair differences for Binet IQ scores (Binet is the name of one standarized IQ test) among four categories. Three important points are apparent from these data:

1. The average difference between identical twins reared together is significantly less than that of any other paired combination.
2. When identical twins are reared apart, the average difference between them increases as an effect directly attributable to environmental influences.
3. There is no significant difference between the scores of fraternal twins and those of any paired sibs of like sex, and the variation between fraternal twins reared together is still greater than that of identical twins reared apart. All of these observations support the existence of a basic genetic component in intelligence.

Another important line of evidence stems from correlation analyses that measure the degree to which variables (paired individuals) vary together. A statistic known as the *correlation coefficient* (r) provides a measure of the intensity of association between variables, such that a value of 1.00 is a perfect correlation, zero indicates no correlation, and values between 1.00 and zero provide a relative measure of the strength of correlation. Table 11-4 lists a summary of correlation data on IQ scores for an array of paired relationships. The table also includes the *coefficient of determination* (r^2) which indicates the proportion of the variation in one pair-member that is accounted for by variation in the other pair-member. For example, an r^2 value of .77 for identical twins reared together means that they have 77% of their variance in common. Identical twins reared apart, with $r^2 = .56$, only have 56% common variation; the difference of 21% must be attributed to environmental differences between the two groups because all pairs consist of identical genotype combinations. All groups other than identical twins have genotypic differences of varying degrees between pair-members in addition to any environmental influences that lower their r^2 values significantly below those of either monozygotic group. Note that, as would be expected, unrelated children reared apart (randomly paired children) have practically no correlation, and the r^2 value is negligible,

Table 11-4. A summary of correlation data on IQ scores for an array of paired groups involving various degrees of genetic and environmental relationships

Pair relationship	Number of groups	Median r value	Range of r values	r² value
Identical twins reared together	15	.88	.76 to .95	.77
Identical twins reared apart	4	.75	.62 to .85	.56
Fraternal twins reared together	21	.53	.38 to .87	.28
Paired sibs reared together	39	.49	.30 to .77	.24
Paired sibs reared apart	3	.46	.34 to .49	.21
Parent-child, reared by parent	13	.52	.22 to .80	.27
Parent—adoptive child	4	.19	.18 to .39	.04
Unrelated children reared together	7	.16	−.17 to .31	.03
Unrelated children reared apart	7	.09	−.04 to .27	.01

(Modified from Guilford, J. P.: The nature of intelligence, New York, 1967, McGraw-Hill Book Company, p. 352. Cited from Erlenmeyer-Kimling, L., and Jarrik, L. F.: Science **142**:1477, 1963.)

meaning that none of their variability is accounted for by the other pair-member either by genetic relationship or similarity of environmental factors. It is readily apparent from these data that high r and r² values are correlated first, with the closeness of the genetic relationship and, secondly, with the similarity of environmental factors, for example, reared together versus reared apart. This interpretation again supports the genetic nature of intelligence and the influence of environmental factors.

The crucial question now arises—how much of an individual's intelligence is dependent on his or her inheritance and how much on environmental influences? In attempts to answer this question, the human geneticist runs into nearly every conceivable problem related to polygenic inheritance, the collection of human genetic data, and the interpretation of these data, in addition to a host of political and cultural pressures and biases related to intelligence and its measurement. Perhaps the most meaningful genetic concept in the nature-nurture controversy is the statistic known as *heritability* (h^2), which, by definition, is a measure of the proportion of phenotypic variation, that is due to heredity, that is, genotypic variation. In its most elementary interpretation (ignoring interaction and covariances), heritability would be the fraction of variation still present among phenotypes in a population, if in some magical way all environments were made equal and all environmental variations were to disappear. The technical measurement of heritability is a number between 0 and 1 that indicates the amount of phenotypic variation that is genetically determined. The problem in calculating heritability is that the environmental variation cannot easily be measured and separated from the genetic component unless very specific experimental crosses are designed and followed over several generations in order to obtain reliable data as input to a sophisticated statistical analysis that ultimately produces an accurate estimate of heritability. Since the prescribed conditions cannot be fully produced to obtain human data, heritability estimates on complex characteristics such as intelligence remain subject to criticism of the data, statistical procedures, and interpretations.

The generally accepted estimate of the heritability of IQ, as calculated almost entirely from data on whites, is .8, which means that about 80% of the variation in IQ is due to genotypic differences generated by the polygenic system that determines each individual's intelligence potential. The remaining 20% of the variation within the population is attributed to the environment.

Perhaps the hardest thing to grasp about heritability is that it is a population phenomenon that indicates the extent to which variation in successive generations is predictable and subject to genetic control (selection). It is a measure of breeding true, useful in predicting how much of a particular trait will be expected in the average offspring of a given set of parents. In general, for polygenic traits (with h^2 less than 1.0) parents who are above or below the population average tend to produce children who approach the average. A substantial body of data shows that the children of *very intelligent* parents tend to be merely *intelligent* and the children of *very dull* parents tend to be merely *dull*, while *average* parents tend to have *average* children. This phenomenon is technically known as *regression toward the mean*. The amount of regression depends on the heritability of the trait—with high heritability, the regression is low, and vice versa. Thus, as long as the heritability of a trait is less than 1.0, there will be a regression effect on that trait within the population.

According to the regression concept we should be able to predict the IQ of the average offspring in a family by the following procedure:

1. Average the IQ of the parents
2. Subtract 100 from the average
3. Multiply the remainder by .8 (the heritability)
4. Add this product to 100

Thus, if both parents had IQs of 125, the average expected in their children would be 120. There would of course be a range from higher (possibly beyond either parent) to much lower for any particular individual, but as a population phenomenon this average would be realized over the long run. If the parents' IQs averaged 75, the expected average of their children would be 80. And average parents (IQs of 100) would be expected to have children with IQs of 100 on the average. However, these predictions for intelligence are only "theoretical" because of one factor—the dependence on IQ measurement as the critical input for the genetic prediction.

Even ignoring the cultural biases, and so forth, associated with IQ measurement, there is a more fundamental nature-nurture aspect to the situation that is absolutely basic to understanding any genetic implications of intelligence based on intelligence measurements. That is the fact that all such devices measure an end product or realized potential, but not the undeveloped potential. Thus, our handle on intelligence, even with accurate heritability estimates, loses much of its predictive power because the environmental factors that influence the realization of intelligence within its genetically determined range are so poorly understood, complex in their interactions, and unequally distributed that no valid statement can be made concerning the "average amount" of intelligence potential that is developed in the "average" person. In other words, the phenotype is not a reliable expression of the genotype. This point is illustrated in Fig. 11-7 which depicts this lack of reliability between the phenotype and genotype. When great disparities such as these exist within entire populations, the nature-nurture problem is difficult to resolve as a population phenomenon, much less to attempt to apply some generalizations at an individual level.

An overall view of the distribution of intelligence (IQ scores) in the population provides a final approach to the nature-nurture relationship. As we have seen in this chapter, polygenic traits tend to be normally distributed within a population. In addition, intelligence tests are constructed in such a way that the distribution of IQ scores is expected to form a bell-shaped distribution. Yet, the actual distribution of intelligence is not quite "normal," but shows certain

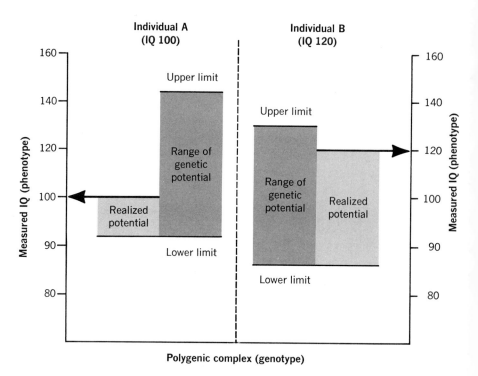

Fig. 11-7. One interpretation of the relationship of genetics to intelligence is that each individual has a fairly wide range of potentialities for intelligence, with the upper and lower limits of this range being determined by heredity. The realization of intelligence, as measured by an IQ score for example, is determined in complex fashion by a variety of factors of the internal and external environment. Thus the performance of individual *A*, with a greater potentiality for intelligent behavior, may fall short of that of individual *B* whose IQ test score may lie much nearer the upper limit of his capacity. (Modified from Srb, A. M., Owen, R. D., and Edgar, R. S.: General genetics, ed. 2, San Francisco, 1965, W. H. Freeman and Company.)

deviations from the expected curve as indicated by arrows *1, 2, 3,* and *4* in Fig. 11-8.

The first deviation, represented by those individuals with IQs below 55 to 60, really represents a range distinct from the rest of the intelligence distribution. Nearly everyone in this category is severely mentally deficient because of physical brain damage or impaired brain development and function caused by specific genetic defects. Physical damage may result from maternal infections, fetal lack of oxygen, premature birth, malnutrition, or brain damage by accident at any point in time. Genetic defects include metabolic disorders such as phenylketonuria and galactosemia, developmental anomalies such as microcephaly, and chromosome aberrations such as "mongolism" which will be discussed in the next chapter. These specific causes override all other genotypic and environmental factors that influence intelligence and throw the afflicted individual out of the "normal" distribution into this severely defective category. Less than 1% of the total population falls into this range. On the other hand, the more mild category of mental retardation, the so-called feebleminded, with IQs ranging from 55 to 75, appears to represent the very lower end of the intelligence scale according to its polygenic distribution. This category is made up mainly of those individuals who inherited the lowest intelligence potentials, the so-called "familial mentally retarded." The distinction is supported by the fact that individuals in this latter group are usually physically indistinguishable from persons in the higher ranges, whereas individuals in the severely retarded category are commonly abnormal in physical appearance or bodily functions which accompany the specific causes of their condition. The strongest evidence to indicate that the severely retarded group is distinct from the normal distribution, whose low end is represented by the mildly retarded individuals (IQs of 55 to 75), comes from the comparison of sibling distributions from both groups. The mildly retarded parents produce children who also range at the low end of the intelligence scale. There is regression toward the mean observed, but the distribution is just about what would be expected in a polygenic

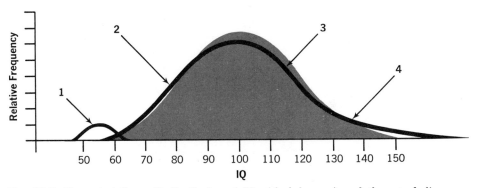

Fig. 11-8. Theoretical "normal" distribution of IQs (shaded curve) and the actual distribution in the population (heavy line), with the lower hump exaggerated for explanatory purposes. See text for explanation. (Redrawn from Jensen, A. R.: How much can we boost IQ and scholastic achievement? Harvard Educational Review **39**:1, 1969.)

model. However, children of parents in the severely retarded category tend toward a nearly normal IQ range and, on the average, surpass the children of the more mildly retarded parents. This sort of distribution is just what would be expected from parents who belong in the normal distribution of the polygenic system but were displaced downward by a specific cause. Except when the specific defect recurs, the children from these parents represent a typical cross section of the population. This phenomenon is illustrated in Fig. 11-9 which shows the IQ distribution of 562 siblings of severely or mildly retarded parents.

The second deviation of the actual IQ distribution from the expected noted in Fig. 11-8 concerning an "excess" of individuals in the 70 to 90 range and the third deviation concerning a "deficiency" in the 90 to 130 range are interrelated. The 70 to 90 bulge is caused in large measure by the effects of environmental disadvantages that place persons at lower levels than their genetic potentials would normally warrant. These individuals, and a certain portion of the severely retarded group previously discussed, come from the 90 to 130 IQ range, thereby making the 90 to 130 category deficient.

The fourth deviation has not yet been adequately explained or accounted for. This is the "excess" of individuals at the high end of the IQ scale (130 and up). Major gene effects that produce superior intelligence have been postulated but seem unlikely. More likely is an environmental explanation that states that truly superior genotypes for intellectual development are often recognized at an early age and these individuals are encouraged and cultivated to still greater superiority by being provided with environmental interactions that allow very high development of their intelligence potential. It is also possible, but only speculative, that positive assortative mating (like marrying like) is strong at the upper levels of intelligence in such a way that the degree of resemblance of parents is greater at the high end of the scale than it is throughout the remainder of the distribution. This would reduce the regression toward the mean

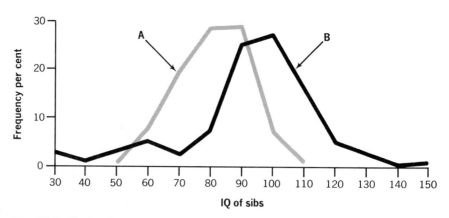

Fig. 11-9. IQ distributions of siblings produced by **A,** mildly retarded parents, and **B,** severely retarded parents. (Data from Roberts, J. A. Fraser: The genetics of mental deficiency, Eugenics Review **44**:71, 1952.)

at the very high end and increase the number of individuals in this portion of the distribution; in other words, this would be a selective process.

In summary, the general conclusion that can be made from the data presented here is that both heredity and environment contribute important conditions in the determination of an individual's intelligence status as measured by intelligence tests. Both heredity and environment establish upper limits for intellectual development and rarely, if ever, does any individual reach either limit. The status actually achieved is determined by either hereditary or environmental factors; whichever is lower will set the actual limit of realized potential for any particular person. The nature-nurture relationship in intelligence really defines *one* complex set of interactions that collectively produce an end product—intelligence—which can only be comprehended, like "life" itself, as a whole greater than the sum of its parts (at least greater than the parts into which man can presently subdivide the phenomenon).

12

The effects of abnormal chromosomes

In our considerations of genetic transmission and the various modes of inheritance, the assumption was always implicit that the chromosomes that carry the genes behaved normally during meiosis (and mitosis). However, the chromosomes are subject to accidents in their maneuvers and to alterations in their structure which produce an array of abnormalities in man. These chromosomal aberrations constitute the second broad category of mutation (recall that point mutations were discussed in Chapter 6). The study of the relationship of the microscopic appearance of the chromosomes and their behavior to the genotype and phenotype of the individual is known as *cytogenetics*. Human cytogenetics is a relatively recent field that has greatly expanded the scope of medical genetics and plays an important role in much genetic counseling.

Until 1956 the human chromosome number was thought to be 48, but in that year improved techniques established the normal number at 46, and by 1959 a variety of chromosomal aberrations was demonstrated in man and linked to abnormal conditions that were heretofore poorly understood. The normal chromosome complement was described in Chapter 4, and at this point the classification system (Table 4-1) and normal female and male karyotypes (Figs. 4-3 and 4-4) should be reviewed and used as references for the discussions in this chapter.

The genetic system of man, and most animals, is delicately balanced and adapted to functioning in the diploid condition where two doses of all genes, except those on the sex chromosomes, represent the norm. Any alteration in chromosome number or gross structural change that disrupts this genetic balance usually produces developmental abnormalities with profound phenotypic effects characterized by many physical defects and accompanied by mental deficiency. The accumulated set of abnormalities so produced is called a *syndrome*, and most syndromes are named after the person, or persons, who first described the conditions. In this chapter we shall first consider the more serious and more commonly detected syndromes that are caused by changes in the number of chromosomes. This will be followed by a description of the types of structural alterations that may occur, including a few examples of these less severe, and lesser known, aberrations.

CHANGES IN CHROMOSOME NUMBER

Since even the smallest chromosomes (with the possible exception of the Y chromosome) carry a large number of genes, a change in chromosome number represents a very drastic disruption in genetic balance. It is very unusual to find any surviving individual with more than one extra or one missing chromosome. If there is one extra chromosome present, this means that some particular chromosome is represented *three* times instead of the usual paired diploid arrangement. This condition is known as *trisomy* (literally three bodies) and is symbolized as $2n+1 = 47$ (recall that n represents the haploid —gametic—condition and 2n the normal diploid—zygotic—condition). When one chromosome is missing from the diploid complement, only one representative of a particular pair is present, producing a condition known as *monosomy* symbolized as $2n-1 = 45$. Trisomics and monosomics are collectively known as *aneuploids* (not true multiples), as contrasted to *euploids* (true multiples) such as diploids (2n) or abnormally higher multiples of the basic haploid (n) set known as *polyploids* (3n, 4n, and so forth).

Theoretically 24 different trisomics and 24 different monosomics would be expected to occur—there would be one trisomic and one monosomic for each of the 22 autosomes plus an X and a Y of each type. Of the 48 possibilities, only the eight listed in Table 12-1, six trisomics and two monosomics, have been observed in live-born children; of these only five are known to survive infancy with any regularity. The reason others are not found is not that they do not occur but that the genetic imbalance is so severe that development is arrested long before birth. This has been confirmed through chromosome

Table 12-1. Aneuploids observed in live-born children

Syndrome name	Abnormal chromosomal constitution	Year of discovery	Frequency
	Trisomics		
Patau's	13, 13, 13	1960	1 in 500 births
Edwards'	18, 18, 18	1960	1 in 3,000 births
Down's	21, 21, 21	1959	1 in 600 births (s)*
Triplo-X	X, X, X	1961	1 in 1,200 female births (s)
Klinefelter's	X, X, Y	1959	1 in 400 male births (s)
Jacobs'	X, Y, Y	1961	1 in 300 male births (s)
	Monosomics		
Al-Aish's	21, O	1967	Rare (about 3 known)
Turner's	X, O	1959	1 in 3,000 female births (s)

(s) indicates those that regularly survive beyond infancy. Note that sex chromosome syndromes will tend toward maleness whenever a Y is present and toward femaleness in the absence of a Y.

analyses performed on spontaneously aborted fetuses of which a large proportion are found to be aneuploids. The trisomics and monosomics that do survive until birth, therefore, represent the least severe developmental assaults but are nevertheless serious departures from normality in most cases. It is not unexpected that the mildest and most common forms of aneuploidy involve the sex chromosomes which, according to the Lyon-Russell principle (Chapter 9), have a dosage compensation mechanism that allows greater latitude in chromosome composition than can be tolerated by the autosomes. Among the autosomes, only one of the very smallest (number 21), presumably containing fewer genes than the larger chromosomes, survives to adulthood in the trisomic condition, and none in the monosomic condition live that long.

The principal cause of aneuploidy is an accident in meiosis that leads to an unequal distribution of a chromosome pair. A lack of separation, known as *meiotic nondisjunction,* results in one daughter cell receiving both members of a particular chromosome pair, while the other cell receives neither. Nondisjunction may occur during Meiosis I or Meiosis II. In either case the basic result is the same although when it occurs in the first meiotic sequence, the proportions of abnormal gametes are greater. These two types of meiotic nondisjunction are compared to normal meiosis in Fig. 12-1. If a 24-chromosome gamete unites with a normal one, a trisomic is produced. If a 22-chromosome gamete is involved in a union with a normal gamete, the new individual will be a monosomic. For example, let us consider the implications of meiotic nondisjunction for the sex chromosomes: in oogenesis two kinds of abnormal eggs can be produced, XX and O (O means that the egg is lacking the X chromosome). In spermatogenesis four kinds of sperm can be produced, XY and O sperm result when nondisjunction occurs at Meiosis I, and XX and YY types (plus O types) can result when nondisjunction occurs at Meiosis II. Since nondisjunction occurs infrequently, these abnormal gametes generally

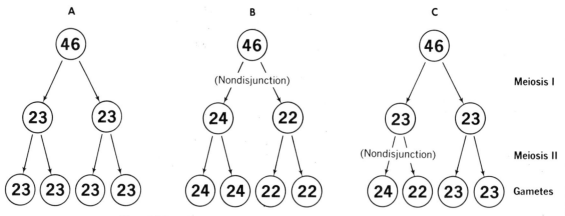

Fig. 12-1. The effect of meiotic nondisjunction on gametic chromosome number: **A,** normal meiosis, **B,** nondisjunction during the first meiotic sequence, **C,** nondisjunction during the second meiotic sequence; affecting only one of the products of Meiosis I.

mate with normal gametes; Table 12-2 lists the outcomes of these unions. Note that these account for all of the sex chromosome aneuploids listed in Table 12-1.

Although sex chromosome aneuploids appear to be the least severe, all suffer physical defects and mental retardation to varying degrees, and all but triplo-X are sterile. These are also the most easily detected chromosomal abnormalities because a simple microscopic analysis (known as a buccal smear, after the technical name of the mouth lining, buccal mucosa) of a few cells scraped from the lining of the mouth to determine Barr body number is diagnostic for X chromosome constitution. Recall that all X chromosomes in excess of one form Barr bodies (refer to Chapter 9); normal males have no Barr bodies, and normal females have one. Deviations from this are indicative of most sex chromosome syndromes as listed in Table 12-3. Note that the table includes several unusual types that have been infrequently observed. Fig. 12-2 illustrates two abnormal buccal smears; *B* is that of a triplo-X female, and *C* is from an unusual male of XXXXY chromosome constitution.

Down's syndrome, the only autosomal aneuploid to belong to the group that survives through infancy with regularity, is also the most severe of the group. For many years this condition was known only as Mongolian idiocy because of the often severe mental retardation accompanied by a physical defect that makes the eyes of Caucasians resemble the almond-shaped eyes of the Mongolian race (although anatomically they are different). Mongolians feel that children of their race who have trisomy 21 look like Caucasians. Now we see why Down's syndrome is the preferred term!

Down's syndrome individuals are characterized by a small, round head and common facial features that include a flat profile, open mouth with protruding furrowed tongue, flat nasal bridge, and speckled irises (Brushfield spots), in addition to the almond-shaped eyes. Their hair is straight and sparse; their skin is dry, and their ears are small, low-set, and misshapen. Their hands and

Table 12-2. Outcomes possible in matings involving nondisjunction of the sex chromosomes during oogenesis or spermatogenesis*

	Abnormal gamete	Normal gamete	Zygote (genotype)	Phenotype
	Eggs	*Sperms*		
	XX	X	XXX	Triplo-X
Nondisjunction in	XX	Y	XXY	Klinefelter's
oogenesis	O	X	XO	Turner's
	O	Y	YO	Inviable (never observed)
	Sperms	*Eggs*		
	XY	X	XXY	Klinefelter's
Nondisjunction in	O	X	XO	Turner's
spermatogenesis	XX	X	XXX	Triplo-X
	YY	X	XYY	Jacobs'

Note that all sex chromosome syndromes except Jacobs' can arise from nondisjunction in either sex.

Table 12-3. Barr body numbers observed in buccal cells of individuals of sex chromosome aneuploids observed in man

Phenotype (apparent sex)	Sex chromosome constitution	Number of Barr bodies in buccal cells
Females		
Normal	XX	1
Turner's	XO	0
Triplo-X	XXX	2
Unusual types	XXXX and XXXXX	3 and 4
Males		
Normal	XY	0
Klinefelter's	XXY	1
Jacobs'	XYY	0
Unusual types	XXXY and XXXXY	2 and 3
	XXYY and XXXYY	1 and 2
	XYYY and XXYYY	0 and 1

feet are broad with webbing of fingers and toes, extra digits, and distinctive dermatoglyphic patterns (palm and sole ridges and creases). Umbilical hernia, congenital heart disease, and intestinal disorders are also typical. Many of these characteristics are apparent in the child shown in Fig. 12-3. These individuals constitute about 10% of mentally retarded patients.

Not only is Down's syndrome a relatively frequent and severe abnormality that often places great social and psychological (and sometimes financial) stresses upon a family, but there is also a maternal age effect that has placed Down's syndrome in the forefront of many genetic counseling controversies. It is well established that the risk of having a Down's syndrome child increases dramatically with the age of the mother from 1 in 3,000 chances for women under 30, to 1 in 40 chances for women over age 45 (Table 12-4). As we have seen, nondisjunction may occur during oogenesis or spermatogenesis. However, the association between incidence and increased maternal age, coupled with observations that Down's syndrome children resemble the blood types of their mother more than would be expected by chance, leads to an assumption that this particular syndrome results primarily from nondisjunction in oogenesis. There are at least two possibilities, based on human reproductive biology, which suggest possible explanations of this maternal effect. First, all the potential eggs a female will ever produce are present from birth, arrested at Prophase I of meiosis (Chapter 4). This means that the eggs released later in life have been exposed to all sorts of possible damage from drugs, X-rays, and so forth, for longer and longer periods of time. Thus, the maternal age factor could actually be an egg age factor. Secondly, and perhaps more likely, is the possibility that among the millions of sperms that compete to fertilize a single egg, abnormal ones are at a distinct disadvantage and, therefore, very few are successful in fertilization. No similar type of competition would operate on a single abnormal egg. In addition, prior to fertilization the egg is much more

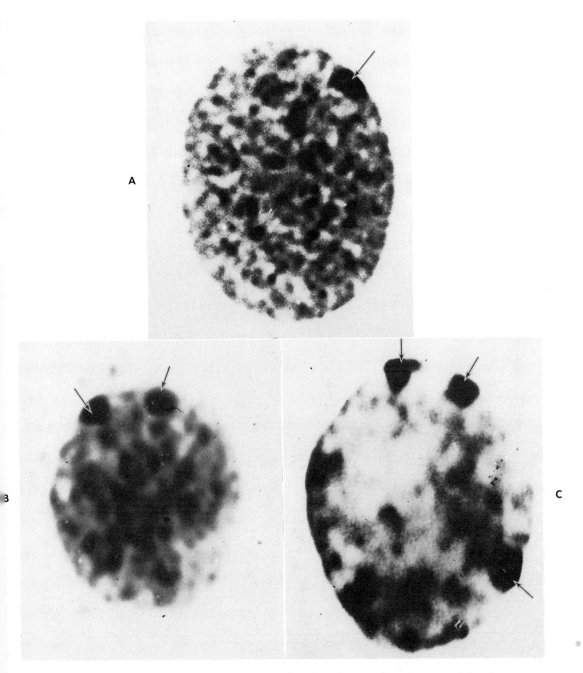

Fig. 12-2. Barr bodies (as indicated by arrows) in buccal smears from **A,** a normal female, **B,** a triplo-X female, and **C,** an XXXXY male. (From Redding, A., and Hirshhorn, K.: Guide to human chromosome defects. In Bergsma, D., editor: Birth defects, Original Article Series, vol. 4, 1968, The National Foundation.

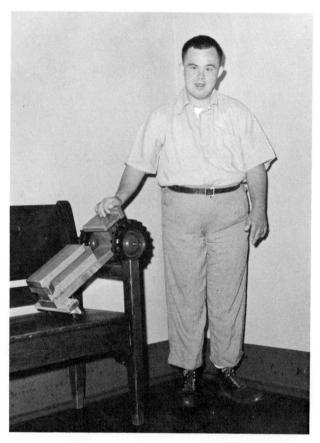

Fig. 12-3. A 25 year old Downs' syndrome male, with a Stanford-Binet IQ of 26. Note the physical characteristics described in the text. (From Moody, A.: Genetics of man, New York, 1967, W. W. Norton.)

Table 12-4. Risk of producing a Down's syndrome child according to the age of the mother

Age of mother	Risk of Down's syndrome child
Less than 29	1 in 3,000
30 to 34	1 in 600
35 to 39	1 in 280
40 to 44	1 in 70
45 to 49	1 in 40
All mothers (Average)	1 in 600

(Data from Porter, Ian H.: Heredity and disease, New York, 1968, McGraw-Hill Book Company, p. 45.)

physiologically passive than sperms that must swim to the oviduct by generating and expending a relatively large amount of energy. This activity might very well be adversely affected by the genetic imbalance of an extra or a missing chromosome, thereby making sperms more susceptible to haploid "trisomy" or "monosomy" than eggs.

About 30% of all Down's syndrome births occur in women over 40 years old. If these births could be prevented, some 1,900 of the more than 6,000 Down's syndrome babies expected annually in the United States could be eliminated. With the advent of amniocentesis (see Chapter 5) it is possible to analyze the chromosome complement of a fetus at an early enough stage to perform an abortion, should a severe abnormality be detected. Because of the well known maternal age effect for Down's syndrome, it is becoming common to justify this procedure when older women seek genetic counseling. If the fetal karyotype should show trisomy 21, like that depicted in Fig. 12-4, the expectant mother is faced with the decision of a therapeutic abortion or the

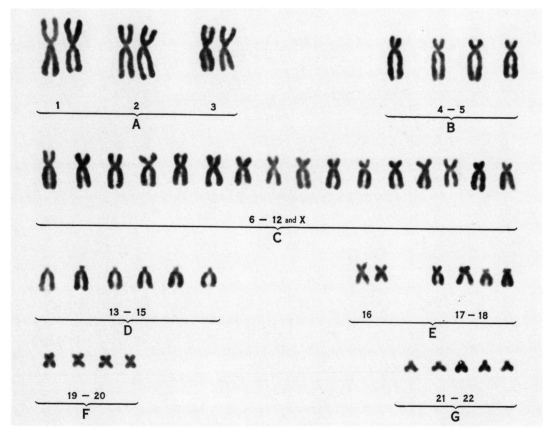

Fig. 12-4. Karyotype of a female exhibiting trisomy 21 (Down's syndrome). (Courtesy Dr. D. S. Borgaonkar, Johns Hopkins School of Medicine)

prospect of giving birth to a Down's syndrome child. Such information places the genetic counselor in an entirely new position, for now the parents (or expectant mother) are not being informed of "the odds" that their child will have a particular genetic defect but are given a clearly positive or negative diagnosis to decide upon. This creates a whole new set of moral and ethical problems for the individuals involved, and for the whole of society, regarding what is now commonly referred to as the cost of our genetic knowledge.

When we can test the fetus by karyotyping, or definitive biochemical tests, to determine if the baby will be "normal," we assume that the standards for normalcy are well defined—but are they? Amniocentesis for detection of chromosomal abnormalities creates another quandry thrust upon society for which outmoded standards are not sufficient guidelines. For example, are 46 chromosomes "normal" and any other number "abnormal" to the extent that any fetus not conforming should be destroyed before birth? Even among children with an affliction as severe as Down's syndrome, some develop IQs above 70 and can be trained to perform simple tasks, and most are delightful fun-loving babies, especially responsive to music. And what decisions should be made concerning the normalcy of fetuses detected with the milder sex chromosome syndromes? Triplo-X females are usually very near normality, including full fertility. But their IQ range is below normal and, theoretically at least, they would produce both X-bearing and XX-bearing eggs; the latter, if fertilized by a normal X-bearing sperm, would produce more triplo-X females and by a Y-bearing sperm, would produce Klinefelter's males (XXY). Klinefelter's syndrome males are sterile and subnormal in intelligence—should they be aborted when detected? Perhaps the most perplexing fetal karyotype detected is Jacobs' syndrome (XYY). When first discovered, it was widely publicized in the scientific literature and the public press as a supermasculine chromosome complement associated with tall stature and behavioral abnormalities such as strong criminal and antisocial impulses. Although the height of these males still seems to be above average, the abnormal behavior has not been substantiated, and it appears that many, if not most, XYY men live rather normal lives, including reproduction.

Amniocentesis and abortion are viewed by many as a new freedom from the burden of producing children with genetic defects, especially chromosomal aberrations. But the question has already been raised as to how long it will be until that freedom becomes a responsibility to abort certain fetuses, whether we want to or not, for the good of society. Societal attitudes toward parents "burdened" with a Down's syndrome or Klinefelter's syndrome child may very well change from sympathetic to indignant when it is known that they *chose* to have *that* child. When individual freedoms conflict with societal goals, such as public health priorities, the pressures from society often result in laws for the mutual benefit and protection of its citizens. In 1972 the Massachusetts legislators passed a law requiring sickle cell screening of all children entering school. The question now is what do we (society) do with such data? Will genetic defects become legally *reportable* diseases, such as VD, which may then label an individual as "inferior," or will some be considered *quarantinable* to prevent the reproduction of individuals so defined? The prospective increase in hemophilia caused by treatment with AHF, discussed in Chapter 9, would be classified in the latter category. These are very serious problems that the so-called biological revolution has created. Let us hope that society will have the collective wisdom to resolve them in concordance with and respect for the dignity of mankind.

All forms of aneuploidy are not as clear-cut as those discussed previously. It is also possible for mixtures of aneuploid and normal cell lines to exist within the same person. These are known as *chromosomal mosaics,* and the degree of "abnormality" of such individuals may range from severe to negligible. Mosaicism may arise in one of two ways. The most common mechanism is thought to be *mitotic nondisjunction* that occurs after conception and, presumably, at an early stage of embryonic development. (Late-stage nondisjunction would not produce a very large amount of abnormal tissue and would likely go undetected.) If nondisjunction occurred during the first mitotic reproduction of the zygote, an individual consisting of essentially half trisomic and half monosomic cells would result. If a chromosome pair failed to separate at some later time, there would be three cell types produced—namely, normal, trisomic, and monosomic—in proportions dependent upon the time of the mitotic accident. Fig. 12-5 diagrammatically compares these two types of mitotic nondisjunction.

The second manner in which chromosomal mosaicism may be produced is by the loss of one chromosome as it moves toward the pole during mitosis. This is known as *anaphase lag* and results in one cell line with normal chromosome constitution and one with monosomy for the lost chromosome. Table 12-3 listed some observed mosaics for the sex chromosomes. Autosomal mosaics may

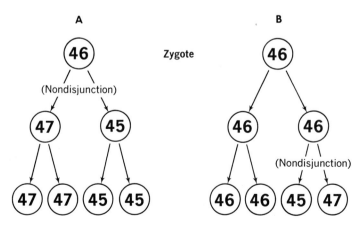

Fig. 12-5. Chromosomal mosaicism as a result of mitotic nondisjunction. **A,** Zygotic nondisjunction, producing trisomic and monosomic cell lines. **B,** Postzygotic nondisjunction, producing three different cell lines.

also occur; the most common example of this is Down's mosaicism in which normal and trisomy 21 cell lines are observed. In these individuals the physical symptoms are less apparent and mental retardation is less severe. Knowledge of these types of aberrations raises another genetic question: should fetuses detected with chromosomal mosaicisms be aborted?

Polyploidy (euploids) in man is extremely rare. A *triploid* cell (3n) would contain 69 chromosomes, and a *tetraploid* cell (4n) 92; such genetic imbalance is apparently intolerable in man and most animals although quite common in the plant kingdom. Nearly all polyploidy observed in man has been in malignant cells and in spontaneous abortions where it accounts for about 10% of all those that occur during the first half of pregnancy. Only eight live-born individuals have been recorded. All but one were mosaics, having a normal diploid cell line in addition to a polyploid line, but in the exception the only tissue examined was blood. One had a tetraploid cell line, and all others were triploid. Polyploidy can arise from fertilization involving unreduced gametes (essentially complete nondisjunction) containing 46 instead of 23 chromosomes, through dispermy (fertilization by two sperms), or by the failure of duplicated chromosomes to separate into two daughter cells during mitosis—this is evidently how the observed live-born mosaics were produced. A few full triploids have been born prematurely, but none survived more than several hours.

CHANGES IN CHROMOSOME STRUCTURE

Chromosomes are also prone to accidents that break and alter their individual structure, often with no change in chromosome number per se. Until very recently (1970) the identification of specific chromosomes was not precise (group placement, A to G, was accurate, but within the group classification was inaccurate), and many structural aberrations were difficult to detect. With the advent of fluorometric techniques the identification of specific chromosomes has been greatly enhanced because, when specially stained and viewed under ultraviolet light, each chromosome displays a characteristic fluorescence pattern. A normal male karyotype prepared by the fluorometric technique is shown in Fig. 12-6; compare this to Fig. 4-4. This new technique will undoubtedly add considerably to our knowledge of structural defects in chromosomes. Coupled with computer programs already established for rapid, automatic and extremely accurate chromosome analysis, it may open the way to chromosome screening on a mass scale in the very near future, which may become as routine as blood analyses are today. This will again raise the critical questions of what is normal and what abnormal and how society will cope with this new technology.

The two most common types of structural abnormalities observed to date in man are *deletions* and *translocations*. Other types of defects have been detected but, until the discovery of fluorescence patterns, these abnormalities were very difficult to identify in human cells. They will be briefly discussed after deletions and translocations are examined.

A deletion is the loss of a portion of a chromosome and, in effect, represents *partial monosomy*. A piece of chromosome that is broken off and lacking a centromere is easily lost during cell reproduction. Breakage may occur by any of a number of agents such as irradiation, chemicals, drugs, and viral infections. We have already seen that full monosomy is extremely rare, presumably because the disruption in genetic balance cannot be tolerated. Deletions add circumstantial evidence to support this conclusion because, although partial monosomy should logically be milder than full monosomy, very few individuals with deletions seem to survive and those who do usually evidence quite severe physical and mental defects.

The most frequently observed and studied deletion is one that occurs in the short arm of chromosome number 5. Since its discovery in 1963, nearly 100 cases have been studied. The most striking feature of the deletion-5 syndrome is the characteristic baby's cry that sounds like the forlorn mewing of a cat for which it was named the *cri du chat* syndrome by its French discoverers. This seemingly small loss of genetic material from one member of a chromosome pair, as seen in Fig. 12-7, produces physical anomalies including low birth weight, round face, squint eyes with oblique "antimongoloid" fissures of the

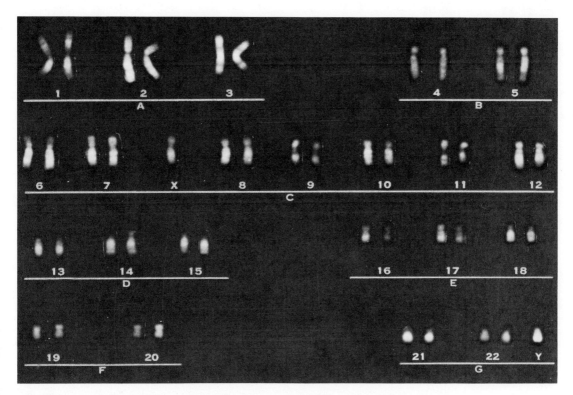

Fig. 12-6. A normal male karyotype displaying fluorescent banding patterns. Note that the fluorometric technique makes it possible to identify each specific homologous pair within the groups. (Courtesy Dr. D. S. Borgaonkar, Johns Hopkins School of Medicine)

eyelids, broad-based nose, abnormal low-set ears, short neck, and microcephaly. The life span of these individuals is not greatly reduced, but mental retardation is profound. Their usual IQ range is 20 and below, although one individual has been reported with an IQ of 56.

A deletion of the short arm of chromosome 4 has recently been recognized as clinically distinct from that of chromosome 5. Although many of the physical features are similar to deletion-5, the cat-like cry is absent, and additional anomalies such as a beaky nose with triangular-shaped nostrils, cleft palate, congenital heart disease, and undescended testes in males are characteristic. These individuals do not respond to stimuli, appear to be deaf, and seem to be more mentally retarded than deletion-5 individuals. Even the majority of older patients, ages 9 to 15, show no signs of mental activity, having IQs below 10 or essentially zero.

Other deletions have been observed in the D, E, and G chromosome groups, including one in the G group called the Philadelphia chromosome that is associ-

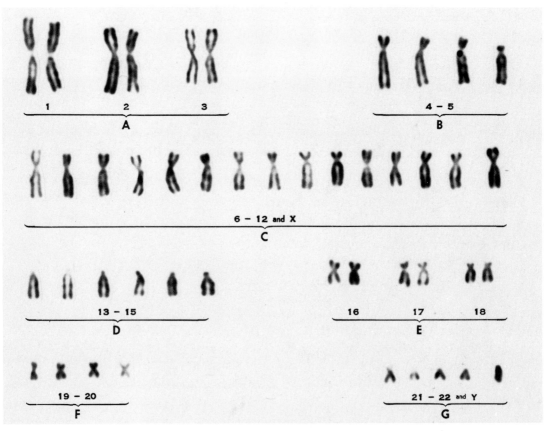

Fig. 12-7. Karyotype of a male exhibiting the *cri-du-chat* syndrome. Note that one member of pair number 5 has a portion of its short arm deleted. (Courtesy Dr. D. S. Borgaonkar, Johns Hopkins School of Medicine)

ated with a particular form of chronic leukemia. Fluorometric studies have revealed that the Philadelphia chromosome is not the same G group member as the chromosome involved in Down's syndrome and, since Down's syndrome is conventionally called trisomy 21, the deletion chromosome is number 22.

Occasionally pieces may break off both ends of a chromosome and be lost, leaving the remainder to join at the broken ends to form a *ring chromosome*. Ring chromosomes associated with congenital abnormalities have been observed in the X chromosome and several autosomes.

Translocation, the other class of structural abnormalities commonly found in man, involves the transfer of a portion of one chromosome to another, usually nonhomologous, chromosome. This occurs when a break in each chromosome is followed by repair involving reciprocal exchange of the broken portions, as in the manner illustrated by Fig. 12-8. Although the typical result of translocation is *two* abnormal chromosomes with no change in total chromosome number, the most common type found in man is a modified version known as *centric fusion*. This involves two acrocentric chromosomes in which the breaks occur in the extremely short arms at or near the centromere, followed by rejoining of the two large portions and the loss of the two small fragments of the short arms, as illustrated in Fig. 12-9 (compare this to Fig. 12-8).

The most thoroughly studied translocation in man is a D/G (a member of the D group and the G group involved) centric fusion, specifically between chromosomes 15 and 21. The presence of a translocation does not necessarily lead to an abnormal phenotype. For example, if an individual possesses a 15/21 centric fusion in which only a small amount of satellite material was lost from each chromosome during the translocation, this *one* chromosome will be genetically equivalent to one number 15 and one number 21. If this individual also possesses one normal number 15 and one normal number 21 chromosome, he or she will be phenotypically normal but will possess a total of only 45 chromosomes. In this case 45 chromosomes is *not* a monosomic; such individuals are called *translocation carriers*.

The presence of a translocation chromosome upsets gametogenesis because normal segregation of the chromosomes cannot occur during meiosis. The result in a translocation carrier is the production of *balanced* and *unbalanced* gametes. It is assumed that the independent chromosomes 15 and 21 will pair normally at their homologous segments of the 15/21 chromosome at Metaphase I. Since it is the centromeres that direct chromosome movement at Anaphase I (when homologues separate to opposite poles), the direction in which these three chromosomes segregate will be determined by how the centromere of the centric fusion member of this triad acts. It may, at least theoretically, function in one of three ways—as a double, as a number 15 only, or as a number 21 only—each of which would produce different kinds of gametes. If these gametes were to be involved in fertilization, an array of normal and abnormal phenotypes, including in our example Down's syndrome,

Fig. 12-8. Diagrammatic interpretation of the formation of a reciprocal translocation, shown in three stages from top to bottom.

Fig. 12-9. Diagrammatic interpretation of the origin of centric fusions, shown in three stages from top to bottom.

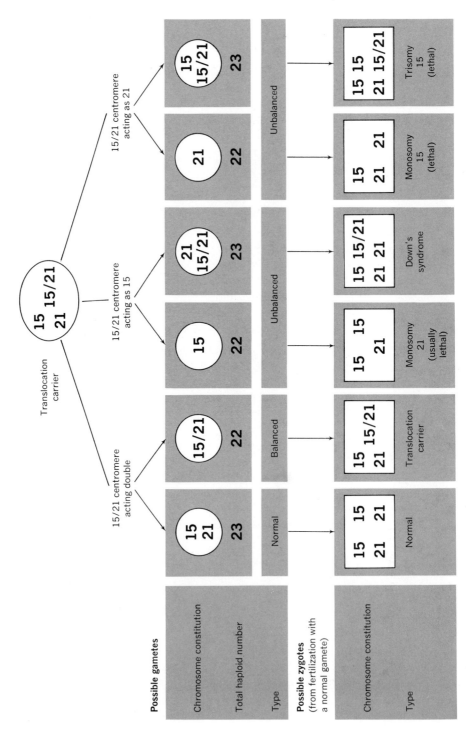

Fig. 12.10. Summary of gametogenesis in a 15/21 translocation carrier and the outcome of fertilization with a normal gamete. Note that of the zygotes that would usually survive to birth, one in three would have Down's syndrome.

would be expected. Fig. 12-10 presents a summary of this example, which is also the general mechanism and genetic implications for any centric fusion. About 3% of Down's syndrome babies are estimated to arise from a "normal" translocation carrier parent. In these cases trisomy 21 is "inherited," and the risk of recurrence in future children is quite high (1 in 3) compared to the average recurrent risk of 1 in 200 in all Down's syndrome cases. A karyotype of a centric fusion Down's syndrome male is shown in Fig. 12-11; compare this to Fig. 12-4. Other observed translocations in man, nearly all centric fusions, include A/B, B/D, B/E, C/X, D/D, and G/G.

Centric fusions create additional problems in the prospect of fetal karyotype monitoring. We might again ponder the question of normalcy in relation to the detection of a 45-chromosome translocation carrier who will be phenotypically "normal" but in reproduction have abnormal gametogenesis, as described in Fig. 12-10. Should such fetuses be aborted? Should such a disease be reportable, and then genetic counseling made either available or mandatory? Or should this be a quarantinable disease?

Several other anomalies of chromosome structure, which are well known in other animals and in the plant kingdom, have been only rarely detected in man. However, as mentioned earlier, the fluorometric technique and computer analysis offer the prospect of significant advancement of our knowledge in this area. Three additional types of abnormalities, *inversions, isochromosomes,* and *duplications* will be mentioned.

An *inversion* involves two breaks in one chromosome, followed by repair during which the broken out section inverts. Inversions interfere with normal pairing during meiosis and may cause nondisjunction of the improperly paired chromosomes. They also suppress the effect of crossing over within the inverted section and retain groups of genes intact, producing what have been called supergenes that can evolve (be selected for or against) as units. Depending on whether or not the centromere is included within the inverted section, two kinds of inversions can be distinguished, one called a *pericentric inversion* when the centromere is included, and one called a *paracentric inversion* when it is outside the inverted section. As shown in Fig. 12-12, pericentric inversions nearly always displace the position of the centromere with respect to a normal homologue (unless the breaks should be equidistant from the centromere). It is this type that has been found in man occasionally. However, with the refined resolution provided by fluorescence patterns, paracentric inversions should be more easily detectable and, if inversions are significant in human variation or disease, the new techniques should soon reveal it.

Isochromosomes are perfectly metacentric chromosomes. They arise during cellular reproduction (mitosis or meiosis) when the centromere splits transversely instead of along the normal longitudinal plane (see Fig. 12-13). Such chromosomes are partly duplications and partly deletions of the genes carried on the normal chromosome. Any gamete produced in this manner and involved in fertilization with a normal gamete will produce a zygote that is partially trisomic and partially monosomic. Isochromosomes have been observed for the X chromosome with some regularity, but they are virtually unknown in autosomes, except possibly for chromosome 21 in a few cases of Down's syndrome.

The final category of structural defects in chromosomes which we shall consider is that of *duplications*. We have already seen that translocations and isochromsomes produce duplication of chromosome segments and that the duplication of genetic material is more common and less harmful than deletions, for example trisomy versus monosomy. Thus, it is not surprising that small duplications involving only a few genes, called *repeats*, can be tolerated. In fact, such duplications are thought to be an important evolutionary mechanism for the origin of "new" genes. The presence of repeats of genes for the same polypeptide production makes it possible for one to mutate and experiment, so to speak, in new directions while the critical function of that particu-

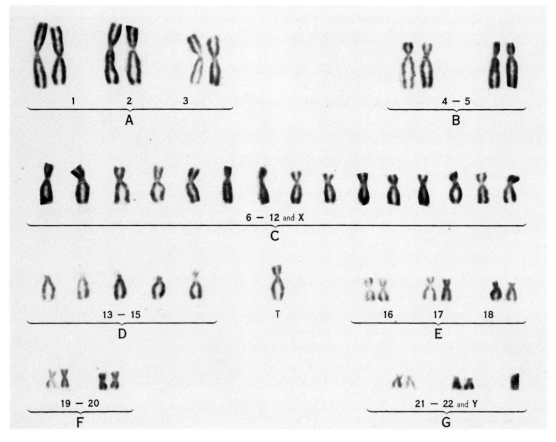

Fig. 12-11. Karyotype of a male with trisomy 21 (Down's syndrome) caused by a 15/21 centric fusion (translocation). The abnormal chromosome, labeled *T*, consists of the second number 15 chromosome, giving a normal group D complement, plus a third number 21 chromosome which produces the trisomy. (Courtesy Dr. D. S. Borgaonkar, Johns Hopkins School of Medicine)

lar gene is maintained by the other copy. The different polypeptide chains in the structure of hemoglobin very likely evolved from duplicate genes.

We have seen that most chromosomal aberrations represent such a drastic assault on the developmental process that they are lethal, usually causing quite early natural abortions. Of those fetuses that survive through development, most exhibit severe congenital defects, with mental retardation a common characteristic. Yet, chromosomal variations may also be detected in some apparently normal people, such as translocation carriers. With amniocentesis, fluorometric technique, and computer analysis, our knowledge (and problems) will

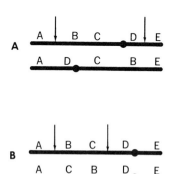

Fig. 12-12. Diagrammatic interpretation of the formation of inversions. **A,** Pericentric, **B,** paracentric. Note that in **A** the centromere is displaced, but in **B** no gross change is detectable. Arrows indicate the break points, and letters indicate gene sequence simply to illustrate structural change.

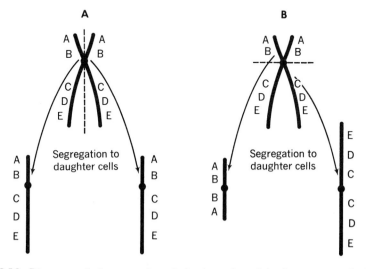

Fig. 12-13. Diagrammatic interpretation of the formation of isochromosomes. **A,** Normal longitudinal splitting of the centromere (dotted line). **B,** Abnormal transverse splitting of the centromere (dotted line) forming isochromosomes.

undoubtedly be increased. It is currently estimated that among "normal" people in the population:

About 3% have structural anomalies of an autosome, particularly groups D, E, and G.

About 0.5% have structural rearrangements such as inversions and reciprocal translocations.

About 3% of the men have variation in the length of the Y chromosome.

Numbers of cells with 45 chromosomes increase with age—in women because of the loss of an X and in men the loss of a Y.

13

The action of genes in populations

Although family involvement presents the most immediate and personal concern with the implications of heredity, it should not be overlooked that families are segments of a larger unit, the population. This, you will recall, is the unit of evolution and the reservoir of genetic variation necessary for the survival of the species. A population may be viewed as a pool of genes in which each individual represents some very small packaged portion and each family serves as a funnel through which a small sample of genes from the gene pool is channeled to produce the individual package. The predictions we can make concerning the sample of genes within a funnel (family predictions) are based on different principles than predictions made for the population as a whole. For example, predicting the probability that an individual will be a carrier for a recessive genetic disease depends on parental genotypes within families and on gene frequencies within the population as a whole. Yet, the composition of the gene pool significantly influences the specific kinds of genes that may be included in any particular family sample. Failure to understand the most elementary concepts of *population genetics*, as this specialty within the field of genetics is called, often leads to confusion and misunderstanding of population phenomena, some of which have important societal ramifications. It is simply not sufficient to treat a population as *one* large family.

In this chapter we shall discuss the foundation of population genetics and attempt to develop the rudiments of population thinking, as opposed to familial thinking, in our consideration of the implications of a few basic genetic situations at the population level. We shall also discuss the concept of race and the definition of racial distinctions in terms of population genetics.

THE FOUNDATION OF POPULATION GENETICS

Within a large population we do not expect to find all dominant traits in a 3:1 ratio to recessive traits or all genotypes for alternate alleles in a ratio of $1AA:2Aa:1aa$. These are only ratios that are expected within *certain kinds* of families, specifically those where both parents are heterozygous at the locus

273

concerned. As we are well aware, if the genotypes of the parents are different, the phenotypic and genotypic ratios expected in their offspring will be different also. A population consists of all these different family groups, and the proportion of phenotypes and genotypes in the whole population will be the sum of the offspring of all these families. Considering the simplest case of two alleles at a single locus with A dominant to a, there are nine different marriage combinations possible, each with a specific genetic expectation among the offspring (family predictions). These are listed in Table 13-1. Only if all marriage combinations are equal in occurrence can the progeny in Table 13-1 be summarized to represent the constitution of the next generation in the proper proportions; this would literally never be expected. It is therefore necessary to determine the frequency of each marriage combination in order to place the proper weight on its contribution to the progeny of the next generation.

Regardless of the proportion of genotypes in a population at any particular point in time, only two kinds of gametes can be produced by either sex—A-bearing and a-bearing—and their frequency will reflect the proportion of each allele in the gene pool. In large populations, mating is usually random with respect to any particular locus, that is, the choice of a mate is not determined by, or a result of, the genotype at one specific locus. This means that the probability that any two gametes will meet to form a zygote (offspring) is the product of their independent frequencies. Thus, if a population is viewed as a gene pool, the *frequency* of the alleles at any particular locus will determine the proportion of genotypes. As long as random mating continues and other factors, such as mutation and selection, that change gene frequency are absent, low, or in balance, the constitution of the population will remain virtually unchanged indefinitely. For example, if 60% of the alleles in the gene pool are A and the other 40% are a, the proportion of AA offspring expected in the population will be

$$A \times A = A^2 = .6 \times .6 = .36$$

Table 13-1. Possible marriage combinations and genetic expectations among the offspring with respect to one locus with two alleles (A and a)

Marriage combinations Parental genotypes Female Male	Offspring Progeny genotypes		
	AA	**Aa**	**aa**
$AA \times AA$	1	0	0
$AA \times Aa$	$\frac{1}{2}$	$\frac{1}{2}$	0
$AA \times aa$	0	1	0
$Aa \times AA$	$\frac{1}{2}$	$\frac{1}{2}$	0
$Aa \times Aa$	$\frac{1}{4}$	$\frac{1}{2}$	$\frac{1}{4}$
$Aa \times aa$	0	$\frac{1}{2}$	$\frac{1}{2}$
$aa \times AA$	0	1	0
$aa \times Aa$	0	$\frac{1}{2}$	$\frac{1}{2}$
$aa \times aa$	0	0	1

And the proportion of *aa* offspring expected will be

$$a \times a = a^2 = .4 \times .4 = .16$$

This accounts for 52% of the genotypes, leaving 48% of the population as heterozygotes. To calculate this remainder, we must account for the fact that the heterozygous genotype can be formed in *two* different ways (*A*-sperm + *a*-egg or *A*-egg + *a*-sperm) and is therefore twice as likely to occur as either homozygote. The proportion of *Aa* offspring expected will be

$$2 \times A \times a = 2Aa = 2 \times .6 \times .4 = .48$$

Thus, a randomly mating population unaffected by mutational or selectional forces and consisting of 60% *A* and 40% *a* alleles is expected to have genotypic proportions of

$$36\% \ AA + 48\% \ Aa + 16\% \ aa = 100\% \ \text{(the entire population)}$$

and phenotypically be 84% dominant types *(AA + Aa)* and 16% recessive types *(aa)*.

The frequencies of the *A* and *a* alleles in the population (the gene pool) can be simply calculated from the genotypic proportions. Since *AA* genotypes are all *A* alleles and *Aa* genotypes are half *A* alleles, the frequency of *A* in the population is

$$A = AA + \tfrac{1}{2}Aa = .36 + \tfrac{1}{2} \ (.48) = .36 + .24 = .60$$

In the same manner the frequency of *a* is

$$a = aa + \tfrac{1}{2}Aa = .16 + \tfrac{1}{2} \ (.48) = .16 + .24 = .40$$

These also represent the expected gametic frequencies to produce the next generation and, since they are the same as the previous generation, the genotypic proportions in the population will remain constant over time as long as random mating persists. Only if some force, such as mutation or selection, should alter the gametic frequencies would the population be expected to change.

A "working proof" of this population equilibrium is given in Table 13-2. This table is comparable to Table 13-1 except that the proportion of each marriage combination is considered (the product of the genotype frequencies) and this "weight" is multiplied through the expected progeny proportions (Table 13-1) for each combination to account for their contribution to the next generation. In contrast to Table 13-1, the progeny in Table 13-2 *can* be summarized to represent the constitution of the next generation in the proper proportions. As the table indicates, the new generation would have the same genotypic (and phenotypic) proportions as its parental generation and would, in turn, produce a new generation like itself in a similar manner, and so on. As long as the proper conditions exist (random mating, and so forth), a stable population structure, dependent on the frequency of the alleles, will be established.

This principle of genetic stability of a population is true for any combination

Table 13-2. The frequency of marriage combinations and progeny proportions expected from the random mating of individuals in a population consisting of 36% *AA* + 48% *Aa* + 16% *aa* genotypes*

Marriage combinations Parental genotypes and proportions		Proportion of combinations in the population	Offspring Progeny genotypes and proportions		
Female	Male		AA	Aa	aa
.36 *AA* ×	.36 *AA*	.1296	.1296	0	0
.36 *AA* ×	.48 *Aa*	.1728	.0864	.0864	0
.36 *AA* ×	.16 *aa*	.0576	0	.0576	0
.48 *Aa* ×	.36 *AA*	.1728	.0864	.0864	0
.48 *Aa* ×	.48 *Aa*	.2304	.0576	.1152	.0576
.48 *Aa* ×	.16 *aa*	.0768	0	.0384	.0384
.16 *aa* ×	.36 *AA*	.0576	0	.0576	0
.16 *aa* ×	.48 *Aa*	.0768	0	.0384	.0384
.16 *aa* ×	.16 *aa*	.0256	0	0	.0256
Total		1.0000	.3600	.4800	.1600
		All possible marriages	Constitution of "new" generation		

Note that the constitution of the "new" generation (sum of progeny) is identical to that of the parental generation and the population remains unchanged.

of allelic frequencies in the gene pool and can be generalized as follows: If the proportion of *A* alleles is called p and the proportion of *a* alleles q, then all of the gametes will be produced in these proportions so that

$$pA + qa = 1$$

and the proportion of *AA* genotypes expected in the population will be

$$A \times A = A^2 = p \times p = p^2$$

the proportion of *aa* genotypes expected will be

$$a \times a = a^2 = q \times q = q^2$$

and the proportion of *Aa* genotypes expected will be

$$2 \times A \times a = 2Aa = 2 \times p \times q = 2 \ pq$$

Thus, the genotypic constitution of the population will be

$$p^2AA + 2pqAa + q^2aa = 1 \text{ (the entire population)}$$

This general formula is known as the *Hardy-Weinberg equilibrium equation*, named after the English mathematician, G. H. Hardy, and the German physician, W. Weinberg, who, in 1908, independently developed this basic mechanism of the action of genes in populations. This formula provides the key to making genetic predictions at the population level and forms the foundation for the mathematical interpretation of natural selection and the processes of evolution. We shall have only a brief encounter with this mathematically sophisticated area of human genetics in the remainder of this chapter.

The Hardy-Weinberg principle explains why traits such as eye color, hair color, and blood groups, or abnormalities such as polydactyly are maintained from generation to generation in relatively constant proportions. Some common errors in applying familial thinking to population phenomena are to assume that dominant traits will swamp out recessive ones or to conclude that traits that are very common in the population must be dominant and those that are rare must be recessive. Dominance and recessiveness are relationships that refer to the phenotypic expression of alleles in *individuals* possessing them. On the other hand, phenotypic proportions in a population are a function of the frequencies of the controlling alleles in the *gene pool.*

This point can be illustrated by applying the Hardy-Weinberg equilibrium formula to a pair of alleles where the dominant form is lower in frequency than the recessive form. If we consider a gene pool where B alleles constitute 20% (p = .2) and b the other 80% (q = .8), random mating would produce genotypic proportions of

$$p^2BB + 2pqBb + q^2bb = 4\% \ BB + 32\% \ Bb + 64\% \ bb$$

Phenotypically the dominant trait *(B___)* would only appear in 36% of the population, and the recessive trait would be the prevalent one (64%). The gametic frequencies expected in this population would be

$$B = BB + \tfrac{1}{2}Bb = .04 + \tfrac{1}{2} \ (.32) = .04 + .16 = .20 = p$$

and

$$b = bb + \tfrac{1}{2} \ Bb = .64 + \tfrac{1}{2} \ (.32) = .64 + .16 = .80 = q$$

These are the same gametic frequencies as those of the previous generation and, as long as random mating persists and no other forces change gametic frequencies, the population will be in equilibrium at this locus. Thus, there is not an inherent tendency for dominant alleles to increase in a population simply because they are dominant. The frequency of any allele is not determined by its being either dominant or recessive but mainly by its selective value in the population through the action of natural selection. When allelic frequencies change over time, it is usually because of a change in their selective value as the population adapts to change in its environment.

Many human traits are in an apparent Hardy-Weinberg equilibrium perhaps because they are adjusted to the present environment and not changing or because cultural evolution (medical technology, sanitation, and so forth) has eliminated strong selective differentials between many alleles. Such genetic variation has possibly become a "relic of the past" which is now nearly neutral with respect to natural selection, in balance with mutation rates, and in conformity to random mating within the population.

A human blood group known as the *MN system* appears to fit perfectly into this situation and, because the alleles in this system are codominant, it provides an ideal example. The gene locus for this system is called L (after Landsteiner and Levine who discovered it in 1927), and the two codominant alleles are designated as L^M and L^N. These alleles produce antigens called M and N respectively, which react with the appropriate antibodies, anti-M and anti-N, in a fashion similar to the ABO system. Table 13-3 summarizes the pertinent information on the MN system. Note that *three* phenotypes, corresponding to the three different genotypes, can be distinguished according to their reactions with the antibodies (this is similar to blood types, A, B, and AB).

In one extensive study of the MN system, the genotypic proportions among 2,320 parents were found to be

$$.311\ L^ML^M + .492\ L^ML^N + .197\ L^NL^N$$

The gametic frequencies expected in the "population" would be

$$L^M = .311 + \tfrac{1}{2}\ (.492) = .311 + .246 = .557 = \mathrm{p}$$

and

$$L^N = .197 + \tfrac{1}{2}\ (.492) = .197 + .246 = .443 = \mathrm{q}$$

These frequencies would describe a Hardy-Weinberg equilibrium of

$$\mathrm{p}^2 = L^ML^M = .310$$
$$2\mathrm{pq} = L^ML^N = .494$$
$$\mathrm{q}^2 = L^NL^N = .196$$

which is very close to that described for the parental group examined and should be the same for the next generation. An examination of 2,734 children of the parental group under study was also made, and the proportions of genotypes among them were

$$.310\ L^ML^M + .495\ L^ML^N + .196\ L^NL^N$$

Thus, we see that except for extremely small deviations (the result of chance sampling error in a finite sample), the population under study is in Hardy-Weinberg equilibrium and predictions about the composition of subsequent generations can be made.

Even so, within any particular family, typical "familial predictions" can be made according to parental genotypes; for example, if both parents are of MN blood types, the expectations among *their* children are $\tfrac{1}{4}$ type M: $\tfrac{2}{4}$ type MN: $\tfrac{1}{4}$ type N. Families of this composition constitute only a certain "segment" of the entire population however. In this case that segment ($L^ML^N \times L^ML^N$ marriage combinations) would be expected to form about 24% of the marriage

Table 13-3. Summary of the MN blood group system

Possible genotypes	Antigens produced	Reactions with		Phenotype (Blood type)
		anti-M	anti-N	
L^ML^M	M	+	−	M
L^NL^N	N	−	+	N
L^ML^N	M and N	+	+	MN

combinations, as calculated from the expected (or actual) frequency of the *MN* genotype $(.494 \times .494)$.

For many genetic traits, especially diseases, it is not always possible, or practical, to identify carriers of recessive alleles by biochemical tests. (Mass screening for certain diseases may change this situation in the future, however.) In these cases the Hardy-Weinberg equation can be used to estimate the frequency of heterozygotes in the population. For example, in an earlier discussion (Chapter 7) of the inheritance of *phenylketonuria* (PKU) it was stated that the incidence of this recessively conditioned disease is 1 in 10,000 births. This means that

$$pp = 1/10{,}000 = .0001 = q^2 \text{ (frequency of the recessive genotype)}$$

and q, the frequency of the *p* allele, can be estimated as the square root of q^2 or

$$q = \sqrt{.0001} = .01 \text{ (frequency of the } p \text{ allele)}$$

since $p + q = 1$, $pP + qp = 1$

$$pP = 1 - qp = 1 - .01 = .99 \text{ (frequency of the } P \text{ allele)}$$

Thus, the frequency of heterozygotes (*Pp* genotypes) can be estimated according to the formula as

$$2pq = 2 \times .99 \times .01 = .0198$$

This means that about 2% of the population, or 1 in 50, is expected to be carriers of PKU. It can also be predicted that the probability that two parents at random from the population, with nothing known of their genetic backgrounds with respect to PKU, might *both* be carriers of PKU is $\frac{1}{50} \times \frac{1}{50} = \frac{1}{2{,}500}$ (product of the independent probabilities). If they are both carriers *(Pp × Pp)*, the chance of having an affected child is 1 in 4; this gives a total probability for two persons at random in the population producing a child with PKU of $\frac{1}{2{,}500} \times \frac{1}{4} = \frac{1}{10{,}000}$. This *is* the observed incidence, as expected, because virtually all PKU children are born to "normal" parents (carriers).

Another serious genetic disease that has recently come to public attention is *Tay-Sachs disease*. This recessively conditioned disease, like sickle cell anemia, is fatal to homozygotes at an early age, lacks any medical treatment, and is conducive to mass screening for the detection of carriers. It is also prevalent in a minority group in the United States, the Jewish people. In 1881, Dr. Warren Tay, a British ophthalmologist, first described the disease noting a characteristic large white patch in the retina of the eye at the center of which is a brownish-red spot, known as the cherry-red spot. Shortly after, in 1896, the first pathological description was made by Dr. Bernard Sachs, an American neurologist. He described the abnormal structure and degeneration of the brain and nervous system which lead to paralysis, severe mental impairment, and blindness. Noting its tendency to run in families, he named the disease *amaurotic* (associated with blindness) *familial idiocy;* Tay-Sachs disease is now the preferred designation. The symptoms result from the accumulation of a specific lipid (fat compound) and a cholesterol compound in the brain and spinal cord. This abnormal storage appears to result from the inability to break down certain fatty compounds because of the inactivity of a particular enzyme (a hexosaminidase), that is, a normal gene product is lacking. This abnormal buildup of compounds that normally circulate in the body produces disastrous consequences. By 6 months of age the afflicted child's purposeful movements decrease, and its facial expression becomes vacuous. As the disease progresses, megalencephaly (abnormal brain enlargement) develops, convulsions become frequent, and the child lays in a froglike position. Blindness develops by the end of the first year; by 2 years of age the child is merely vegetating, and death usually occurs before the age of 5.

Detection of Tay-Sachs carriers (heterozygotes) can now be accomplished by a hexosaminidase assay from the blood serum. This test for activity of the enzyme is specific for the "Tay-Sachs gene" and very dependable because there appears to be no overlap in the amount of activity between heterozygotes and homozygous normal individuals. The lack of activity in afflicted homozygotes can be detected in fetuses after the sixteenth week of pregnancy by amniocentesis and hexosaminidase assay. In one group of pregnant women at risk for having a Tay-Sachs child (both parents carriers), six of fifteen were prenatally diagnosed to be carrying Tay-Sachs fetuses. The accuracy of the diagnoses was confirmed in five fetuses after therapeutic abortions and in the other after its birth and subsequent development of the symptoms.

Should mass screening to detect Tay-Sachs carriers be conducted? It is estimated that 1 in 6,000 Jewish births is a Tay-Sachs baby, whereas the incidence among non-Jewish births is 1 in 500,000. Even among Jews the incidence is greatest in those of eastern Polish ancestry known as the Ashkenazic Jews. It is presumed that the mutation first arose in this group of people centuries ago—just why the mutation would have had a selective advantage to account for its maintenance is not understood. The frequency of the Tay-Sachs gene *(t)* in the Jewish population can be estimated as

$$tt = 1/6,000 = .000167 = q^2$$
$$q = \sqrt{.000167} \approx .013$$

and the frequency of the normal allele *(T)* as

$$p = 1 - q = 1 - .013 = .987$$

The predicted proportion of carriers in the Jewish population is

$$Tt = 2pq = 2 \times .987 \times .013 = .0257$$

Thus, about 2.5%, or 1 in every 40 Jews, carries this deleterious gene. Similar calculations for non-Jews predict that 1 in 380 is heterozygous.

We can see the value of carrier detection for such a severe genetic disease as this, especially within the Jewish population where 1 in every 1,600 marriage combinations ($\frac{1}{40} \times \frac{1}{40}$) is expected to be at risk (both carriers) if random mating occurs. Now society must decide how to handle this genetic technology. Should two carriers be prohibited by law from marrying? Should a Tay-Sachs fetus, confirmed by amniocentesis and hexosaminidase assay, be aborted? Does society have the right to *require by law* that everyone be tested to detect carriers of diseases such as this (or sickle cell anemia)? If so, does it then have the right to *require* the testing of all fetuses of parents at risk with *mandatory* abortion of all those confirmed to be afflicted? These and many more disturbing and perplexing problems face individual parents and the whole of society as our genetic knowledge advances and is applied to man. In the next chapter we shall consider some of the more far-reaching and radical effects of genetic technology on the future of the human species.

Predictions can also be made for sex-linked traits. In these cases the genotypic (and phenotypic) proportions are not expected to be the same for both sexes, however. The hemizygous male genotypes (and phenotypes) are a *direct* reflection of the allelic frequencies in the gene pool of an equilibrium population, whereas the expected proportions of female genotypes still conform to the Hardy-Weinberg formula. These expectations are:

$$\text{females:} \quad p^2 \, AA + 2pq \, Aa + q^2 \, aa$$
$$\text{males:} \quad p \, AY + q \, aY$$

For a severe sex-linked recessive disease, such as hemophilia, which appears almost exclusively in males, it is possible to estimate the proportion of female carriers who are the potential mothers of afflicted sons. In our discussion of hemophilia in Chapter 9, its incidence was given as 1 in 10,000 male births. This represents q, the frequency of the allele for hemophilia *(h)*. Thus, if $q = \frac{1}{10,000} = .0001$, then p (the frequency of H) = .9999. Female carriers *(Hh)* would be expected to constitute 2pq of all females, or $2 \times .9999 \times .0001 = .000199 = .02\%$ or about 1 in 5,000 females. The proportion of female hemophiliacs expected would be $q^2 = (.0001)^2 = .00000001$ or 1 in one hundred million female births or roughly 1 in every 2 hundred million members of the population (females and males). However, since it takes a hemophiliac father to produce an afflicted female and these men do not reproduce at a normal rate, the expected frequency of female hemophiliacs is further reduced by the male reproductive disadvantage. Assuming a reproductive disadvantage five times less than normal would lower our prediction of hemophiliac females by $\frac{1}{5}$ or only 1 in a billion births (1 in 500 million female births).

WHY YOU SHOULDN'T MARRY YOUR FIRST COUSIN

From time to time in previous chapters the genetic hazards of *consanguinity* have been alluded to. The most common form of consanguinity in the human population is cousin marriages. Brother-sister and parent-child combinations are the very closest genetic relationships possible, but these matings are rare and are viewed as unacceptable in the mores of nearly every society.

As members of the same species, all human beings are obviously related to some extent—we all share in the same gene pool. Consanguinity, or *inbreeding,* refers to marriage between close relatives, where closeness is defined by the probability that two individuals might both possess a particular allele that is *identical by descent* through their immediate family. Such alleles originate as replicas of the very same allele in a common ancestor of some previous generation. The offspring produced by parents who have a recent common ancestor (within two or three generations) have a much higher probability of becoming homozygous for a particular allele than offspring of parents who are "unrelated" (no recent common ancestor). Homozygosity in the latter case requires independent lines, or distantly related ancestors, as the sources of the alleles. These alleles are said to be *alike in state,* as contrasted to identical by descent.

Random mating, or the mating of "unrelated" individuals, means that the probability is nil that any alleles in the genotypes of the parents are identical by descent and the probability of having any allele in common, alike in state, is solely dictated by the frequency of the particular allele in the gene pool (as predicted by the Hardy-Weinberg formula). Inbreeding represents nonrandom mating because the related individuals may possess alleles in common, which are identical by descent and not simply a product of their frequency in the population as a whole. The proportion of alleles that are expected to be identical by descent in any two individuals defines the "closeness" of their relationship. This is known as the *coefficient of relationship* (r) and can be calculated as the probability of transmitting any particular allele along a pedigree pathway that connects the individuals in question. Fig. 13-1 illustrates the calculation of the coefficient of relationship (which is really a correlation coefficient identical to the r discussed in Chapter 11) between parent and child *(A)* and grandparent and grandchild *(B).* Note that the gray line in the figure traces the pedigree pathway. The probability that one particular allele will be transmitted from a parent to a child anywhere along the pathway is always $\frac{1}{2}$, and the total probability of transmission through a pathway is the product of each independent "step" along the way, or $(\frac{1}{2})^n$ where n equals the number of steps (transmissions)—this procedure is known as the method of path coefficients. The coefficients of relationship in Fig. 13-1 mean that a child has half of its genes in common with one of its parents, while a grandchild has, on the average, one fourth of its genes in common with one of its grandparents. In other relationships there are two different pathways (through each of two common ancestors) by which two individuals may receive identical alleles; in these cases r is the sum of the separate path probabilities. The calculation of r values for three close relationships, including first cousins, is illustrated in Fig. 13-2. Thus, we see that, on the average, first cousins are expected to have one eighth of their alleles identical by descent.

It is well substantiated that all persons are heterozygous at many loci and that we all carry deleterious recessive alleles of various kinds. For example, it has been estimated that every person carries about four *lethal equivalents.*

This means that, on the average, each person possesses some combination of recessive alleles which, if homozygous, could cause death in four different combinations (full lethals) or give a 50% probability of death in eight different combinations (partial lethals). There is also the possibility that in some other ways these alleles could add up to the equivalent of four possibilities of genetically caused death. These and all other masked defects carried within the whole

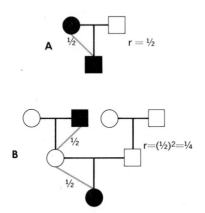

Fig. 13-1. Calculations of the coefficients of relationship *(r)* for **A,** a parent-child relationship and **B,** a grandparent-grandchild relationship. The shaded symbols indicate the individuals between whom the relationship is estimated.

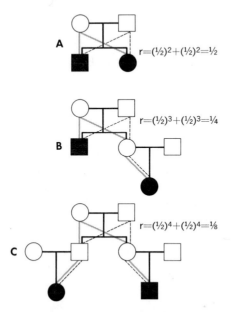

Fig. 13-2. Calculations of the coefficients of relationship *(r)* between **A** two sibs, **B** uncle-niece (aunt-nephew), and **C** first cousins. Each of these relationships consists of two pathways through each of two common ancestors; the heavy gray line indicates one path and the light gray line indicates the other; *r* symbolizes the sum of both paths.

population form what is called the *genetic load,* much of which is generated by recurrent mutations. Given the tremendously large number of loci in the total human complement, the likelihood that two unrelated individuals will possess the *same* deleterious alleles, for example, lethal equivalents, is quite low. On the other hand, inbreeding increases homozygosity of identical alleles, thereby increasing the chances of exposing these deleterious mutations—this is called *inbreeding depression.* It is therefore valuable to have a relative measure of inbreeding to estimate the amount of homozygosity expected in the offspring of consanguineous marriages.

The proportion of homozygosity expected in an inbred offspring can be calculated as the probability that two alleles at a given locus in a given individual are identical by descent (that is, are derived from the same ancestor). This probability is known as the *coefficient of inbreeding* (F). It can be computed by the method of path coefficients as the sum of the probabilities of all pathways connecting the offspring from one parent back to the other parent through each ancester common to both parents. Fig. 13-3 illustrates the calculation of F for the offspring of an uncle-niece (aunt-nephew) mating *(A)* and that of a first cousin mating *(B).* It should also be noted that F can be calculated from the coefficient of relationship as $F = r \times \frac{1}{2}$. This follows because if one parent transmits a particular allele to the offspring, the probability the other related parent has the identical allele is r, and the probability that that parent will transmit it to the same offspring is $\frac{1}{2}$. Thus, for an uncle-niece mating $F = \frac{1}{4} \times \frac{1}{2} = \frac{1}{8}$, and for a first cousin mating $F = \frac{1}{8} \times \frac{1}{2} = \frac{1}{16}$.

In a first cousin mating, if one member *is* a carrier of a particular deleterious allele, the probability that the other member also carries the same allele, identical by descent, is *always* $\frac{1}{8}$, r, and the chance that their child will be homozygous for that allele is *always* $\frac{1}{8}$ (the probability that one first cousin is carrier when the other one definitely is) $\times \frac{1}{4}$ (the probability that two carriers will have a homozygous recessive child) $= 1$ in 32. Note that this is exactly half the inbreeding coefficient because F measures homozygosity for

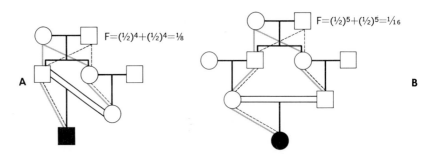

Fig. 13-3. Calculation of the coefficients of inbreeding *(F)* for the offspring of **A,** an uncle-niece mating and **B,** a first cousin mating. The shaded symbols indicate the individuals for whom *F* is estimated.

any allele. Since the normal allele in each carrier first cousin has the same probability of being identical by descent and becoming homozygous in the offspring as does the recessive deleterious allele, $F = \frac{1}{32} + \frac{1}{32} = \frac{1}{16}$. Thus, we can say that in any consanguineous mating where one member *is* definitely a carrier of one particular allele, such as a deleterious recessive, the probability that a child who is homozygous for that allele will be produced is $\frac{1}{2}F$. However, if either carrier were to mate with someone from the population at random (unrelated), the probability that that person would be a carrier for the same allele, alike in state, would be 2pq. In other words depending on the frequency of the allele in the gene pool, the probability that this mating would produce a child homozygous for the allele in question would be 2pq $\times$ $\frac{1}{4}$ (the probability that two carriers will have a homozygous recessive child). If the allele is rare or relatively low in frequency in the population, 2pq will be much less than $\frac{1}{8}$ (r for first cousins) and the probability of having a child homozygous for the allele will be much less than 1 in 32.

Fig. 13-4 illustrates the difference between a carrier for a rare recessive (III-2), such as the allele for PKU, marrying his first cousin (III-1) as opposed to marrying an unrelated woman representing a random selection from the gene pool with respect to the PKU allele (III-3). Earlier in the chapter it was calculated that the frequency of carriers of PKU in the population was estimated at 1 in 50 (2pq); thus, the probability that child IV-1 will have PKU is 1 in 32 ($\frac{1}{8} \times \frac{1}{4}$), while the probability that child IV-2 will be afflicted with the disease is 1 in 200 ($\frac{1}{50} \times \frac{1}{4}$). Remember that in both cases III-2 *is* a carrier, so the probability in each case is influenced by this fact.

If we were to start from scratch to consider the probabilities of a PKU child from a random mating of two individuals with no pedigree information available or influencing the prediction, the best prediction would be 2pq $\times$

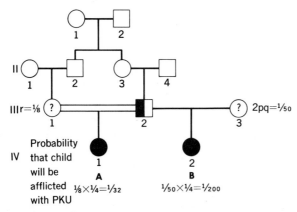

Fig. 13-4. Comparison of the probabilities of producing a child with PKU disease when a carrier (III-2) marries **A,** his first cousin or **B,** an unrelated woman at random from the population.

2pq (the probabilities that both are carriers) $\times$ ¼ (chance of PKU child) $=$ ⅟₅₀ $\times$ ⅟₅₀ $\times$ ¼ $=$ ⅟₁₀,₀₀₀ (which is q^2 of the Hardy-Weinberg formula). But, if the marriage involves consanguinity, then the effect of inbreeding must be considered in addition to the frequency of the allele in the gene pool. For consanguineous marriages the formula (which will not be derived but simply stated for our use) $q^2(1-F) + qF$, where q is the frequency of the allele in question and F is the inbreeding coefficient for the child in question, gives the probability of having a homozygous recessive child when neither mate is known to be a carrier and pedigree information does not influence carrier probability. In our example q $=$ .01 or ⅟₁₀₀ (see discussion of PKU earlier in the chapter), and F is ⅟₁₆ for a child of a first cousin mating. Thus:

$$q^2 (1-F) + qF = (1/100)^2 (1-1/16) + 1/100 (1/16) =$$
$$1/10,000 (15/16) + 1/1,600 = 15/160,000 + 1/1,600 =$$
$$15/160,000 + 100/160,000 = 115/160,000, \text{ or approximately } 7/10,000$$

This represents a seven-fold increase in the risk (⁷⁄₁₀,₀₀₀ : ⅟₁₀,₀₀₀) of producing a PKU child from a first cousin mating as compared to random mating, even when no factor other than the frequency of the deleterious allele in the population is considered.

Considering another example previously discussed in this chapter, Tay-Sachs disease, where q $=$ .013 and 2pq $=$ approximately .025 in the Jewish population, the probabilities of producing an affected child in the situations described above are as follows:

When one parent is a carrier and the parents are unrelated, the probability of a Tay-Sachs baby is 2pq $\times$ ¼ $=$ .025 $\times$ .25 $=$ .00625 or 1 in 160. But if the parents are first cousins, the probability is r $\times$ ¼ $=$ ⅛ $\times$ ¼ $=$ 1 in 32. This is a five-fold increase in risk over random mating.

In the Jewish population at random the probability of a Tay-Sachs baby from the mating of two unrelated individuals is 2pq $\times$ 2pq $\times$ ¼ $=$ .025 $\times$ .025 $\times$ .25 $=$ .00015625 or 1 in 6,000. But if the mating is between first cousins, the probability is $q^2 (1-F) + qF = (.013)^2 (1-⅟₁₆) + (.013) (⅟₁₆) = .000169 (.9375) + (.013) (.0625) = .000158 + .000813 = .000971$ or approximately 1 in 1,000. This represents a six-fold increase in risk over random mating.

These calculations concerning consanguinity have been made under the same assumption used in analyzing pedigrees when the trait in question has a low frequency in the population, that is, it is assumed that the allele entered a pedigree only once. With rare alleles this simplifying assumption is nearly always valid. Hence, for rare recessive traits, which comprise the majority of genetic diseases and abnormalities, the probabilities derived above are reasonably accurate. From this it is quite evident that consanguinity significantly increases the probability of uncovering rare recessive alleles and exposing the progeny of such matings to greater genetic risks than those of the randomly mating portion of the population. Although some families assume that "closeness" is the best way to preserve the family line or keep hidden certain skeletons in their closets, the genetic skeletons in our family closets are best kept masked by random mating!

One sensitive problem that has been at issue in society for centuries and continues to plague the social fabric of our culture is the question of race. What is a race, and what is the significance of racial diversity? Emotion-laden debates occur today over issues such as "racial equality" and the "intelligence of races." It seems appropriate at this time to place the biological, or genetic, concept of race into the arena in the hope that this will shed some light on the problem and aid in the development of a rational approach to this difficult social issue.

The concept of race is a biological phenomenon; however, it is also a category of classification used to reduce the array of phenotypic diversity in the human species in manageable groupings for identification. Overemphasis of the latter use without recognition of the former has lead to much misunderstanding and abuse of the significance of racial distinctions. It is readily apparent that groups of people native to different parts of the world are often phenotypically different enough from each other to be easily distinguished. Thus, races as categories of classification are simply a natural outgrowth of the human tendency to systematize and label similar things into groupings that make diversity more comprehensible. It has been pointed out that if races did not exist, they would have to be invented. Where would anyone start in sorting out the more than three and a half billion people of the world other than with some obvious physical attributes that characterize them as individuals coming from similar geographical regions and similar or related cultures? We do not have to invent races; what is important, however, is to understand why and how racial distinctions exist. Only then can we ponder the implications of racial differences.

Races are broadly defined as subpopulations of a species that differ in allelic frequencies at certain loci. Races are not separate species isolated by reproductive incompatabilities (Chapter 2). Rather, they represent "geographic isolates" in which natural selection has produced phenotypic differences that enhance local adaptation to prevailing conditions. It is essential to understand that it is allelic frequencies and not gene differences which characterize races, and that a race is an interbreeding population consisting of genetically different individuals. A race is not a single genotype or genetically identical group of people; races are compounds of individual differences.

The differences in allelic frequencies that characterize races are considered to be the products or by-products (pleiotropic effects) of natural selection. As we observe racial differences today, some make "evolutionary sense" and others do not. In other words, we can trace some, but not all, differences in allelic frequencies between populations to particular selection pressures that account for the prevalence of a particular allele in one geographical area (in one race) and its low frequency in some other locale (in another race).

The classic example of this is skin color, the most common physical attribute traditionally used in race classification. We have already seen that this is a polygenic trait that produces a wide array of phenotypes, even within racial groups (Chapter 11). Yet, we can distinguish groups such as a "white" race and a "black" race. The difference between these groups lies in the frequency of the pigment-inducing alleles. But why should the frequencies be different?

In general dark-skinned populations are native to the areas of the world where sunlight is most abundant and intense. Light-skinned populations are native to areas where sunlight is less abundant and usually seasonally limited. Thus, these two geographic isolates were exposed to quite different conditions of sunlight to which the polygenic system governing skin color had to adapt. It is an old idea that heavy pigmentation developed as protection against the sun; however, an acquired tan in light-skinned people also provides effective protection against sunburn. Simple sunburn protection does not appear to be a sufficiently strong selection pressure to account for allelic differences, but the addition of two other pieces of evidence lends support to differential action of natural selection on pigment-inducing alleles.

The first involves vitamin D, the "sunshine vitamin," which is essential for normal calcium and phosphorus metabolism. It can be synthesized in the skin in the presence of sunlight; this was probably its major source for our ancestors. Its deficiency causes soft, fragile bones (rickets in children), whereas excessive amounts can result in the calcification (hardening) of soft tissues and kidney damage. Light pigmentation allows sunlight to penetrate the skin easily, which facilitates vitamin D production. It appears quite plausible that light pigmentation was adaptive to "sunlight deficient" populations for vitamin D production, especially in the most deficient season (winter) when the skin would be the lightest. Conversely, dark pigmentation provides protection against excessive amounts of the vitamin by keeping the production rate down.

There is also evidence that, although an induced tan can provide protection against sunburn, the incidence of skin cancer is higher in lightly pigmented individuals than in those with heavier pigmentation for comparable amounts of exposure to sunlight. This would provide another selective advantage favoring pigment-inducing alleles in sunlight abundant populations.

Another racial distinction for which differential selection pressures in geographically separated populations is well known concerns the allele for sickle cell anemia. Its high frequency in the black population stems from the fact that carriers (heterozygotes) have a selective advantage for survival in the tropical environment to which the black race is native. In this area where malaria (long a major killer of mankind) is prevalent, carriers of the sickle cell allele are more resistant to the disease than are normal homozygotes. Such heterozygote "superiority" produces a genetic load in the population in the form of recessive homozygotes who die of sickle cell anemia. But overall, it provides adaptation to the malarial environment; therefore, the sickle cell allele has a selective advantage and its frequency is established at a high level. On the other hand, populations in nonmalarial areas of the world suffer only the selective disadvantage of the recessive homozygote; because of this negative selection pressure, the frequency of the sickle cell allele is kept quite low, essentially at the level of its recurrent mutation rate. Now that malaria is fairly well under control and large black populations are established outside the malarial range, the detrimental genetic load of afflicted homozygotes is a prime genetic disease that is desirable to eliminate. Not only has the carrier lost the selective advantage of malarial resistance, but sickle cell trait can be considered disadvantageous in modern society.

The same circumstances, including malarial resistance of carriers, explain the high frequency of Cooley's anemia (thalassemia) in Mediterraneans (see Chapter 10). However, the high frequency of an allele like that responsible for Tay-Sachs disease is more difficult to explain because this so-called "ethnic disease" developed a high frequency in one group of people (Jews) within a larger population of a particular geographical area (eastern Europe). A general environmental factor such as malaria cannot explain the differential allelic frequencies between Jews and Gentiles in the "same" environment. To produce such a high frequency of a detrimental recessive allele, a selective advantage for the heterozygote would be expected. But how did the Ashkenazic Jews achieve their genetic individuality in the midst of all other eastern Europeans? It has been suggested that the restriction of Jews to ghettoes where they suffered a multitude of diseases as well as malnutrition set the "ecological factors" in which the heterozygote was superior. The close interbreeding within these communities could have contained the allelic frequency to its ethnic distribution, and inbreeding could have hastened its increase. This is still a speculative attempt to produce an explanation of an otherwise obscure variation in allelic frequencies between populations.

Racial diversity of blood group systems is also well known. We saw in Table 8-1 that the frequency of ABO blood types varies between the black and white populations of the United States. Among the Chinese the frequencies, which differ from both American populations, are: 25% A, 34% B, 31% O, and 10% AB. Selective advantage of one blood type over another has been correlated with certain pathological conditions (for example, duodenal ulcers), and dietary habits (for example, high fat diets or predominantly carbohydrate diets), and disease immunity (for example, resistance to smallpox or syphilis), but no convincing evidence is yet available to explain this genetic diversity.

Understanding the formation of races does not answer the critical question of the magnitude of genetic differences between races. It has long been stressed by many biologists that, although races are distinguished by conspicuous physical features such as skin color, hair texture, and facial structure, all mankind shares a common gene pool in which racial diversity represents only minor fluctuations of allelic frequencies in an otherwise very similar, but quite variable, genetic system. Only recently has other than circumstantial data in support of this contention become available. The stumbling block in the past had been the inability to collect data on a random sample of many loci of "average genes" for comparison between races. In other words, how much difference is there in a random sample of genes rather than a selected set of obvious traits.

A technique, called electrophoresis, is capable of detecting subtle differences in protein structure (gene product) by differential migration through an electrically charged field. This makes it possible to obtain allelic frequency data for a random sample of protein loci. Two or more "charged forms" of a protein produced by amino acid substitutions in the polypeptide chain distinguish alternate alleles at a particular locus. In one study involving 28 different protein loci sampled in the American black, American white, and Japanese populations, the following observations were made:

1. Sixty-six percent of the loci were monomorphic (no allelic variation), and the fixed allele was always the same for all three populations. Thus, there was no complete gene substitution at any locus.

2. The remaining loci were polymorphic (two or more allelic forms detectable). Only this portion can contribute to genetic variability within or between races. However, when allelic frequencies were used to estimate the expected proportion of different alleles between two randomly chosen individuals, it was found that the difference between individuals from the same population was nearly as great as that between individuals of different populations. It is readily apparent from the data given in Table 13-4 that the net difference between populations is quite small compared to the differences between individuals within any population.

The conclusion drawn from these data, and supported by studies of various blood group frequencies, is that the genes in these three racial groups are remarkably similar, despite conspicuous phenotypic differences. These data also reiterate the well established fact that there is a great deal of genetic variation *within* races. Of course when complex traits, such as intelligence, are considered, we must include physical environmental effects on the polygenic genetic system that governs the trait, plus the environmental effects of cultural differences that influence any measurement of the trait, before it is even possible to

Table 13-4. Estimates of the expected proportion of different alleles between two randomly chosen individuals within and between three racially distinct populations derived from 28 common protein loci*

Population	American black	American white	Japanese
American black	12.0%	1.6%	1.5%
American white		14.6%	0.5%
Japanese			12.0%

Figures on the diagonal are within population differences (in percent of alleles), and those above the diagonal are net differences (over and above within population variation) between populations.
(Data source: Nei, M., and Roychoudhury, A. K.: Gene differences between Caucasian, Negro and Japanese populations, Science **177**:434, 1972.)

allege significant genetic differences. It is commonly stated that the average IQ of American blacks is lower than the average for American whites living in the same cities or states. However, this difference cannot be solely a genetic distinction unless all environmental factors are at least relatively equal, but they are not. No one can deny that there is great variation *within* races for intelligence and a broad overlap *between* races; thus, the net genetic variation may be as small as that observed for random protein loci in Table 13-4. A genetic identity between two different races is not really expected, but neither is a significant genetic difference. Even if some portion of the measured intelligence difference between racial groups is genetic, because of the broad overlap and wide variation, there is no rationalization for treating a member of any particular race as some uniform representative of that race. Nor is there any reason on this basis to declare a person "better" or "worse" than all members of some other race with a slightly different population mean for a polygenic trait.

In France a positive correlation has been found between stature and intelligence. It cannot be ascertained to what extent the correlation between these two polygenic traits is genetically or environmentally caused. Therefore, can it be argued with any validity that tall Frenchmen are genetically superior to short Frenchmen and thereby justify social and economic advantages to tall people while depriving others because of their shortness?

Of course racial distinctions disappear when reproductive barriers, which are a by-product of geographical isolation, break down. As mentioned in Chapter 11, this does not reduce genetic variation unless new selection pressures emerge to eliminate certain alleles. The net result is simply that the variation no longer characterizes specific racial groups. Current estimates place the proportion of alleles of Caucasian origin in the American black population at somewhat more than 20%. With the physical release from geographic isolation and the social release from reproductive barriers, racial and ethnic distinctions are likely to wane in the United States and perhaps worldwide in future generations. Someday the entire world may be able to regard itself as one cosmopolitan human race, with emphasis on the "human."

14

Biosocioprolepsis—the future

The title of this chapter is taken from a word coined by Albert Rosenfeld in his book *The Second Genesis: The Coming Control of Life*. He offers the word *biosocioprolepsis* as a term to describe the various possibilities that are, or soon will be, at hand for manipulating our biological future. Prolepsis means anticipation and is used to describe a method by which objections are anticipated in order to weaken their force; proleptics is the art or science of *prognosis*. Biosocioprolepsis, then, is the anticipation of biology's impact on society. Thus, by projecting present potentialities into future prospects, it may be possible to anticipate dangers and problems so that they can be avoided by our foresight rather than bewailed through our hindsight.

Biosocioprolepsis encompasses a wide range of biological potentialities that will have an impact on man and society. The effects of air and water pollution, drugs and psychotherapy, modern surgery, and research into the prolongation of life are facets of this "biological revolution" which cannot be adequately treated in the context of this book; indeed entire books are available on all of these subjects. The objective in this last chapter of our consideration of human heredity is to discuss the major genetic potentialities that may have an impact on our biological and social future.

The potentials for genetic manipulation can be sorted out for consideration at two levels of social impact. The first involves the prospects for the treatment of hereditary diseases on an individual basis, that is, treatment for the benefit of the afflicted person. Genetic manipulation at this level is called *euphenics* and represents new forms of treatment that essentially conform to the traditional goals of medical science for the relief of human suffering.

The second, and undoubtedly more controversial, level of social impact at which genetic manipulation can be considered is that dealing with the potentials for designing and shaping the very biological composition of our future generations through goal-oriented processes of selective breeding. This is known as *eugenics*.

PROSPECTS FOR TREATMENT OF HEREDITARY DISEASES

Over 1,500 hereditary diseases are known to afflict man, and, although most are rare, collectively they form a formidable challenge to medical science. About 100 of these diseases are understood at the functional level, that is, a specific enzyme deficiency (gene product) has been identified. For the remaining vast majority much less is known.

There is a very basic difference between hereditary diseases and other types (infectious or contagious), which must be recognized before our discussion proceeds. Although it is common to think in terms of *curing* disease, not a single human hereditary disease has ever been cured! For nongenetic diseases, such as attacks on our systems by micro-organisms, a cure rids the afflicted person of the infection. However, in the case of hereditary diseases, the source of the affliction, an abnormal gene, cannot be cured or corrected. Instead, a *treatment* to override or by-pass the malfunction may be able to relieve, or even eliminate, the symptoms for the individual concerned, but the defective gene persists and can be passed on to offspring. Hence, hereditary diseases cannot be cured in the conventional sense. Furthermore, their treatment may, in fact, produce an increase in the disease in future generations. This counterproductive aspect of medical "advancement" in the treatment of hereditary diseases creates one of the serious dilemmas of genetic engineering.

Past and present treatment of hereditary diseases has been accomplished in one of three ways, (1) dietary therapy, (2) product therapy, or (3) drug therapy. For example, phenylketonuria (PKU) and galactosemia are treated by dietary restrictions to lower or eliminate the compounds that cannot be handled normally by the body—phenylalanine and lactose, respectively—thereby avoiding the buildup of toxic substances. In other instances a product missing or inactive because of a defective gene, such as insulin in a person with diabetes mellitus, is supplied as the treatment. Missing or inactive enzymes (direct gene products) are not easily substituted, however, because the body either digests the enzyme or forms antibodies against it since it represents a foreign protein in the afflicted person's system. Finally, drugs have been used with some success to block or reduce the accumulation of toxic products and/or by-products from a malfunctioning metabolic pathway. The drug allopurinal helps reduce the accumulation of uric acid associated with the Lesch-Nyhan syndrome and gout by inhibiting an enzyme (xanthine oxidase) that produces the abnormal buildup. Once we understand the biochemical details of more genetic diseases, it may be possible to use drug therapy more extensively. Future prospects for this mode of treatment will depend on the development of drugs with specificity to correct the metabolic abnormality without eliciting undesirable side effects.

Future approaches to the treatment of hereditary diseases go beyond diets, products, and drugs and focus on the source, the defective gene. Sophisticated genetic manipulations, known as *gene therapy,* offer the possibility of more effective and efficient treatment, but the prospects for cures remain elusive. Gene therapy involves the insertion of a "good," that is, functional, gene (DNA) to replace a defective one in a person suffering from an inherited disease. Ideally the therapy would be permanent and heritable (transmitted by mitosis to replacement cells) and, if germinal tissue were included, it could theoretically become a true cure. However, current approaches are aimed toward treatment of a specific tissue or organ where the critical gene function occurs rather than "head to toe" treatment in hope of replacing the defective gene in every cell of the body. (After all, most genes are inactive in all cells except those differentiated to perform the specific function in question—see gene regulation, Chapter 6.) Several approaches to gene therapy seem feasible in light of current research and theory. These involve three modes of transferring the "good" genes to recipient cells and two sources, with variations, to supply the donor genetic material.

The most common mode of gene transfer to date has been through *viral infection*. Some viruses infect their host cell without killing it. These are called temperate viruses, as opposed to virulent types that cause illness and kill the cells they infect. Instead of multiplying in its host cell, a temperate virus inserts its genetic material into the genetic material (chromosomes) of the host. Here the viral genes are replicated with host chromosomes and passed on to daughter cells (by mitosis) as though they were normal components of the genetic complement.

Once the DNA of a temperate virus is inserted into the chromosome of the host cell, a combination of controls regulates its replication and expression. It is well established that at least some viral genes so inserted can function in the host cell, even if the host is man. The first evidence in man was discovered by accident in the early 1960s when several laboratory technicians at Oak Ridge National Laboratories, who had been working with a rabbit virus called *Shope papilloma*, became infected. When this virus infected rabbit cells, it induced the production of an enzyme, arginase, presumably the product of a functioning viral gene in the rabbit host. The infected laboratory technicians had abnormally low amounts of arginine (the amino acid metabolized by arginase) in their blood (with no ill effects), presumably because the viral gene was functioning to produce excess enzyme in their bodies.

In 1970 the first attempt at gene therapy in human patients was made. Two sisters in West Germany were discovered to be suffering from a very rare genetic defect, called hyperargininemia, by which they were unable to manufacture the enzyme arginase. The lack of this enzyme resulted in the abnormal buildup of the amino acid arginine in their blood to toxic levels, leading to progressively worsening mental retardation, epileptic seizures, and paraplegia. One sister, aged 5 years, was already in serious difficulty, and the other, aged 18 months, was deteriorating in the same way as her older sister. These two girls were purposely infected with the Oak Ridge Shope papilloma virus in the hope that arginase production could be induced in them as it had been earlier in the accidently infected laboratory technicians. If arginase production could be induced, the prospect was that the progression of ill effects, especially in the younger sister who was still in the early stages of the symptoms, could be arrested. Unfortunately, this landmark event in medical and genetic technology was unsuccessful. It is not known whether the dose was insufficient, the vaccine too impure, or just what went wrong, but the principle of gene therapy has been established as a distinct prospect in the human species.

Viruses are not only capable of inserting their own genes into host cells, but they can also serve as vehicles to carry nonviral DNA to the cells they infect. This occurs when a temperate virus becomes infectious again. The controls that regulate the stability of viral DNA in the host chromosome can be upset, inducing the temperate virus to replicate rapidly in the production of many new viruses. These viruses kill the host cell and are released to infect other cells. In the process, the host chromosome fragments and genes from the host may become incorporated into some of the newly formed viruses. If such a host-gene–carrying virus infects another cell, the nonviral genes may be inserted into the new host cell. These genes may then be integrated into the chromosome of the new host and begin to function.

Recent experiments have demonstrated that human cells, along with those of other animals and of plants, can be genetically modified in this manner. For example, it has been shown that a selected bacterial gene can be transmitted via virus to human cells (in tissue culture) to take up a function lacking in the human tissue. A virus, called *lambda,* that infects the common intestinal bacteria *Escherichia coli* was used to infect and pick up genetic material from its bacterial host. Lambda viruses have been extensively studied, and it is possible to select specific types of lambda according to which particular bacterial genes they are carrying. In one experiment, viruses carrying the bacterial gene that manufactures the same galactose-metabolizing enzyme (galactose-1-phosphate uridyl [GPU] transferase) that is present in normal humans but lacking in persons suffering from galactosemia was chosen to infect cultured human skin cells (fibroblast cells) from a patient with galactosemia. Infection of the skin cells resulted in the appearance of GPU transferase activity. This experiment represented the first successful directed transplant of bacterial genes into living human tissue.

It is also technically possible to construct infectious virus-like particles by enclosing pieces of foreign DNA within the protein coat of normal viruses that have had their own DNA removed. These particles, called *pseudovirions,* can be used to deliver specified genes to host cells. It has already been demonstrated that pseudovirions of an animal virus (protein coat of the polyoma virus) containing DNA from mouse cells can deliver this DNA inside the nuclei of human embryo cells. Recent developments make it possible to isolate particular genes from bacterial cells or to artificially synthesize a gene in the laboratory. Because of this, it has become increasingly possible that pseudovirions specific for infecting a particular tissue and carrying a selected "good" gene to replace a malfunctioning one might be used in the practice of gene therapy in the near future.

In addition to viral infection, two other modes of genetic transfer are plausible to achieve gene therapy. One, known as *transformation*, involves the direct transfer and integration of "naked" DNA into a host cell. The other mode, known as *cell fusion*, produces genetic transfer by the coupling of two cells with differing genetic complements. Each of these modes of gene transfer will be briefly described.

It has been known for many years that bacterial cells can be genetically modified by direct treatment with exogenous DNA, that is, DNA extracted from other cells. More recently, experiments with mutant strains of the fruit fly *Drosophila* indicate that the genetic properties of their eggs can be modified by treatment with DNA extracted from other strains of flies. And laboratory cultured human cells have been modified by exposure to exogenous DNA added to their growth medium. Successful modification to normal functioning has been achieved in immature human red blood cells (reticulocytes) that were synthesizing abnormal hemoglobin and in cultures of human cells lacking the purine enzyme HGPRT (hypoxanthineguanine phosphoribosyltransferase). These successes have not been easily repeated, perhaps because bulk DNA extracted from cells does not contain sufficient quantities of the particular gene desired. Gene isolation or artificial synthesis of a particular "good" gene in large quantities may enhance the reliability of transformation by exogenous DNA in the future.

New genes may also be introduced into particular cells by the fusion of two different kinds of cells (donor and recipient) grown together in tissue culture. When two cells fuse together, the "hybrid" cell formed contains two nuclei. However, whichever nucleus undergoes mitosis first becomes the controlling nucleus of the cell; this selection is primarily determined by the nucleus closest to undergoing mitosis at the time fusion occurs. The other nucleus may be forced to reproduce prematurely, causing abnormal condensation of its chromosomes, which results in their fragmentation. As the chromosomes of the non-surviving nucleus break down, fragments representing only a few genes may be incorporated into the genetic complement of the surviving nucleus. The integration of this genetic material is presumably accomplished in the same manner as viral-delivered or exogenous DNA.

Cell fusion was shown to be a feasible prospect for gene therapy when, in 1971, the successful transfer of a functional gene from chicken cells to mouse cells was reported. The mouse cells had a genetic deficiency expressed as the inability to manufacture the enzyme inosinic acid pyrophosphorylase (IAP); this enzyme is also deficient in humans with the Lesch-Nyhan syndrome. After cell fusions with chicken cells capable of producing IAP, cells controlled by mouse nuclei were obtained which could manufacture the enzyme. There was no mass transfer of genetic material but apparently a selective incorporation of the gene providing the function that was lacking in the recipient (mouse) cells. If such selectivity proves to be characteristic of cell fusions, this technique would indeed be very promising for clinical application to man. Human-mouse cell fusions have also been accomplished, but it is usually the mouse nucleus that takes control. In one experiment a laboratory culture of mouse cells deficient in a particular enzyme (dipeptidase-2) demonstrated enzyme production in hybrid cells produced from fusions with human skin cells.

Regardless of the source of the "good" gene (natural or synthetic) or the mode of transfer (viral infection, exogenous DNA transformation, or cell fusion), the accuracy and reliability of the current methods of gene transfer are not yet high enough to allow the clinical use of gene therapy as common medical practice. It has been suggested that culturing and monitoring techniques might be necessary to avoid the hit-or-miss inaccuracy of attempts at direct genetic modification of intact body cells. For example, if a genetic defect causes the lack of a particular liver enzyme, it might be possible to establish a laboratory culture of the patient's cells which can be subjected to gene therapy. The cultured cells can then be monitored, and any cells that are successfully modified can be selected, recultured to produce sufficient numbers, then implanted back into the patient's liver. These implants should be highly successful and would not be rejected because the cells would be essentially the patient's own. Such reprogrammed liver cells could provide a normal set of liver enzymes to permanently treat deficiencies such as those in a child with phenylketonuria or galactosemia.

PROSPECTS FOR SHAPING OUR GENETIC COMPOSITION

Our understanding of evolutionary, reproductive, and hereditary processes has reached a level of technological know-how that leaves no doubt that the human species can, if it wishes, undertake to design its very genetic composition. It is not nearly so certain, however, that we have developed the *wisdom*, that is, the knowledge of how to use our knowledge, to proceed with a directed program for the "improvement" of mankind. Indeed there is no unanimity of objectives, goals, or even desire for directed "improvement." Nevertheless, the genetic technology now exists, and the prospects are real. It, therefore, seems imperative that the possibilities of the application of this technology to the human population be considered. It should be noted at the onset that every research development to be considered here has important therapeutic implications for various individual circumstances (euphenic), and the eugenic potential is a possible by-product. Although it is true that the potential for abuse or mismanagement of eugenic measures must be seriously guarded against, it is not true that this genetic technology is being developed by "mad" scientists bent on creating Frankenstein monsters or weird forms of suprahumans for a surrealistic future setting. Quite to the contrary, the eugenic implications arise from research aimed at correcting sterility, improving animal breeding for food supplies, treating genetic disease, regenerating human tissue for repair and replacement, or alleviating individual human suffering in some way. As with all human technological advancements, the solution to the potential dangers and abuses is not to stop the basic research but to use our knowledge wisely. We must understand that advances in technology do not automatically define the criteria for their use—man must make the decisions on the limits of application. History warns us to be cautious, for the abuses of technology to the detriment of life and social harmony are manifold. Let us then consider the current areas of research most conducive to extensions into eugenic programs with the objective, and hope, that a clearer understanding of the prospects will help in the formulation of rational uses of and limits to genetic engineering in our own species.

In the previous chapter we saw that through amniocentesis, Tay-Sachs disease can be diagnosed in a fetus at an early enough stage to perform a therapeutic abortion. Chromosomal abnormalities, such as Down's syndrome, and several other genetic diseases, can also be detected at early stages of fetal development. Thus, *fetal monitoring* for genetic defects, now utilized in high risk situations for euphenic purposes, has the potential, especially as we learn to detect more and more fetal abnormalities, to be used in a program of *negative eugenics*. This refers to the elimination of "undesirable" or "unfit" individuals from the population. The basic concept of negative eugenics is to improve the population by relieving its genetic burdens when they occur. At any level beyond birth, negative eugenics is morally unacceptable to most individuals and societies. Even if selective elimination of "undesirables" were performed *before* birth, through fetal monitoring and abortion, it is doubtful that any large scale, directed program of negative eugenics could be effectively instituted or morally accepted by any modern society. Although it appears that the negative eugenic potential of fetal monitoring is unlikely to materialize, its euphenic value in marriages with high risk of producing genetically defective children has merit. The availability of fetal monitoring as a component of genetic counseling on a voluntary basis would have beneficial side effects for society by actually lowering its genetic burdens through the choices made by the individuals concerned. A well informed public would undoubtedly avail itself of clinical analyses for carrier detection of severe genetic defects, and most married couples at risk would probably choose therapeutic abortion of a "hopeless" fetus to try again for success in a future pregnancy. The evidence to date supports this contention. Recall in the previous chapter that five out of six couples diagnosed with a Tay-Sachs fetus being carried by the mother chose to have a therapeutic abortion. The chances are encouraging for such couples that a subsequent pregnancy will result in a normal child (at each pregnancy the odds favor normal over abnormal by 3 to 1).

Any time that improvement of the population is attempted by increasing the proportion of desired types or raising the level of a particular trait in the population, for example, intelligence, some form of selective breeding must be instituted. This mode of population improvement is called *positive eugenics.* Any measure that increases the genetic contribution of selected types into subsequent generations is a positive eugenic device. The sterilization of genetically caused mentally retarded individuals or of AHF-treated hemophiliacs would be positive eugenic measures because in actuality "normal" individuals are being selected to contribute to the next generation in greater proportions than if "normals" and "defectives" both reproduced at equal rates. But, we have seen that most genetically defective children are produced by "normal" parents who are carriers of recessive genes, possessors of newly arisen dominant mutations in their gametes, or bearers of gametes in which chromosomal accidents have occurred. Sterilization programs for genetically defective individuals produce little overall "improvement" but do prevent deterioration of the gene pool by increased frequency of defective alleles. The eugenic aspect of *selective sterilization* becomes an increasingly more important prospect, however, as medical science develops the treatment for more and more genetic diseases. These, in turn, would lead to increased frequencies of the detrimental alleles in any cases where reproductive potential is reduced when the disease lacks a treatment, but is restored by the treatment. Such restoration of "normal" fertility in effect increases the fitness of the treated individuals, and their increased genetic contribution to the next generation raises the frequency of the detrimental gene they bear (usually as homozygotes).

Much more dramatic, and controversial, techniques of selective breeding, conducive to very high degrees of "selectivity" among breeding types, are now possible as a result of recent advances in genetic and reproductive technologies. There are three major modes of modified reproduction that have considerable positive eugenic potential. These are: (1) artificial insemination and/or artificial inovulation, (2) in vitro fertilization and embryo transplantation, and (3) asexual reproduction, known as cloning.

Artificial insemination simply involves the collection of semen, which is quite easily done by masturbation, followed by its introduction directly through the vagina to the cervix of the uterus by a syringe. This process has been used for many years to achieve pregnancy in marriages where the husband is subfertile or infertile. In the former case the husband's own semen is used (providing a shorter, less acid journey to the egg for low numbers of sperm with below normal mobility), but in the latter case "donor" semen is required. It is estimated that for the past decade in the United States about 10,000 babies per year have been produced by artificial insemination. A large portion of these resulted from AID (artificial insemination by a donor). Donors, so far, have not represented any selected ideal, nor has any eugenic AID program been established—most donors are medical students who receive a fee for their contribution.

Donor semen must be used for insemination soon after it is ejaculated (use within 30 minutes is recommended). Under these conditions a eugenic program matching "desirable" donor sperm to selected females would be difficult to manage and would become quite inconvenient and taxing on the donors if a large number of pregnancies involving many "mothers" were consummated to increase the frequency of the donor genes in the population. The potential for using selected donor sperm for the production of innumerable pregnancies in selected "mothers" anywhere at any time is greatly enhanced by the freezing and storage of sperm. This technique, long practiced in animal breeding, not only is applicable to man, but has become a commercial enterprise.

The first commercial sperm bank was established in the United States in 1970. Sperm banking consists of freezing the semen, mixed with glycerol and egg albumen, in liquid nitrogen at −321 F and storing it in thin plastic "straws" (a dozen per ejaculate). Most deposits are made to provide "insurance" for men undergoing vasectomies; however, these "frozen assets" can be used for AID. Although only recently made available to the public, frozen sperms have been used in artificial insemination for 20 years, accounting for over 400 births in the United States. Sperm banks have begun to expand their business by supplying frozen AID stock. To maintain this stock, donors are paid for semen deposits (just as blood donors are by commercial blood banks). The sperm banks code certain genetic characteristics for matching purposes, and as more and more genetic information about the donor becomes available, greater selectivity on the part of the recipient will undoubtedly be exercised— this in itself would represent a minor form of positive eugenics. The whole system of sperm banking and supplying donor sperm represents a distinct potential for organized programs, perhaps government sponsored or perhaps sponsored by particular religious, ethnic, or racial groups with private funds, to "push" certain genetic qualities by using frozen semen from "desirable" donors to enhance the probability of producing "improvement" in the progeny. There are presently no state or federal statutes governing the operation of sperm banks. It is even possible that, as with prize bulls, the frozen sperms of "prize men" can continue to fertilize females and produce offspring long after the death of the "father" for as long as the frozen stock holds out (assuming deterioration of frozen stocks is negligible). Sperm banks are here and their assets are growing; it is now time to consider their role in human society.

Artificial inovulation is also a possibility. It would involve collecting eggs and placing them in the oviduct for fertilization and subsequent development in the normal fashion. This technique has euphenic application for women whose sterility is caused by blocked oviducts. Such women produce eggs that cannot be reached by sperm for fertilization nor can they pass to the uterus for implantation. By means of a simple surgical operation (laparoscopy), eggs (released by drugs that stimulate multiple ovulation) can be removed and transferred to the lower portion of the oviduct, thus detouring around the blockage. Fertilization can then occur through normal intercourse or by artificial insemination with the husband's sperm.

It is not difficult to project eugenic potentials from this point. Eggs for inovulation could be supplied by selected donors. In fact, if donors with normal oviducts are used, eggs produced by induced multiple ovulation can literally be flushed out of the oviducts with a pipette, eliminating any surgical procedures. Thus, a selected female donor could make contributions of genetic material to the next generation at frequencies unheard of through traditional reproductive measures. In the 9 months a woman would spend in pregnancy with one child (developing one of her eggs), a donor, stimulated for multiple ovulation, could supply perhaps 100 eggs for "surrogate mothers" to incubate. This represents a phenomenal increase in reproductive potential which, in an evolutionary sense, represents high Darwinian fitness and would produce a change in gene frequencies comparable to the changes possible through AID.

The combination of using selected donor eggs and sperms to increase the genetic potential of a child so conceived is not only a potent eugenic possibility but a serious threat to our traditional marriage-family relationship and structure. When investigating such eugenic measures as these, we must seriously evaluate the social costs along with the biological gains, and if the total is not a significant plus or if the gain occurs through social bankruptcy, our endeavors may all be in vain.

The second major mode of modified reproduction, *in vitro fertilization* and *embryo transplantation*, is really an extension of the eugenic potential of artificial insemination and inovulation. In vitro (literally, in glass) fertilization refers to the union of the egg and sperm outside the body, in the laboratory. This feat is not nearly as simple as just pouring some semen on top of eggs in a test tube. It involves removal of eggs from a hormonally treated female donor at precisely the right time plus a properly balanced liquid medium (including fluid from the egg follicle, mucus from the cervix of the uterus, and calf or human blood serum) to permit fertilization to be accomplished.

The first substantiated report of successful in vitro fertilization came in 1969 from the laboratory of Drs. Robert Edwards and Patrick Steptoe in England. In subsequent experiments (1970) they reported 53 successful fertilizations out of 182 attempts, with cellular reproduction (mitosis) initiated in 40 cases. In a few instances development to the blastocyst stage (60 to 100 cells) occurred. These apparently normal blastocysts (embryos), if transplanted to a receptive womb (uterus with endometrium in the implantation condition, naturally or artificially induced), would most likely have implanted and developed normally. Such embryo transplants have been successfully accomplished in mammals, such as lambs, with no apparent harm to the offspring, but, to date, a successful full term human embryo transplant has not been reported.

This mode of reproduction can overcome female infertility caused by blocked oviducts and would probably be easier, safer, and much more successful than artificial inovulation and insemination in utero. Thus, there is a clear therapeutic value to this technique for particular individuals—and this is the objective of the current research in this area. However, the eugenic potential of in vitro fertilization and embryo transplantation is much more dramatic and controversial than that of artificial insemination and inovulation. The reason for this is that tests and analyses can be performed on the test-tube embryo to ensure its "acceptability" before in utero development (transplantation to a mother) even begins. This presents serious moral and ethical questions that are different from any others we have ever had to face. At first glance, the problems to resolve may appear to be the same as those regarding therapeutic abortion of "unfit" fetuses identified by amniocentesis. However, there is a major difference. Therapeutic abortions are performed on seriously abnormal and defective fetuses; whereas, certain test-tube embryos that are part of a goal-oriented eugenic program might be rejected simply because they do not meet the highest expectations of the desired type. In a highly selective program perhaps the majority of embryos would be rejected. Even the euphenic use of this technique poses the problems of deliberately letting some test-tube embryos die. Since test-tube fertilization will probably never be 100% efficient, in order to achieve a successful transplant pregnancy, a dozen or more eggs might have to be used to ensure three or four fertilizations. This means that after transplanting one embryo, for instance to a female with blocked oviducts who supplied her own eggs, the other two or three embryos would have to be disposed of. Is this morally acceptable to society? Would it be a medically ethical procedure for a doctor to perform? Has a *human* life (or lives) been eliminated? Is it murder?

Engineering the genetic composition of future generations would indeed be possible through either of the two modes of modified reproduction described previously. However, because both methods are still sexual processes, that is, they involve the union of two haploid gametes to produce the zygote, eugenic progress toward any particular goal is not ensured in every child produced. The genes of selected parents still get shuffled by meiosis and crossing over, and each parent still supplies only a random half of his or her diploid genetic complement to the offspring. Thus, even with artificial insemination and inovulation, all children would not be expected to be as outstanding as their parents. In fact, although the population would advance in the direction of the selected characteristics on the average, most children would individually fall below their selected parents phenotypic expression (most "desirable" traits to be improved would probably be polygenic and exhibit regression toward the mean). This regression might be minimized, or even eradicated, with in vitro fertilization followed by the development of rigid "test-tube tests" to be passed by the embryo before transplantation. However, this raises the previous problems related to elimination of "undesirable" but not seriously abnormal in vitro embryos.

Many of these problems are overcome by the third major mode of modified reproduction, *cloning*, which avoids the chance shuffling of selected genetic material (selected individuals) by eliminating the sexual process altogether. On the other hand, this represents the most radical departure from traditional reproduction to be considered. As such, it presents a whole array of new problems.

Cloning, or asexual reproduction, is a process in which no genetic shuffling occurs. The significant feature of this mode of reproduction is that all the offspring formed are identical in genetic make-up to their single parent. Cloning is not a simple procedure. The problem is that the highly differentiated cells making up the tissues of the adult form of complex animals, such as mammals, lose the capacity to undergo the whole cycle of development again even though the total genetic complement is still present in their nuclei. The trick is to unlock the controls that irreversibly turn off the genes that govern embryonic development so that the sequence can be repeated by a single body cell stimulated to act like a newly formed zygote. Serious consideration has been given to animal cloning since 1966 when Dr. J. B. Gurdon of Oxford University reported the production of clonal frogs from intestine cells of tadpoles. To achieve development, Gurdon and his colleagues transplanted the nucleus from an intestinal cell into an enucleated frog egg (nucleus destroyed by ultraviolet radiation). This was achieved by inserting the diploid intestinal cell nucleus into the egg cytoplasm through a micropipette (tiny hollow needle) used to puncture the egg membrane. In the cytoplasm of the egg, the genetic system controlling development was "turned on," and the transplanted nucleus directed the entire epigenetic process, resulting in a new individual. The resulting frog was a genetic replica of its single parent—an identical twin.

Transplantation of diploid nuclei to human eggs by micropipette insertion does not appear promising because the human egg is much smaller than that of the frog, which makes the surgical procedure very difficult, as well as increasing the likelihood of irreversible damage. However, the more recent development of techniques for fusing animal cells has increased the possibility for cloning man. The cell-fusion method involves the actual union of two cells into one as they grow in laboratory cultures, under the stimulation of certain viruses. When an egg cell is fused with a diploid body cell, the genetic material of the egg becomes inactive and the diploid nucleus of the body cell assumes genetic control. Under the influence of the egg cytoplasm, the diploid nucleus is "turned on" in its embryonic state and mitosis leads to blastocyst formation. The transplantation of such a blastocyst to a receptive uterus should then lead to implantation and normal fetal development. Once the technique of embryo transplantation becomes subject to routine application, the source of the blastocyst could be an embryonically induced body cell rather than a test-tube fertilized egg.

The eugenic potential of cloning lies primarily in increasing the proportion of selected phenotypes in the population by producing exact replicates in quantity. In addition, it should be possible to provide clonal progeny with even better environmental conditions than those in which their single parent developed. This would increase their possibility of surpassing the genetic potential realized by the selected parent for a characteristic such as intelligence. It has been suggested, for instance, that selected individuals might spend the latter years of their lives educating their own clones. Such a teacher-student relationship (essentially identical twins) might allow for almost telepathic communication, thereby accelerating the educative process.

Even more astonishing is the fact that cell-fusion can be accomplished between cells from different species altogether. Such cell hybridization has already established cultures of man-ape and man-mouse cells. In 1972, the first organisms ever to be produced by cell hybridization were reported by a research group from the Brookhaven National Laboratory in New York. The organisms were hybrid tobacco plants produced from two different tobacco species, *Nicotiana glauca* and *Nicotiana langsdorffii*. These hybrids came from the fusion of two leaf cells and perfectly resembled known hybrids produced by standard sexual techniques of cross fertilization in plants. This new method has been called *parasexual hybridization*.

The first cloning from a specialized cell was also accomplished among plants (with carrots) in 1963, and only three years later Gurdon had cloned frogs. It may also be a relatively short time until animal parasexual hybridization is accomplished. This includes the potential for the development of man-animal hybrids. Parasexual hybridization holds great promise for the production of new crop varieties to increase food production, and perhaps animal breeders may someday utilize the technique for the same purpose, but the inclusion of some human genetic material would create a moral and ethical dilemma unlike anything ever encountered by humanity.

Cloning and parasexual hybridization may never be used in a human eugenic program and neither may artificial insemination/inovulation or in vitro fertilization and embryo transplantation. However, knowledge of these possibilities may prepare society to avoid abuses of these techniques while recognizing the euphenic value to our species and the potential for plant and animal breeding that directly benefits man. Genetics will become increasingly important in human affairs in the future. The material in this book will hopefully allow the reader to more fully understand and appreciate the implications of advances in genetics.

Suggestions for further reading on hereditary processes

Baer, Adela S., editor. 1973. Heredity and society: readings in social genetics.* Macmillan Inc., New York. A set of readings collected with the purpose of addressing the genetic aspects of an array of social problems including environmental hazards, agricultural genetics, behavior, intelligence, and population problems.

Bresler, Jack B., editor. 1973. Genetics and society.* Addison-Wesley Publishing Co., Inc., Reading, Massachusetts. This collection of papers considers how the human gene pool affects society and

*Denotes paperback

how society affects the gene pool. The emphasis is on population rather than individual characteristics and includes topics such as chromosomal abnormalities, ethnic characteristics, genetic counseling, and genetic engineering.

Hamilton, Michael P., editor. 1972. The new genetics and the future of man.* Wm. B. Eerdmans Publishing Company, Grand Rapids, Michigan. This book consists of three sections: "New Beginnings of Life," "Genetic Therapy," and "Pollution and Health." In each section a prominent scientist in his field sets forth the current state of knowledge, future possibilities, and moral and social problems arising from this knowledge. Then three leaders from other professions (lawyers, theologians, and philosophers) discuss and criticize the technology and its implications. The result is a stimulating and thought provoking experience in the new genetics.

King, James C. 1971. The biology of race.* Harcourt Brace Jovanovich, Inc., New York. This book brings together in a concise and comprehensive manner the pertinent biological information on the concept of race.

Leach, Gerald. 1972. The biocrats: implications of medical progress.* Revised edition. Penguin Books, Baltimore. Mainly intended for the layman, this book lucidly discusses the so-called biological revolution and its implications. It contains provocative sections on population control, selective breeding, and fetal medicine.

Levine, Louis. 1973. Biology of the gene. The C. V. Mosby Company. St. Louis. An up-to-date basic genetics textbook recommended to the student who wishes to extend her or his study of genetics beyond the human realm.

Levitan, Max and Ashley Montagu. 1971. Textbook of human genetics. Oxford University Press, New York. This is a comprehensive, technically advanced human genetics textbook containing a wealth of examples and data on human inherited traits.

Scheinfeld, Amram. 1972. Heredity in humans. J. B. Lippincott Company, Philadelphia. Written for the layman, this book discusses human genetics in a very elementary manner and contains many interesting and fascinating points of information.

Wallace, Bruce. 1972. Essays in social biology, Volume 2, Genetics, Evolution, Race, Radiation Biology.* Prentice-Hall, Englewood Cliffs, New Jersey. This second volume of a three-volume series represents a set of valuable essays of particular concern to anyone interested in human genetics.

Winchester, A. M. 1971. Human genetics.* Charles E. Merrill Publishing Company, Columbus, Ohio. This book provides a basic introduction to the field of human genetics for the student with no previous background in biology.

Glossary of selected terms

The terms in this glossary are defined in the context of the material presented in the text. No attempt was made to give a full "dictionary" definition; rather, a definition that will be helpful in understanding the terminology and vocabulary herein was my objective. The student is encouraged to consult this glossary frequently to supplement his or her study. I hope that I have included most of the terms that one might desire to look up—if something has been omitted, consult the index.

ABO blood group A major blood antigen system important in compatibility for transfusions and in pregnancy. Individuals are typed as A, B, O, or AB, depending upon the presence or absence of the A- and B-antigens produced by a multiple allelic mode of inheritance. See *antigen* and *multiple alleles.*

acrocentric Refers to a chromosome having its centromere very close to one end, forms an I-shaped chromosome during anaphase. See *centromere.*

adaptation Any characteristic of an organism that improves its chances of survival and reproduction in comparison to the chances of similar organisms occupying the same environment but lacking in the particular characteristic. See *natural selection.*

adaptive norm The predominant, best-adapted phenotypes that fall within a range close to the mean in a normal (bell-shaped) distribution of phenotypes representing a population.

adaptive radiation The relatively rapid evolutionary divergence of the members of a single major lineage into a number of genera and species occupying a variety of ecological niches. See *niche.*

adenine One of the nitrogenous bases, a purine, found in DNA and RNA.

agglutinogen See *antigen.*

albinism Absence of pigment (melanin) in the eyes, hair, and skin. Inherited as an autosomal recessive.

allele Any of two or more alternative forms of the same basic gene. A normal gene and its various mutant forms are alleles, for example, *A* and *a.* See *multiple alleles* and *point mutation.*

amaurotic familial idiocy See *Tay-Sachs disease.*

amino acid Any one of a class of organic chemical compounds characterized by the presence of an amino group (NH_2) and a carboxyl group (COOH) attached to either side of a central carbon atom. They are the primary building blocks of proteins; twenty major types are found.

amniocentesis Technique of removing a small amount of amniotic fluid and fetal cells by means of a slender needle inserted through the abdomen and into the uterus of a pregnant woman. The cells are cultured and analyzed for detection of various abnormalities.

amnion The inner membrane enclosing the embryo in its surrounding liquid medium, the amniotic fluid.

amphoteric Possessing both positively and negatively charged sites on the same macromolecule. Proteins are amphoteric.

anaphase The stage in mitosis and meiosis during which the chromosomes (or chromatids) move from the metaphase plate to the poles on the cell. See *telophase*.

androgen General name for any substance with male sex hormone activity. See *testosterone*.

anemia A condition characterized by a decreased oxygen-carrying capacity of the red blood cells because of a reduced number of cells, too little hemoglobin, or malfunctioning hemoglobin. See *sickle cell anemia* and *thalassemia*.

aneuploid A chromosomal aberration in which the chromosome number has been increased or decreased by a number other than some exact multiple of the haploid (n) or basic number. Usually an increase of one (2n + 1) or a decrease of one (2n − 1). See *monosomy, trisomy, euploid,* and *nondisjunction.*

anthropoid A member of the primate suborder Anthropoidea (in the shape of man), includes monkeys, apes, and man. See *primate, catarrhine,* and *platyrrhine.*

antibody Protein synthesized in the body which attacks and neutralizes some specific "foreign protein" introduced into the body. See *antigen* and *immune response.*

anticodon A sequence of three unpaired nitrogenous bases found in transfer RNA which by complementarity to messenger RNA codons provides the mechanism for tRNA—mRNA recognition during polypeptide synthesis. See *codon* and *transfer RNA.*

antigen A substance, usually a protein, that elicits antibody formation and/or attack when it enters the body as a "foreign intruder." See *antibody* and *immune response.*

arboreal Adapted to life in the trees; tree-dwelling.

asexual reproduction Any process that does not involve the union of two haploid sex cells (gametes) in progeny production, the offspring are genetically iden-tical to their single parent. See *cloning* and *sexual reproduction.*

assortment See *independent assortment.*

Australopithecus The "South African ape-man," earliest erect hominid and tool user in the human family (Hominidae), succeeding *Ramapithecus* and preceding *Homo erectus.*

autosome Any chromosome other than the sex chromosomes (X and Y). Humans have 22 pairs of autosomes.

autotrophs Organisms that are capable of manufacturing organic nutrients from inorganic raw materials (as in photosynthesis); literally, self-feeders. Most plants are autotrophic, but animals are not. See *heterotrophs* and *photosynthesis.*

baldness The common type, loss of hair from the head in a pattern in which the sides and lower back regions retain varying amounts; also called pattern baldness. Inherited as a sex-influenced trait acting dominant in males. See *sex-influenced.*

Barr body Deeply staining mass found attached to the inner surface of the nuclear membrane in the somatic cells of a normal female but not in a normal male. Inactivated X chromosome. See *Lyon-Russell principle.*

base See *nitrogenous base.*

binomial expansion The exponential multiplication of an expression consisting of two terms connected by a + or − sign, such as $(a + b)^n$.

biogenesis The biological rule that a living organism can originate only from a parent or parents similar to itself, life only from pre-existing life. See *biopoiesis.*

biopoiesis The process leading to the origin of living from nonliving material; spontaneous generation. See *biogenesis.*

bivalent A pair of synapsed homologous chromosomes during meiosis. See *synapsis* and *univalent.*

blastocyst A ball of 60 to 100 cells with a central fluid-filled cavity formed from the zygote. The blastocyst implants into the uterus to complete development. See *implantation.*

camptodactyly A genetic abnormality characterized by permanently bent and stiff little fingers. Inherited as an autosomal dominant with incomplete penetrance. See *penetrance.*

carrier An individual apparently normal, but possessing a single dose of a recessive gene obscured by a dominant allele; a heterozygote. See *heterozygous.*

catalyst A substance that speeds up a chemical reaction without itself being changed. See *enzyme.*

cataract An eye abnormality in which the lens becomes opaque, causing blindness.

catarrhine A member of the primate infraorder Catarrhini (hooknosed), characterized by a narrow septum between comma-shaped nostrils. Includes the more advanced anthropoids such as the Old World monkeys, gibbons, apes and man. See *anthropoid* and *platyrrhine.*

centric fusion A chromosomal aberration in which two acrocentric chromosomes undergo translocation when breaks occur in the very short arm of each and the large portions fuse into one large chromosome while the two small fragments are lost; a modified translocation. See *translocation.*

centriole A tiny body outside the nucleus of a cell, concerned with establishment of the poles and formation of the spindle in mitosis and meiosis.

centromere A specialized region (primary constriction) of a chromosome that holds the two chromatids of a duplicated chromosome together and which is involved in directing chromosome movements during cellular reproduction. Its location along the length of a chromosome is a characteristic used in classification and identification. See *acrocentric, metacentric* and *submetacentric.*

chiasma (plural: **chiasmata**) A crosslike pattern seen late in Prophase I of meiosis when two nonsister chromatids that have undergone crossing over begin to relax from tight synapsis. See *crossing over* and *synapsis.*

chorion The outer embryonic membrane that surrounds the amnion and the embryo. Chorionic villi implant the embryo in the uterine wall and establish the placenta. See *placenta.*

chorionic gonadotrophin (CG) The pregnancy hormone produced by the chorion to take over stimulation of the corpus luteum from LH and keep the uterine wall in its glandular condition. It is the basis for pregnancy tests.

Christmas disease Hemophilia B. See *hemophilia.*

chromatid One of the two identical units resulting from the duplication of a chromosome prior to cellular reproduction

(mitosis or meiosis); the precursor of a chromosome.

chromosomal aberration An abnormality of chromosome structure or number.

chromosomes Deeply staining bodies found in the nucleus of cells, composed of DNA and proteins. The carriers of the genetic information. Each species has a characteristic chromosome number; the human number is 46.

cloning The production of progeny by asexual means from a single parent; formation of exact genetic replicas. See *asexual reproduction.*

coacervate A protein aggregate of varying degrees of complexity and stability, formed by coalescence in a colloidal suspension. See *colloid.*

code See *genetic code.*

codominance The condition in which both members of an allelic pair are expressed approximately equally, resulting in the heterozygote exhibiting a phenotype intermediate to and readily distinguished from either homozygote. It is the lack of dominance in allelic interactions. See *dominance.*

codon A sequence of three nitrogenous bases in DNA or messenger RNA that represents the "code word" for a specific amino acid. Messenger RNA codons are the base complement to the DNA codons from which they are formed. See *anticodon* and *nitrogenous base.*

coefficient of inbreeding (F) The probability that an individual has received both alleles of a pair from an identical ancestral source.

coefficient of relationship (r) The proportion of alleles in any two individuals that have been inherited from common ancestors.

colloid A substance divided into particles that fall within a certain size range (10^{-7} to 5×10^{-5} cm) that is larger than one of a true solution but smaller than one in a course suspension or mixture. A colloidal system contains such particles in some dispersion medium (usually water), as a semipermanent suspension.

colorblindness The inability to distinguish colors normally, several types are recognized, inherited as an X-linked recessive.

congenital Present at birth, may or may not be genetically caused.

consanguinity Relationship by descent from a common ancestor. Marriage between individuals more closely related than two persons selected at random from the population. See *inbreeding.*

control genes Genes that control the function of other genes. See *operator gene* and *regulator gene.*

convergent speciation See *secondary speciation.*

Cooley's anemia See *thalassemia.*

corpus luteum A structure in the ovary derived from the remnants of a ruptured follicle. It secretes both estrogen and progesterone.

coupling With reference to two pairs of genes located on the same chromosome, the situation of having both dominant alleles on one chromosome and both recessive alleles on the homologous chromosome, for example, *(AB)* and *(ab).* See *repulsion.*

cri du chat syndrome An abnormality caused by a deletion in the short arm of chromosome number 5, characterized by mental retardation and an array of physical anomalies. See *deletion* and *syndrome.*

Cro-Magnon man See *Homo sapiens.*

crossing over The exchange of segments between homologous chromosomes (nonsister chromatids) during synapsis in meiosis. See *synapsis.*

cytogenetics The branch of genetics that deals with the relationship to heredity of structures and functions within the cell, especially the chromosomes.

cytokinesis Cytoplasmic division and the splitting of one cell into two, following the events of cellular reproduction (mitosis or meiosis).

cytoplasm The portion of the cell outside and surrounding the nucleus. It contains cell organelles, such as mitochondria and ribosomes.

cytosine One of the nitrogenous bases, a pyrimidine, found in DNA and RNA.

deletion A chromosomal aberration involving the loss of a portion of a chromosome.

diabetes mellitus An inherited disease in which the body cannot utilize sugar normally, inheritance probably polygenic.

differentiation Change in structure and/or function of cells progressively toward a more specialized or mature state during embryonic development.

dihybrid An individual heterozygous for two pairs of genes under consideration; A double heterozygote. See *heterozygous.*

diploid (2n) Having two of each kind of chromosome (homologous pairs); the normal condition in cells except for the gametes which are haploid. The diploid number is 46 in humans, 23 homologous pairs. See *haploid* and *homologous chromosomes.*

disruptive selection Natural selection that operates on two or more different sets of adaptive types in different environments, thereby establishing two or more adaptive norms and splitting the population into distinctive subpopulations. See *adaptive norm* and *natural selection.*

directional selection Natural selection that occurs during or after environmental change producing a progressive shift of the adaptive norm. See *adaptive norm* and *natural selection.*

divergent speciation See *primary speciation.*

dizygotic (DZ) twins Refers to twins who originate from two different eggs (multiple ovulation) fertilized by two different sperms (two zygotes); such fraternal twins are genetically no more similar than any other sibs. See *monozygotic twins* and *sibs.*

DNA (deoxyribonucleic acid) The nucleic acid that is the material of the gene. The molecule is a double-stranded helix consisting of two sugar (deoxyribose) and phosphate backbones connected by pairs of the four nitrogenous bases, adenine, cytosine, guanine, and thymine, arranged in specific (complementary) combinations of A-T (T-A) and C-G (G-C). A gene consists of a specific linear sequence of these bases (a segment of DNA). See *gene, genetic code,* and *nucleic acid.*

dominance The condition in which one form of a gene (dominant allele) suppresses or overrides the effect of another form of the gene (recessive allele) on the phenotype of a heterozygote. See *codominance, heterozygous,* and *recessive.*

dominance lacking See *codominance.*

dosage compensation See *Lyon-Russell principle.*

Down's syndrome An abnormality caused by the presence of an extra chromosome number 21 characterized by mental retardation and an array of physical anomalies; previously known as Mongolian idiocy. Technically, trisomy-21. See *syndrome* and *trisomy.*

Dryopithecus A member of the dryopithecine ("oak ape") group from which the family of great apes and the human family diverged about 25 million years ago.

Duchenne's muscular dystrophy An X-linked recessive condition in which the muscles progressively degenerate, culminating in death by the late teens or soon after, generally found in males. See *X-linked.*

duplication A chromosomal aberration in which a portion of a chromosome is present more than once, may involve whole genes, parts of genes, or series of genes. Also called repeats.

egg The female gamete, characterized as a large cell containing an abundant nutrient supply to support initial embryonic development if fertilization occurs. See *gamete.*

electron The elementary unit of negative electric charge which orbits around the nucleus of an atom. See *proton.*

electrophoresis The differential migration of amphoteric substances, such as proteins and enzymes, in an electric field. It allows different molecular forms to be separated and identified. See *amphoteric.*

elliptocytosis A form of anemia characterized by the red blood cells becoming oval (elliptical) in shape, rather than their normal circular condition; inherited as an autosomal dominant.

embryo The developing organism from the formation of the zygote until the end of the eighth week; characterized by the establishment of the rudiments of all the adult organs. See *fetus.*

endometrium The glandular mucous membrane that lines the inside of the uterus into which an embryo implants. See *implantation* and *uterus.*

enzyme Organic catalyst that speeds up a chemical reaction in a living system. Usually the reaction would not occur effectively under the conditions of the body without the enzyme. See *catalyst.*

epididymus The highly coiled portion of the sperm duct adjacent to the testis; mature sperms are stored here. See *sperm duct.*

epigenetic Referring to the developmental process, the principle that development occurs as a building, not simply a growing, process from which new structures differentiate from patterns set down in a previous stage; the entire phenomenon is under genetic control. See *differentiation.*

epistasis Interaction of nonallelic genes in which the action of one gene masks or alters the action of a gene or genes at other loci; distinguished from dominance which refers to the interaction of one allele with another at the same locus. See *dominance* and *locus.*

erythroblastosis fetalis A condition in which the red blood cells of the fetus are destroyed, causing severe anemia and other serious, often lethal, problems such as heart failure and brain damage; caused by Rh blood incompatibilities between an Rh-negative mother (sensitized) and an Rh-positive fetus.

erythrocyte A red blood cell; filled mainly with hemoglobin and possessing the blood antigens (ABO, Rh, and so forth) on its surface.

estrogen A female sex hormone produced by the follicle cells and the corpus luteum. Stimulates the development of the female secondary sex characteristics and is involved in the regulation of egg production.

eugenics Genetic manipulations aimed at the overall improvement of the human population through goal-oriented processes of selective breeding. See *euphenics.*

euphenics Genetic manipulations aimed at the treatment of genetic diseases for the benefit of the afflicted individual. See *eugenics.*

euploid Having each of the chromosomes of the haploid (n) or basic number present in exact multiples, such as the normal diploid (2n) condition. See *aneuploid* and *polyploid.*

expressivity Degree to which a particular genotype exhibits a phenotypic effect; for example, with variable expressivity, the trait expressed may range from mild to severe in different individuals. See *penetrance.*

F_1 First filial generation; the first successive generation of descent from a given or prescribed mating.

F_2 Second filial generation; the offspring from the intercrossing of the F_1, generally any successive second generation progeny from a given or prescribed mating. See F_1.

fertilization Fusion of a sperm with an egg; two haploid gametes forming a diploid zygote. See *zygote.*

fetus The developing organism from the ninth week through the remaining time to birth; characterized by the growth and maturation of the rudiments set down in the embryo. See *embryo*.

fitness The relative survival value and reproductive capability of a given genotype in comparison to others of the population. See *adaptation*.

follicle See *primordial follicle*.

follicle-stimulating hormone (FSH) A sex hormone (gonadotrophin), produced by the pituitary gland, involved in the regulation of sperm and egg production. See *gonadotrophin*.

fossil Remains of an organism, or direct evidence of its presence, preserved in rocks; usually found as impressions or casts of the hard parts (bones) that are preserved when mineral deposits from water replace the original material.

fraternal twins See *dizygotic twins*.

frequency See *gene frequency*.

FSH See *follicle-stimulating hormone*.

galactosemia A genetic disease characterized by the inability to convert the sugar galactose to the usable form of glucose; inherited as an autosomal recessive.

gamete The mature reproductive cell, or sex cell, produced by meiosis and possessing the haploid complement of chromosomes; in females the egg, and in males the sperm. See *haploid* and *meiosis*.

gene The basic unit of inheritance, located in a fixed position (locus) on a chromosome; a linear sequence of nitrogenous bases in DNA that serves as the code for synthesis of a particular polypeptide. See *DNA, genetic code,* and *polypeptide*.

gene frequency The proportion of one allele relative to other alleles in a particular population; given as a percent or a fraction.

gene interaction See *epistasis*.

gene pool The sum total of all genes (all alleles) in a particular population at any given time.

gene therapy Treatment of genetic disease by the insertion of a "good" (functional) gene to replace the defective one; a form of euphenics.

genetic code The triplet combinations of nitrogenous bases found in DNA that carry the information for determining specific amino acid sequences in polypeptide synthesis; the nucleic acid instructions for translation into amino acid chains through the mechanisms of gene function. See *codon* and *gene*.

genetic disease An incapacitation of a normal bodily function produced by a malfunctioning (mutant) gene.

genetic drift Change in gene frequency in a small population because of chance fluctuations (sampling error), not a result of selection.

genetic load The detrimental genes (mutations) which for various reasons, such as being masked as recessives or advantageous in the heterozygous but not the homozygous condition, are carried in the gene pool for periods of time.

genotype The actual genetic constitution of an individual, usually used in a limited sense to refer to the allelic combinations (homozygous or heterozygous) at one or more loci under consideration. See *heterozygous, homozygous,* and *phenotype*.

germ cells The cells that give rise to the gametes. See *oogonium* and *spermatogonium*.

germ line (germplasm) General term for the gametes and the cells and tissues from which they arise.

germinal mutation A mutation in a germ cell; the raw material in the generation of variation in a population. See *mutation*.

gonad A sexual gland; the ovaries and testes.

gonadotrophins Hormones produced by the pituitary gland and acting on the gonads; FSH and LH are gonadotrophins.

guanine One of the nitrogenous bases, a purine, found in DNA and RNA.

H-substance A blood antigen that serves as the substrate for the production of the A and B antigens of the ABO system. It is normally found unchanged in individuals with type O blood but is totally absent in individuals with the rare Bombay blood type.

haploid (n) Having one of each kind of chromosome; a single set of chromosomes consisting of one member of each homologous pair, characteristic of gametes. See *diploid*.

Hardy-Weinberg equilibrium equation A general formula describing phenotypic stability in a population over generations when mating is at random and gene frequencies are unchanged; expressed as $p^2AA + 2pqAa + q^2aa$ where p = the frequency of the A allele and q = the frequency of the *a* allele. See *gene frequency* and *random mating*.

Heberden's nodes A genetic abnormality characterized by a nonpainful swelling of the terminal finger joints. Inherited as a sex-influenced trait acting dominant in females. See *sex-influenced*.

helix A spiral-shaped structure. The DNA molecule is a double helix.

hemizygous The condition in which only one allele is present at a given locus in an otherwise diploid zygote. Since males have only one X chromosome, they are hemizygous (half a zygote) for their X-linked genes.

hemoglobin The iron-containing pigment (protein) of the red blood cells responsible for their color, important in the transportation of oxygen to the cells of the body.

hemophilia A genetic disease characterized by the inability of the blood to clot normally. Inherited as an X-linked recessive.

heritability A measure of the degree to which the variation in a polygenic characteristic is caused by the genes, as opposed to environmental influences. See *polygenic inheritance*.

hermaphrodite An individual with both female and male sex organs; connotes functioning as both sexes. See *intersex* and *pseudohermaphrodite*.

heterotrophs Organisms that must obtain raw organic nutrient, along with inorganic materials, from the environment; literally, other-feeders. All animals are heterotrophic, but plants typically are not. See *autotrophs*.

heterozygote An individual heterozygous at a particular locus or loci. See *heterozygous*.

heterozygous The condition in which two different forms of a gene (alleles) make up the genotype at a particular locus, for example, *Aa*. See *homozygous*.

holandric The pattern of inheritance of genes located on the Y chromosome. Females would never show the trait, and a father would pass it to all of his sons. See *Y-linked*.

hominid Specifically, a member of the human family (Hominidae). Evolving out of the dryopithecine group, this family includes *Ramapithecus*, *Australopithecus* and *Homo erectus* in direct succession to *Homo sapiens*, the only surviving member of the family. See *primate* and *hominoid*.

hominidae See *hominid*.

hominoid A member of the Primate superfamily Hominoidea which includes the gibbon family (Hylobatidae), the great ape family (Pongidae), and the human family (Hominidae). See *primate* and *hominid*.

Homo The genus to which we, *Homo sapiens*, and our nearest extinct ancestor, *Homo erectus*, belong. See *hominid*.

Homo erectus "Erect man," the immediate predecessor of our own species, *Homo sapiens*, and successor to *Australopithecus*.

homoiothermic Maintenance of a constant body temperature, usually above the surrounding temperature; "warm-blooded." Characteristic of birds and mammals. See *poikilothermic*.

homologous chromosomes Chromosomes that carry the same gene loci and represent a "matched pair" confirmed by the fact that they are partners in synapsis during meiosis. See *locus* and *synapsis*.

Homo sapiens "Wise man," our own scientific name, includes Neanderthal and Cro-Magnon men as extinct members of our species.

homozygote An individual homozygous at a particular locus or loci. See *homozygous*.

homozygous The condition in which two similar forms of a gene (alleles) make up the genotype at a particular locus; for example, *AA* or *aa*. See *heterozygous*.

hormone A chemical substance produced by an endocrine gland and secreted into the bloodstream in small quantities to circulate and influence functions in other parts of the body. See *gonadotrophins*.

Huntington's chorea A genetic disease characterized by nerve degeneration causing spasmodic movements and loss of mental faculties; inherited as an autosomal dominant with delayed time of onset.

hybridization The interbreeding of members of two different species; sometimes used loosely to describe mating between different prescribed genotypes or members of dissimilar populations within the same species. See *species*.

hypertrichosis pinnae auris "Hairy ears," possibly a Y-linked locus. See *Y-linked*.

hypophosphatemia A genetic disease characterized as a form of vitamin D–resistant rickets. Inherited as an X-linked dominant.

identical twins See *monozygotic twins*.

implantation The attachment of an embryo (blastocyst) to the wall of the uterus.

immune response The reaction of an antibody with an antigen. The body's protective mechanism against "foreign" proteins. See *antibody* and *antigen*.

inbreeding Reproduction involving the mating of closely related individuals. See *consanguinity*.

inbreeding depression The loss of viability and/or fertility resulting from the homozygosity of detrimental recessive genes exposed through inbreeding. See *inbreeding* and *viability*.

independent assortment The random distribution of one member of each gene pair to the gametes during meiosis. The different genes assort independently in producing the haploid gene complement of an egg or sperm; linked genes are an exception. Mendel's second law of inheritance. See *meiosis, linkage,* and *segregation*.

inorganic Refers to chemical molecules that are not organic; that is, matter other than that characteristic of organisms, includes water and minerals. See *organic*.

intelligence quotient (IQ) A measurement of human intelligence derived by dividing the chronological age into the mental age, as determined by some standardized testing procedure. The genetic component inherited as a polygenic trait. See *polygenic inheritance* and *heritability*.

interphase The stage in the life span of a cell when it is carrying on its particular bodily functions and is not actively undergoing the process of cellular reproduction (mitosis or meiosis).

intersex An individual who shows some secondary sexual characteristics of both sexes but does not function in both ways, in fact, often sterile. See *hermaphrodite*

inversion A chromosomal aberration in which a segment of a chromosome becomes reversed with regard to the linear sequence of genes in the normal chromosome. If the centromere is included in the reversed segment, it is called a pericentric inversion; if the centromere lies outside the reversed segment, it is called a paracentric inversion. See *centromere*.

isochromosome A chromosomal aberration in which a chromosome with two arms of equal length and identical loci (in reverse sequence) is formed by crosswise rather than longitudinal division of the centromere during cellular reproduction; such chromosomes are partly duplications and partly deletions of the genes carried on the normal chromosome. See *deletion* and *duplication*.

isolating mechanisms See Table 2-2.

Jacobs' syndrome An abnormality caused by trisomy of the sex chromosomes involving one X and two Y chromosomes (XYY). Individuals are males with above average height and apparently normal fertility. See *syndrome* and *trisomy*.

juvenile cataract A genetic abnormality characterized by cataracts of various patterns that develop in children; inherited as an autosomal dominant with variable expressivity. See *cataract* and *expressivity*.

karyotype A chart made from a photograph of the chromosomes (in metaphase) in which the homologous pairs are matched and arranged in numerical order from the longest to the shortest pair.

Klinefelter's syndrome An abnormality caused by trisomy of the sex chromosomes involving two X chromosomes and one Y (XXY). Individual appears male, but testes are not functional and breasts may develop as in a female. See *syndrome* and *trisomy*.

lack of dominance See *codominance*.

Lesch-Nyhan syndrome A genetic disease characterized by severe nervous deterioration causing mental retardation associated with spasticity and a tendency toward self-mutilation. Inherited as an X-linked recessive.

LH See *luteinizing hormone*.

linkage The association of loci on the same chromosome; linked genes do not assort independently in meiosis. See *independent assortment* and *sex-linked*.

linked genes Nonallelic genes whose loci are on the same chromosome. See *allele* and *linkage*.

locus (plural: **loci**) The site (location) on a chromosome occupied by a particular gene, in all its forms (alleles). Homologous chromosomes have identical loci. See *homologous chromosomes*.

luteinizing hormone (LH) A sex hormone (gonadotrophin) produced by the pituitary gland, involved in the regulation of sperm and egg production. See *gonadotrophin*.

Lyon-Russell principle The phenomenon of the inactivation of all X chromosomes in excess of one in all somatic cells; the method of dosage compensation to balance the content of genes on the X chromosome between the sexes. See *Barr body.*

meiosis A specialized type of cellular reproduction in which the chromosome number is reduced in half (haploid) in the production of the gametes involved in sexual reproduction. See *gamete* and *mitosis.*

melanin A brown pigment (protein) that affects the coloration of the eyes, hair, and skin according to its concentration and deposition. Its production and its distribution are under genetic control.

Mendelian population A local interbreeding population sharing a common gene pool; the primary unit of evolution.

menopause Termination of the reproductive life span of a female, usually occurs between the ages of 45 and 50.

menstruation The cyclic bleeding and sloughing off of the uterine lining (endometrium) following the preparation for a pregnancy that does not materialize.

messenger RNA (mRNA) The type of RNA that carries the transcribed genetic code (codons) from DNA in the nucleus to the ribosomes in the cytoplasm to direct polypeptide synthesis. See *RNA* and *codon.*

metabolism The sum total of all chemical processes carried on in the body, such as digestion, excretion, and the release and utilization of energy.

metacentric Refers to a chromosome having its centromere centrally located; forms a V-shaped chromosome during anaphase. See *centromere.*

metaphase The stage of mitosis or meiosis in which the chromosomes become aligned on the metaphase plate. See *anaphase* and *metaphase plate.*

metaphase plate A characteristic figure in a cell undergoing mitosis or meiosis formed by the alignment of the chromosomes across an equatorial plane of the cell at right angle to the spindle. Viewed from above (polar view), it presents the best opportunity to photograph, count, and study the chromosomes. See *metaphase* and *spindle.*

microcephaly An abnormality characterized by an underdeveloped brain and, consequently, a small head. The genetic form is inherited as an autosomal recessive.

mitochondria Small bodies found in the cytoplasm that provide energy to the cell through oxidative respiration; the "powerhouse" of the cell.

mitosis The process of cellular reproduction that produces exact replicas with respect to the genetic content (chromosomal complement). See *interphase, prophase, metaphase, anaphase, telophase, cytokinesis,* and *meiosis.*

Mongolian idiocy See *Down's syndrome.*

monosomy A chromosomal aberration in which one chromosome of a certain pair is missing. A human monosomic has only 45 chromosomes (2n − 1) in each cell. See *aneuploid* and *trisomy.*

monozygotic (MZ) twins Refers to twins who originate from a single fertilized egg (one zygote) that splits into two parts very early in embryonic development, with each part developing into a complete individual; such identical twins are exact genetic replicas of each other. See *dizygotic twins.*

mosaic Genetically, an individual (or tissue) that contains two or more genetically distinguishable lines; often resulting from nondisjunction, but also in the female from X chromosome inactivation. See *nondisjunction* and *Lyon-Russell principle.*

multiple alleles Refers to the condition in which three or more forms of a gene (alleles) are known to exist in a population, as in the ABO blood system. See *allele.*

muscular dystrophy See *Duchenne's muscular dystrophy.*

mutagen An environmental agent that increases mutation rate beyond that which occurs spontaneously.

mutant A cell or individual that exhibits a change caused by a mutation.

mutation An alteration of the genetic material. See *chromosomal aberration* and *point mutation.*

nail-patella syndrome A genetic abnormality characterized by reduced fingernails and small or missing kneecaps (patella). Inherited as an autosomal dominant.

natural selection Literally, selection by nature in the determination of which individuals shall live and reproduce and which shall not. It results from those individuals (genotypes) that are best adapted to the environment producing more offspring, therefore influencing the genetic constitution of subsequent generations more significantly than the less adapted individuals; essentially differential reproduction. See *adaptation, fitness, directional selection, disruptive selection,* and *stabilizing selection.*

Neanderthal man See *Homo sapiens.*

niche The constellation of environmental factors into which a species, or other taxonomic category, fits; its specific way of utilizing its environment.

nitrogenous base A nitrogen-containing molecule found in DNA and RNA which functions chemically as a base, as opposed to an acid. Two classes are distinguished, purines and pyrimidines. See *adenine, cytosine, guanine, thymine,* and *uracil.*

nonallelic interaction See *epistasis.*

nondisjunction Failure of two members of an homologous chromosome pair (or of two sister chromatids) to separate (disjoin) from each other during anaphase, so that both go to one pole and none to the other, thereby producing aneuploidy. See *anaphase* and *aneuploid.*

nucleic acid A macromolecule composed of complexes of nucleotides linked by sugar-phosphate bonds; the principle types are DNA and RNA. See *nucleotide.*

nucleotide The structural unit of nucleic acids, composed of a phosphate, a sugar, and a nitrogenous base. See *DNA, RNA,* and *nitrogenous base.*

nucleus A relatively large spherical body inside an interphase cell that contains the chromosomes in their uncoiled, thread-like state.

oocyte The cell that undergoes meiosis to produce an egg; the primary oocyte undergoes Meiosis I, and the secondary oocyte undergoes Meiosis II.

oogenesis The production of eggs in the female through the process of meiosis, modified so that only one egg with a large nutrient supply is formed. See *meiosis* and *polar bodies.*

oogonium A female germ cell before meiosis begins. See *germ cells.*

operator gene A gene, or chromosomal region, that controls the functioning of a group of structural genes to which it is adjacent in an operon. See *operon* and *structural gene.*

operon The unit of gene regulation consisting of an operator gene and the structural genes it controls. Interaction of this unit with a repressor substance from a regulator gene is the mechanism for turning genes on or off. See *operator gene, regulator gene, repressor,* and *structural gene.*

organic Refers to chemical molecules containing carbon-to-carbon bonds; pertaining to organisms and life processes in general. See *inorganic.*

ovary The female reproductive gland (gonad) in which meiosis occurs.

oviduct A tube that begins at the ovary and runs to the uterus through which the egg passes; fertilization takes place in the oviduct.

ovulation The discharge of the egg (secondary oocyte) from the ovary.

oxidation Chemical reactions where one or more atoms of oxygen are incorporated into a molecule. In living organisms such reactions are directed by enzymes. See *enzyme.*

p See *Hardy-Weinberg equilibrium equation.*

paracentric inversion See *inversion.*

pedigree A diagrammatic record of inheritance showing the ancestral history of a person (proband) for a particular trait or traits. See *proband.*

penetrance The proportion of individuals with a particular genotype that express the corresponding phenotype. When penetrance is incomplete, some individuals with a particular genotype show the expected phenotype while others do not; an all-or-none effect on the individual. See *expressivity.*

penis The male sex organ used to introduce sperms directly into the female egg tract. An evolutionary adaptation to internal fertilization developed in the first fully terrestrial animals, the reptiles.

peptide bond The chemical bond that holds two amino acids together in a polypeptide, formed between the amino group of one amino acid and the carboxyl group of another. See *amino acid.*

pericentric inversion See *inversion.*

phenocopy The expression of a phenotype that is usually associated with a specific genotype, produced instead by the interaction of some environmental factor with a different genotype; an environmentally induced mimic of a known genetic condition. The drug thalidomide produced phenocopies of phocomelia. See *phocomelia*.

phenotype The appearance or observable nature of an individual as determined by her or his genotype and the influence of the environment. Individuals that appear alike may be genetically different and not breed alike. See *genotype*.

phenylketonuria (PKU) A genetic disease characterized by the inability to metabolize the amino acid phenylalanine. Inherited as an autosomal recessive.

phocomelia An abnormality of limb development in which one or more limbs are extremely shortened to a flipperlike condition; literally, "seal limbs." Genetic form inherited as an autosomal recessive; also environmentally caused, as in thalidomide babies. See *phenocopy*.

photosynthesis A series of chemical reactions by which green plants utilize energy from the sun to manufacture carbohydrates (organic nutrients) from carbon dioxide and water (inorganic raw materials). See *autotrophs*.

phyletic speciation Gradual change over long periods of time producing "modern" species recognized as distinct from "ancient" species but representing a continuum of change that at any particular point in time would appear to be under the influence of stabilizing selection. See *speciation* and *stabilizing selection*.

piebaldness A mottled or spotted condition of the skin produced by pigmented and unpigmented patches; inherited as an autosomal dominant.

pituitary gland The "master" endocrine gland of the body located beneath the brain; it produces several hormones, including the gonadotrophins, which act on other glands. See *gonadotrophins*.

PKU disease See *phenylketonuria*.

placenta An organ formed in the uterine wall during pregnancy from a combination of endometrial tissue and the embryonic membranes. Within it blood vessels of the embryo come in close proximity to those of the mother; it serves to nourish the embryo. See *chorion* and *endometrium*.

platyrrhine A member of the primate infraorder Platyrrhini (broad-nosed) characterized by a broad nasal septum between round-shaped nostrils and a prehensile (seizing or holding) tail. Consists of the less advanced anthropoids, the New World monkeys. See *anthropoid* and *catarrhine*.

pleiotropy The influence of a single gene on several characteristics, because of the involvement of the gene product in several, usually related, processes. Multiple phenotypic effects from one gene.

poikilothermic Pertaining to animals without internal control of body temperature; "cold-blooded." See *homoiothermic*.

point mutation A mutation within a gene, such as a single base change. See *transition* and *transversion*.

polar bodies Minute by-products formed in oogenesis that discard excess chromosomes as the cytoplasm is conserved and concentrated in the production of a single egg cell. See *oogenesis*.

poles (of a cell) The ends of the spindle to which the chromatids (or chromosomes in Meiosis I) move during anaphase of mitosis or meiosis. See *anaphase* and *spindle*.

polygenic inheritance A mode of inheritance that produces continuous variation in a complex characteristic, because of the interaction of many genes at different loci that are cumulative in their effect on the same trait; for example, stature, skin color and intelligence. Also called quantitative inheritance. See *qualitative inheritance*.

polypeptide A chain of amino acids held together by peptide bonds and manufactured at the ribosomes through the genetic code; the basic gene product. One or more polypeptides form an active protein molecule, such as an enzyme. See *amino acid* and *protein*.

polyploid A chromosomal aberration in which the chromosome number is increased beyond the normal diploid (2n) condition by some exact multiple of the haploid (n) or basic number; for example, triploid (3n), tetraploid (4n), pentaploid (5n), hexaploid (6n), and so forth. Very rare in humans and all animals but relatively common in the plant kingdom.

polysome An aggregate of several ribosomes activating the same messenger RNA molecule with each manufacturing a polypeptide. See *messenger RNA* and *ribosomes*.

population Entire group of organisms of one kind (species) sharing a common gene pool; often equivalent to a Mendelian population but may also be larger or smaller depending on the particular usage. See *Mendelian population.*

population genetics The branch of genetics that deals with hereditary principles at the population level; for example, gene frequencies, breeding patterns, and population predictions.

primary speciation The splitting of one species into two or more as the result of directional selection operating independently on geographically isolated segments of a previously united population. (Also theoretically possible through disruptive selection.) See *directional selection, disruptive selection,* and *speciation.*

primate A member of the Primate order of placental mammals adapted to tree-dwelling and characterized by a grasping hand, well-developed vision, and large brain; includes monkeys, apes, and man. See *anthropoid* and *prosimian.*

primordial follicle Cluster of cells within the ovary that contains the potential egg (primary oocyte).

probability The likelihood of the occurrence of any particular form of an event, estimated as the ratio of the number of ways in which that form might occur to the total number of ways the event might occur in any form; the fraction or percent of the time that some one of several alternative events is expected to occur.

proband The individual who first draws attention to a particular trait for pedigree construction; also called propositus.

Proconsul A member of the dryopithecine group of ancestral apes. It appears to be in the ancestral line of the great ape family, not that of modern man. See *Dryopithecus.*

progeny The individuals that arise from a particular mating; offspring. See *sibs.*

progesterone A female sex hormone produced by the corpus luteum, involved in regulation of egg production.

prophase The stage of mitosis or meiosis in which the chromosomes are condensed and readily visible but not yet aligned on the metaphase plate. See *metaphase.*

propositus See *proband.*

prosimian The primative primate suborder; literally, premonkeys. Includes the tree shrews and tarsiers. See *anthropoid* and *primate.*

protein A macromolecule composed of one or more polypeptides. Innumerably many different kinds, including the enzymes, are present in all living organisms. See *enzyme* and *polypeptide.*

proton The elementary unit of positive electric charge found in the nucleus of an atom. See *electron.*

pseudohermaphrodite An individual with physical abnormalities that cause sexual misclassification or an apparent mixture of sexual characteristics. See *hermaphrodite* and *intersex.*

pseudovirion An artificial virus-like particle constructed by enclosing selected DNA within a viral protein coat to be used to deliver genetic material to a host cell by viral infection.

purine One class of nitrogenous base found in DNA and RNA; includes adenine and guanine. See *pyrimidine.*

pyknodysostosis A genetic abnormality characterized by thick, dense, fragile bones and pleiotropic phenotypic effects. Inherited as an autosomal recessive. See *pleiotropy.*

pyrimidine One class of nitrogenous base found in DNA and RNA; includes cytosine, thymine, and uracil. See *purine.*

q See *Hardy-Weinberg equilibrium equation.*

qualitative inheritance A pattern of inheritance that is characterized by sharply contrasting phenotypic alternatives of some particular characteristic, due mainly to the action of alleles at a single locus; for example, dominance-recessive or multiple allelic relationships. See *polygenic inheritance.*

quantitative inheritance See *polygenic inheritance.*

Ramapithecus An early member of the human family (Hominidae) preceding *Australopithecus.*

random mating Mating determined by chance rather than being prescribed; random selection of a mate implies no close relationship. See *inbreeding.*

recessive The condition in which a particular form of a gene (recessive allele) is phenotypically expressed only in the homozygous state; it is masked by a dominant allele in the heterozygous state. See *dominance, heterozygous, homozygous,* and *phenotype.*

reciprocal Corresponding to each other as by being complementary; for example, *AB* and *ab* are reciprocal gametes from a double heterozygote, or $A \male \times a \female$ and $a \female \times A \male$ are reciprocal crosses.

recombinant In crosses involving linkage, the reciprocal gametes produced by crossing over. See *repicrocal* and *recombination.*

recombination The shuffling of the genetic material that takes place in sexual reproduction as a result of meiosis. Also the formation of new allelic combinations within an homologous chromosome **pair** as a result of crossing over; the percentage of crossing over equals the recombination value.

reduction Chemically, the incorporation of hydrogen into a molecule; hydrogenation; often used as the opposite of oxidation. In living organisms such reactions are directed by enzymes.

regulator gene A gene that produces a product, the repressor, that controls structural genes by interacting with the operator gene. See *operon* and *repressor*.

repeats See *duplication*.

replication The reproduction of a copy; a duplication accomplished by copying from a template, such as the formation of chromatids and DNA molecules. See *template*.

repressor The product of a regulator gene, involved in controlling structural genes. See *operon*.

reproductive cell See *gamete*.

reproductive isolation In contrast to geographical (physical) isolation, a situation in which the gene pool of a population is kept separate from all others because no interbreeding takes place between the different populations; the basic species criterion. It is accomplished through an array of biological barriers called isolating mechanisms. See *species* and Table 2-2.

repulsion With reference to two pairs of genes located on the same chromosome, the situation of having the dominant allele of one pair and the recessive allele of the other pair on one chromosome, while the homologous chromosome has the opposite arrangement; for example, *Ab* and *aB*. See *coupling* and *linkage*.

Rh (rhesus factor) A major blood antigen important in compatibility for transfusions and in pregnancy. Individuals are typed as positive (having the antigen) or negative (lacking the antigen). Antigen production (positive) is inherited as an autosomal dominant. See *antigen*.

ribosomal RNA (rRNA) The type of RNA associated with proteins to form the ribosomes. See *ribosomes* and *RNA*.

ribosomes Small spherical bodies composed of RNA and proteins found in the cytoplasm of a cell; the sites of polypeptide synthesis. See *RNA*.

RNA (ribonucleic acid) A nucleic acid synthesized in the nucleus on a DNA template and differing from DNA by possessing ribose instead of deoxyribose in its sugar-phosphate backbone, being single- rather than double-stranded, and replacing the nitrogenous base thymine with uracil. The basic types are distinguished by function; see *messenger RNA*, *ribosomal RNA*, and *transfer RNA*. Also see *nucleic acid*.

secondary speciation The fusion of two species into one through hybridization followed by either directional or stabilizing selection in the establishment of a new adaptive norm. See *adaptive norm* and *hybridization*.

secretor A person possessing the A and/or B antigens of the ABO blood group in a water-soluble form in such a way that they can be detected in bodily fluids, such as the saliva, in addition to their presence on the red blood cells. Secreting is inherited as an autosomal dominant.

segregation Genetically, the separation of alleles during meiosis so that only one member of each pair is incorporated into each haploid gamete; Mendel's first law of inheritance. See *alleles* and *independent assortment*.

selection See *natural selection*.

semen The combination of sperms and fluid from the glands associated with the sperm tract.

sex cell See *gamete*.

sex chromosome A chromosome particularly involved in sex determination; the X and Y chromosomes are sex chromosomes (XX produces females and XY produces males). See *autosomes*.

sex hormone A hormone produced by or acting on the sex glands (gonads). See *gonadotrophin* and *hormone*.

sex-influenced Refers to autosomal genes in which a particular allele acts dominant in one sex but recessive in the opposite sex; for example, the allele for baldness acts dominant in males but recessive (with perhaps reduced penetrance) in females. Compare to *sex-limited* and *sex-linked*.

sex-limited Refers to genes, usually autosomal, that produce a phenotypic effect in one sex or the other but not in both; for example, the alleles for breast shape are only expressed in females, while those governing beard pattern are only expressed in males. Although limited in expression, such genes are transmitted in normal fashion through both sexes. Compare to *sex-influenced* and *sex-linked*.

sex-linked Refers to genes located on the sex chromosomes (usually used as synonymous to X-linked) that produce a pattern of inheritance modified from the typical autosomal mode. See *X-linked* and *Y-linked*.

sex ratio The relative proportion of males to females in a population either at conception (primary ratio), at birth (secondary ratio), or at a particular age after birth (tertiary ratio); usually expressed as the number of males per 100 females.

sexual reproduction Progeny production through the union of two haploid sex cells (gametes) to form a zygote representing a new and unique genetic combination (genotype) resulting from the random shuffling (through meiosis) of the genes of each parent. See *asexual reproduction*.

sibs (siblings) Brothers and sisters of the same two parents born at different times; that is, not twins. Half-sibs have one parent in common. See *progeny*.

sickle cell anemia A genetic disease characterized by defective hemoglobin resulting in severe anemia. Inherited as an autosomal allele codominant to the normal allele; the heterozygote is intermediate with a mild anemia called sickle cell trait. See *codominance*.

somatic cells (somatoplasm) All the cells of the body except those of the germ line. See *germ cells*.

somatic mutation A mutation in a somatic cell; that is, in any cell of the body other than a germ cell. See *mutation*.

speciation Species formation. See *phyletic speciation, primary speciation, reproductive isolation, secondary speciation,* and *species*.

species Populations considered on the basis of various criteria, especially reproductive isolation, to represent a natural group of organisms distinct from all other such groups. See *reproductive isolation* and *speciation*.

sperm The male gamete; characterized by a head containing the haploid chromosome set, a neck containing mitochondria for energy supply, and a long tail to provide motility. See *gamete*.

spermatid The product of meiosis in the male; it matures into a functional sperm. See *sperm*.

spermatocyte The cell that undergoes meiosis to produce sperms; the primary spermatocyte undergoes Meiosis I, and the secondary spermatocyte undergoes Meiosis II.

spermatogenesis The production of sperms in the male through the process of meiosis. See *meiosis*.

spermatogonium A male germ cell before meiosis begins. See *germ cell*.

sperm duct A tube that leads from the testis inside the abdominal cavity to connect with the urethra. In a vasectomy the sperm duct is severed.

spindle An arrangement of longitudinally oriented protein molecules (called spindle fibers) that is spindle-shaped with the poles of the cell established at each end and the metaphase plate located across its equatorial plane. See *metaphase plate* and *poles*.

spontaneous generation See *biopoiesis*.

stabilizing selection Natural selection that eliminates the extreme deviants from a population and maintains a rather constant adaptive norm, occurs under relatively constant environmental conditions. See *adaptive norm* and *natural selection*.

structural gene A gene that produces polypeptides to be incorporated into specific proteins (enzymes, antigens, and so forth) for use in the body; a "typical" gene. See *operon*.

submetacentric Refers to a chromosome having its centromere located off center, but not very close to the end; forms a J-shaped chromosome during anaphase. See *centromere*.

supergene A group of genes "locked" together as one unit such as those contained within the chromosomal segment involved in an inversion; the whole group may be acted upon by natural selection as a single (super) gene. See *inversion*.

synapsis Very intimate (locus-by-locus) pairing of homologous chromosomes during Prophase I of meiosis; bivalent formation. See *homologous chromosomes*.

syndrome A group of symptoms that occur together and characterize a particular disease; for example, Down's syndrome.

Tay-Sachs disease A genetic disease characterized by the inability to breakdown certain fat compounds which instead accumulate in the brain and spinal cord causing blindness, paralysis, and death; inherited as an autosomal recessive.

telophase The stage of mitosis or meiosis at which the chromosomes reach the poles of the cell and cytokinesis begins. See *cytokinesis.*

template A pattern or mold. DNA acts as a template for its own replication and for RNA production. See *replication.*

test cross The cross of an individual with a dominant phenotype (the "testee") to an individual homozygous for the recessive alleles of the gene in question (the "testor") in order to ascertain the genotype of the "testee" (homozygous or heterozygous for the dominant allele) according to the progeny produced. If the "testee" is homozygous, all progeny will have the dominant phenotype; if the "testee" is heterozygous, half the progeny will have the dominant phenotype, and the other half will have the recessive phenotype.

testicular feminization A genetic abnormality characterized by a male genotype (XY) but female appearance because of estrogen rather than androgen secretion from the testes (which are often retained in the abdominal cavity); inherited as a sex-limited recessive. See *sex-limited.*

testis The male reproductive gland (gonad) in which meiosis occurs.

testosterone The primary male sex hormone produced by the interstitial cells of the testes.

thalassemia An inherited form of anemia exhibiting codominance with its normal allele. Abnormal homozygotes exhibit a severe anemia, called thalassemia major, and heterozygotes have a milder form, called thalassemia minor. See *anemia.*

thymine One of the nitrogenous bases, a pyrimidine, found in DNA, but usually replaced by uracil in RNA.

transcription In gene function, the complementary copying (transcribing) of the genetic code from DNA to messenger RNA; nucleic acid–to–nucleic acid exchange. See *translation.*

transfer RNA (tRNA) The type of RNA that transports specific amino acids to the ribosomes for assembly into polypeptide chains. Recognition of tRNA to mRNA on the ribosome is accomplished by base complementarity of anticodon to codon. See *anticodon, codon, ribosome,* and *RNA.*

transformation The direct incorporation of DNA into a host cell followed by its integration and function in the chromosomes of the host.

transition A point mutation within a codon in which one pyrimidine is replaced by another, or one purine is replaced by another. See *transversion.*

translation In gene function, decoding the messenger RNA into an amino acid sequence in the production of a polypeptide; nucleic acid-to-amino acid exchange. See *transcription.*

translocation A chromosomal aberration in which a portion of one chromosome is attached to a nonhomologous chromosome; often a reciprocal exchange of segments between nonhomologous chromosomes. See *centric fusion.*

transversion A point mutation within a codon in which a purine is replaced by a pyrimidine or vice versa. See *point mutation* and *transition.*

triplo-X Trisomy of the X chromosome (XXX); such individuals are very nearly normal females, including normal fertility. See *trisomy.*

trisomy A chromosomal aberration in which one chromosome is present in triplicate instead of the normal double (diploid) condition. A human trisomic has 47 chromosomes $(2n + 1)$ in each cell. See *aneuploid* and *monosomy.*

Turner's syndrome An abnormality caused by monosomy of the sex chromosomes involving the presence of a single X chromosome (XO); such individuals appear as juvenile females with regard to development of the ovaries and secondary sex characteristics; they are sterile. See *syndrome* and *trisomy.*

twins See *dizygotic twins* and *monozygotic twins.*

uniformitarianism A geological concept based on the premise that former processes were the same and acted in the same manner as those active today and that change occurred slowly and gradually over very long spans of time.

univalent An entire duplicated, but unpaired, chromosome during meiosis. See *bivalent* and *synapsis.*

uracil One of the nitrogenous bases, a pyrimidine, found in RNA as a replacement for thymine in DNA.

uterus The organ in which embryonic development takes place; the womb.

vagina Portion of the female egg tract connecting the uterus with the exterior; receives the penis during copulation and sperms upon ejaculation.

viability The capability of living, growing and developing normally.

vulva Collective term for the female external genital organs; includes the labia majora and minora, the clitoris, and the prepuce.

X chromosome One of the sex chromosomes found in males and females; medium-large in size. See *sex chromosome* and *Y chromosome*.

Xg blood group A blood group system inherited as an X-linked trait; X_g^a is dominant for antigen production over the inactive X_g allele. See *X-linked*.

X-linked Refers to a gene located on the X chromosome, usually used synonymously with sex-linked. See *sex-linked*.

XYY syndrome See *Jacobs' syndrome*.

Y chromosome One of the sex chromosomes found only in males; very small in size. See *sex chromosome* and *X chromosome*.

Y-linked Refers to a gene located on the Y chromosome. See *holandric* and *sex-linked*.

zygote The diploid cell produced by the union of an egg and sperm in sexual reproduction. Also used in genetics to refer to the individual produced from such a cell; for example, homozygote or heterozygote.

Index

Breast feeding, prolonged, for birth control, 106

Breeding, selective, techniques of, as mode of population improvement, 299-304

Broom, Robert
  research on *Australopithecus*, 57
  research on *Paranthropus*, 59

Bruno, Giordano, in evolution of evolutionary theory, 19

Brushfield spots in Down's syndrome, 257

Buffon, George in evolution of evolutionary theory, 19

Bulbourethral glands, location and function, 89

C

Calendar rhythm method of birth control, 107

Callosities, ischial, as adaptation for survival in catarrhines, 50

Camptodactyly, incomplete penetrance in, 149

Cancer
  skin, pigmentation of skin in protection against, 288
  as uncontrolled mitosis, 78

Carbon dioxide as by-product of fermentation in heterotrophs, effect on primordial atmosphere, 16

5-Carbon sugar in nucleic acid, 120

Catalysts, definition, 12

Cataracts, juvenile, variable expressivity in, 149-150

Catarrhines as stage of man's descent, 50-51, 52, 55

Cell(s)
  follicle, function of, in oogenesis, 94-95
  germ, in production of gametes, 79
  human, periods in life cycle of, 77-78
  interstitial, location and function, 88

Cell fusion as mode of genetic transfer therapy, 296-297

Centric fusion, 267
  in Down's syndrome, 267, 269, 270
  problems of karyotype monitoring caused by, 169

Centromere
  definition, 72
  in directing movement of chromosomes, 72
  location of, in identification of chromosomes, 72-73

CG; *see* Chorionic gonadotrophin (CG)

Chemical evolution, 4-16

Chemical nature of primordial atmosphere, significance for origin of life, 6, 9

Chiasmata, formation of, in meiotic process, 79

Chloasma, associated with oral contraceptives, 110

Chorion, formation of, 97

Chorionic gonadotrophin (CG)
  in maintenance of pregnancy, 99, 102
  presence of, as basis for pregnancy tests, 100, 102

Chorinonic villi, functions of, 98

Christmas disease, as X-linked recessive abnormality, 200

Chromatid, definition, 72

Chromosomal mosaics, 263

Chromosome(s), 71-74
  aberrations of, definition, 136, 139
  abnormal, effects of, 254-272
  acrocentric, 72
  appearance of, in nonreproducing and reproducing cells, 71-72
  combination of, in products of meiotic process, 81-82
  Denver classification of, 73
  homologous, definition, 73
  human, Denver classification of, 73
  metacentric, 72
  number of, changes in, 255-264
  ring, 267
  sex, 73
    genes on, 195-208
    in genetic control of sex differentiation in man, 191-192
    specific, fluormetric technique in identification of, 264, 265
  structure of
    abnormalities in, types of, 264-271
    changes in, 264-272
  submetacentric, 72
  X, 72; 191-192; *see also* X chromosomes
  Y, 73; *see also* Y chromosomes

Chromosome maps, 182
  tentative, for X chromosome, 204-205

Church in suppression of evolutionary theory, 19

Circumcision, definition, 89

Clitoris, location and function, 93, 94

Cloning as technique of selective breeding, 302-304

Coacervates
  characteristics, 11-12
  definition, 11
  in origin of life, 10-13

Coacervation, definition, 11

Codominant alleles
  autosomal, in thalassemia major and minor, 220-222
  in sickle cell trait, 148

Codons in genetic code, 130, 131

Coefficient
  of determination in analysis of intelligence data, 247
  of inbreeding (f), 284-285
  of relationship (r), 282, 283

Coitus interruptus for birth control, 105

Colloidal system, definition, 11

Colorblindness, red-green, as X-linked recessive trait, 195-196
  pedigrees for, 219-220

Complementarity, base of DNA molecule, 121

Condom for birth control, 106

Congenital deafness, nonallelic gene interaction in, 174